Weed Ecology in Natural and Agricultural Systems

Weed Ecology in Natural and Agricultural Systems

Barbara D. Booth

Department of Plant Agriculture
University of Guelph
Canada

Stephen D. Murphy

Department of Environment and Resource Studies
University of Waterloo
Canada

and

Clarence J. Swanton

Department of Plant Agriculture
University of Guelph
Canada

CABI *Publishing*

CABI *Publishing* is a division of CAB *International*

CABI Publishing
CAB International
Wallingford
Oxon OX10 8DE
UK

Tel: +44 (0)1491 832111
Fax: +44 (0)1491 833508
E-mail: cabi@cabi.org
Web site: www.cabi-publishing.org

CABI Publishing
44 Brattle Street
4th Floor
Cambridge, MA 02138
USA

Tel: +1 617 395 4056
Fax: +1 617 354 6875
E-mail: cabi-nao@cabi.org

A catalogue record for this book is available from the British Library, London, UK.

Library of Congress Cataloging-in-Publication Data

Booth, Barbara D.
Weed ecology in natural and agricultural systems/Barbara D. Booth, Stephen D. Murphy, and Clarence J. Swanton.
p. cm.
Includes bibliographical references (p.).
ISBN 0-85199-528-4 (alk. paper)
1. Weeds—Ecology. 2. Weeds. I. Murphy, Stephen D.
II. Swanton, Clarence J. III. Title.
SB611.B59 2003
632'.5—dc21

2002010884

ISBN 0 85199 528 4

Typeset by Wyvern 21 Ltd, Bristol
Printed and bound in the UK by Biddles Ltd, Guildford and King's Lynn

Contents

Preface

Our goal in writing this book was to describe why weeds occur where they do. We have made no attempt to discuss their management and control: there are excellent texts available for that. Rather, we think that students should understand how and why weeds fit into their environment. This text presents ecological principles as they relate to weeds. Ecology is central to our understanding of how and why weeds invade and yet there are few books that make this connection. That is the niche we hope to fill.

We make no excuses for using the word 'weed', and, since humans decide what species are considered to be a weed, we make no attempt at a detailed definition. We could really have used the word 'plant' throughout the text. We have tried to present a broad array of weed examples, and have therefore selected weed examples from different types of systems – agricultural, managed (e.g. forestry) and natural systems – and from around the world.

The book was designed as a teaching text for a middle year undergraduate course. No ecological background is assumed, although some basic biology is required. We have tried to write it and arrange the material so that it is presented in a clear concise manner. At the beginning of each chapter, we have listed concepts that will be addressed, as an overview of what is to come, and to assist the reader when reviewing the material. At the end of each chapter there is a list of questions, the first of which refers to a weed of your (the student's) choice. It can be a common widespread weed, or it may be a local problem. You will be asked to summarize information that is known about your weed in relation to the material discussed in each chapter. There may be a lot or very little information available to you. The idea behind this is to apply the ecological principles you learn in the chapter to a weed of interest, and to give you practice in researching a topic. Our hope is that by the end of the book, you will have created a 'case history' of your chosen weed.

For the instructor, we designed this book so that the material could be covered in a single-term course by covering approximately one 'content chapter' per week. Chapters 1 and 15 are a brief introduction and conclusion. Two chapters (10 and 14) discuss how ecology 'is done', i.e. methodology, experimental design and basic calculations. These can be used as you see fit. We have tried to keep the writing precise and concise and to include only pertinent information. If we have done our job well, students should be able to read and understand all of the information.

We have used common names throughout the text with Latin names given the first time the species is mentioned in each chapter. We did this because common names are easier to remember when first learning about a species. A species list of common and Latin names is provided at the end of the book.

We thank many people who assisted in the writing and production of this book. David Clements and Jason Cathcart provided detailed comments on many versions of the text. Cheryl Corbett, Sara Mohr and Sheryl Lonsbary read sections or chapters. Of course we accept the responsibility for any errors that occur. We also thank the authors and publishers who allowed us to use their illustrations and Tim Hardwick of CAB *International* who kept us on in spite of many missed deadlines.

Finally, we thank our spouses, David Beattie, Tara Murphy and Josee Lapierre, who probably heard more about 'the book' than they wanted, but kept smiling and nodding their heads anyway. We dedicate this book to them.

1
Ecology of Weeds

Concepts

- The terms 'weed', 'invader', 'colonist', 'exotic', 'non-native' and others are often used in overlapping and conflicting manners.
- Weeds are classified based on their impact on human activities. Therefore, the effect of a weed is difficult to quantify because it depends on our personal biases.
- Definitions and classifications in ecology are often arbitrary and made for purely practical reasons. They do not necessarily reflect any innate structure of nature.
- Ecology can be studied at a variety of levels. In this book, we focus on population and community ecology.
- Weed ecology provides a basic understanding of the distribution and abundance of weeds in natural and managed systems. In the long term, it may change our attitudes and perceptions towards weeds and alter the way we manage them.

Introduction

It may be tempting for you to start this book with Chapter 2. After all, the *real* information doesn't start until then, and exam questions rarely focus on what you learn in Chapter 1. *However*, Chapter 1 is important because it sets the tone for what is to follow. A Shakespearean play or an opera always begins with a prologue. If you walk in after the prologue has finished, you will certainly follow the plot and enjoy the play, but you might not understand the 'why' of the characters' actions. Consider this chapter to be a prologue. You may already know much of what we are about to say, and you may not be tested on it, but it will put what you are about to learn into context.

There are a number of excellent weed science (Radosevich *et al.*, 1997; Zimdahl, 1999a) and plant ecology (Crawley, 1997a; Barbour *et al.*, 1999) texts. We have found, however, that very few texts are devoted entirely to the basic ecology of weeds. A number of books are available on plant invasions; however, they often: (i) assume an in-depth understanding of ecological principles; (ii) focus heavily on the control and management of invasive species; or (iii) provide a detailed description of the biology of

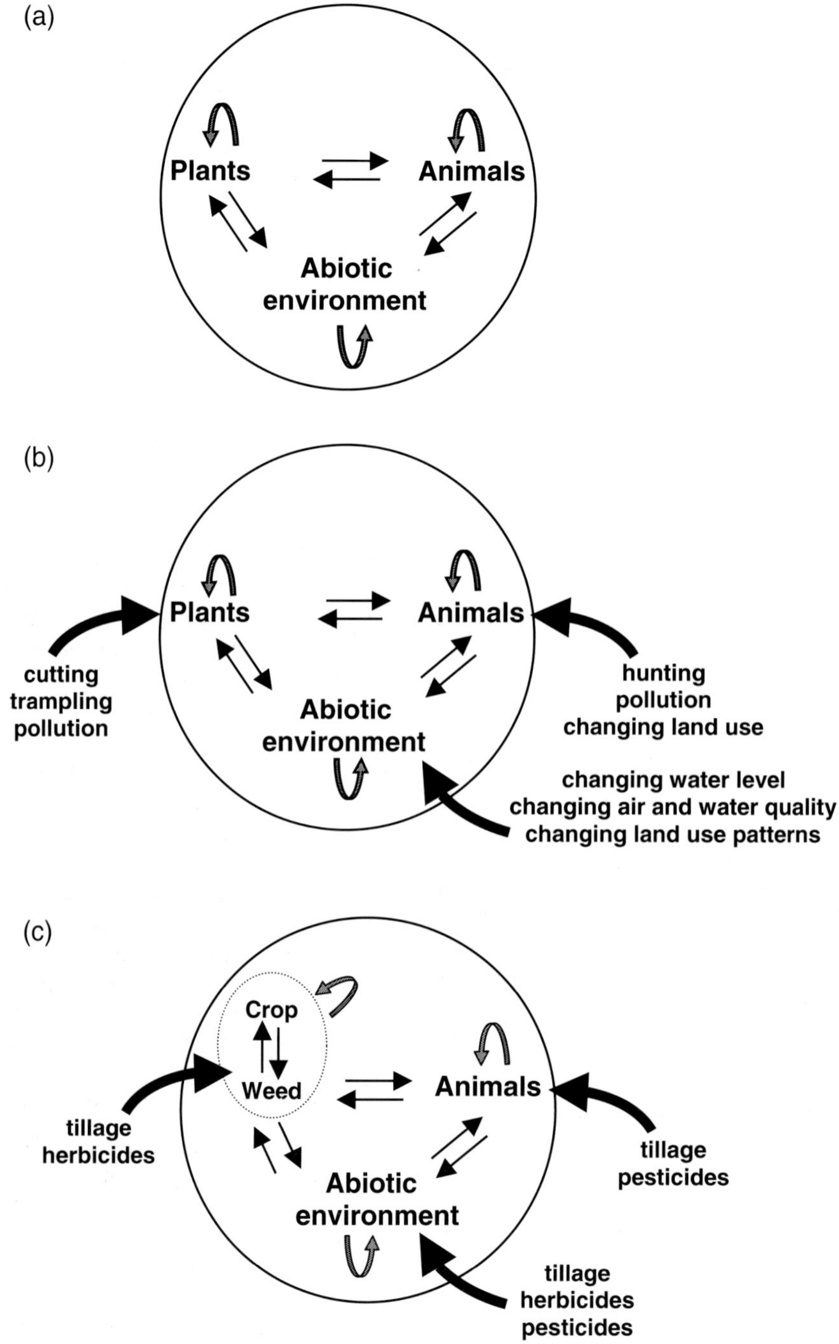

Fig. 1.1. Schematic diagram of three community types: (a) natural community with no human disturbance, (b) natural community with human disturbance and (c) agricultural system.

individual weedy species without providing a broad background. Our goal is to provide you with a link between the fields of weed science, plant invasions and ecology. This book will give you a basic ecological understanding of how plants invade natural, disturbed and agricultural ecosystems.

This book was not designed to replace a good, comprehensive text on basic ecological theory. Rather, we hope to entice readers into exploring such volumes, by presenting an overview of ecology and suggesting ways in which it is useful to applied situations. While ecology texts may seem intimidating and not useful to applied scientists, we hope that, by providing examples of how these concepts are useful in real situations, the importance of ecological theory will become apparent. If we can convince one of you to pick up one of those large, intimidating tomes, then we will have succeeded.

While the focus of this book will be the use of ecological principles to the study of weeds, it is also important to recognize the role that weed science has played in the development of ecology. Several of the earliest ecologists began their careers working on agricultural weeds. The eminent population ecologist John L. Harper began as an agronomist. Early in his career he recognized the importance of ecology to weed management (Harper, 1957). He also developed many of the basic principles of plant population ecology and his 1977 book titled *The Population Biology of Plants* is still a basic text cited in many population ecology papers and texts. Many examples used by him to illustrate ecological principles are weeds. In fact, 'ecologists have far to repay the debt to agriculture for all that they have learned from it' (Trenbath, 1985).

The scope of this book is to examine weeds in systems from highly managed agricultural and grazing land to disturbed or undisturbed natural communities. Is this possible? On the surface, it appears impossible to compare a forest to a field. To the eye, they appear very different in structure and function. However, all types of ecological systems are controlled by the same processes including natural and anthropogenic (human caused) disturbances (fire, construction, tillage) (Fig. 1.1). The human activities that influence natural or managed systems are ultimately biological in nature.

In the three main sections of this chapter we introduce you to weeds, to ecology and finally to weed ecology. In Part I, we present the muddled vocabulary used to describe, define and characterize weeds. In the second section, we describe how ecology is related to other fields of study and how ecology studies can be approached in different ways. In Part III, we integrate the study of weeds with ecology.

Colonizers, Invaders and Weeds: What's in a Name?

Every book on weeds or invasive species must first start with an attempt at defining the terms. Many attempts have been made to define 'weed', 'invasive', 'non-invasive', 'alien', 'naturalization' and other terms describing a species' status, place of origin or population trend (Schwartz, 1997). Pyšek (1995), for example, reviewed definitions of 'invasive' and found it to be described as:

- an alien in a semi-natural habitat (Macdonald *et al.*, 1989);
- a native or alien entering any new habitat (Mack, 1985; Gouyon, 1990);
- a native or alien that is increasing in population size (Joenje, 1987; Mooney and Drake, 1989; Le Floch *et al.*, 1990);
- any alien increasing in population size (Prach and Wade, 1992; Binggeli, 1994; Rejmánek, 1995), or
- any alien species (Kowarik, 1995).

Weeds have typically been defined as 'plants which are a nuisance' (Harper, 1960) or 'a plant where we do not want it' (Salisbury, 1961). Barbour *et al.* (1999) defined a weed as a 'non-native invasive plant' and they distinguished between 'invasive plants' that invade only natural or slightly disturbed habitats, and 'pest plants' that interfere with agricultural or managed natural areas. This definition, however, requires us to further define 'non-native' and 'invasive', and to separate natural from disturbed habitats. The Weed Science Society of America defines a weed as 'any plant that is objectionable or interferes with the activities and welfare of humans'. These definitions are based on our perceptions of the impact of the plant. Thus, the term 'weed' is more a convenient classification than a grouping of plants with common biological characteristics.

Crawley (1997b) recognized the difficulties of defining weeds, and suggested that for a plant to be considered a weed (a problem plant), its abundance must be above a specific level and someone must be concerned. This refines the definition somewhat because it suggests that a plant is only a weed if it is present above a specific abundance; however, it introduced the problem of determining what that threshold level is. This definition recognizes that a weed is only a weed under specific circumstances, that the inclusion of a plant into this category is arbitrarily based on human perceptions and that a specific plant species will not always be considered a weed.

The terms weed, invader and colonizer have often been used in a conflicting manner. The distinctions between them are quite subtle and result from differing viewpoints. According to Rejmánek (1995), *weeds* interfere with human land use; *colonizers* are successful at establishing following disturbance; and *invaders* are species introduced into their non-native habitat. There is substantial overlap among these terms. A plant may be considered as only one of these, or it may be included in all of these categories (Fig. 1.2).

Clearly, we will not definitively solve the problem of 'what is a weed' in this text and it is not necessary to do so. Here, we take a general, all-inclusive view of the term 'weed'. To us, a weed is a native or introduced (alien) species that has a perceived negative ecological or economic effect on agricultural or natural systems.

The traditional approach to the study of weeds is to examine their control or management rather than study their effect on the community. Our focus is on the latter. Whether a weed is in a natural community or a highly managed farm, the underlying questions and principles will be the same. The first part to weed management is to understand why weeds exist and why they have an impact. We leave the bulk of the discussion of weed management to others (Luken and Thieret, 1997; Radosevich *et al.*, 1997; Zimdahl, 1999a).

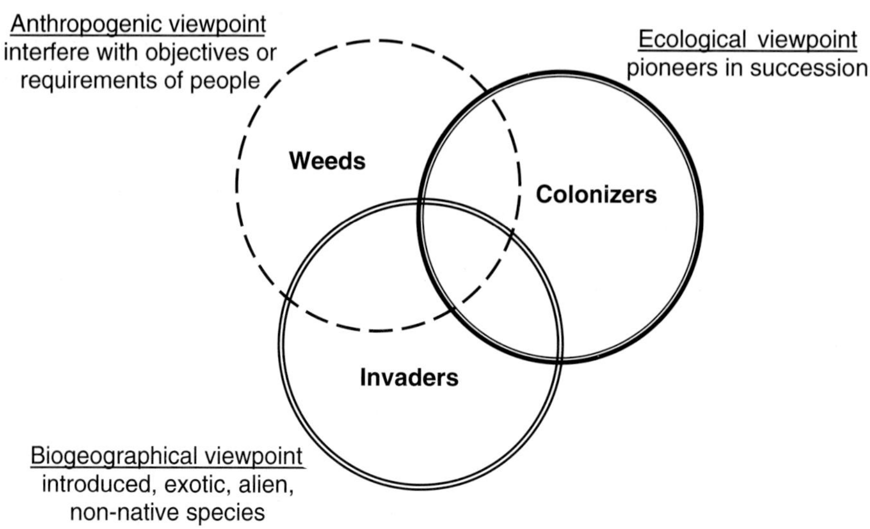

Fig. 1.2. Weeds, colonizers and invaders are similar concepts but result from differing viewpoints (redrawn from Rejmánek, 1995).

Table 1.1. Classification of weeds based on habitat type (based on Holzner, 1982).

Classification	Explanation
Agrestals	Weeds of agricultural systems, e.g. cereal/root crops, orchards, gardens, plantations
Ruderals	Weeds of waste/human disturbed sites (ruderal sites), e.g. roadsides, railway lines, ditches
Grassland weeds	e.g. pasture, meadows, lawns
Water weeds	Weeds that affect water systems, e.g. affect navigation, recreational use
Forestry weeds	e.g. tree nurseries, afforestation sites
Environmental weeds	Suppress native vegetation

Types of weeds

One common way to categorize weeds is based on the habitat they invade. Holzner (1982) divided weeds into agrestrals, ruderals, grassland weeds, water weeds, forestry weeds and environmental weeds (Table 1.1). Environmental weeds have often been called invasive species. There is a tendency to use the word 'invasive' when considering natural habitats, and 'weed' for managed habitats; however, there is a gradient between natural and managed systems, and some apparently natural systems are managed.

Weed characteristics

There have been many attempts to list characteristics associated with weeds. Baker (1965, 1974) summarized weed characteristics based on adaptations (Box 1.1). A species with more of these characteristics is more likely to be a successful weed. Baker (1965) said that a plant possessing all of the traits would be 'a formidable weed, indeed'. A weed will not necessarily possess all (or even any) of these characteristics, and conversely, a plant possessing some (or all) of these characteristics will not necessarily be a weed. A weed may require certain characteristics to invade, but a community must be invasible in order for the invasion to be successful.

A list of a species' characteristics cannot necessarily be used to predict its weediness or invasion success. Weed characteristics, community characteristics, the interaction between the community and the potential weed, as well as timing and chance will determine whether an introduced species is successful (Lodge, 1993; Hobbs and Humphries, 1995). Furthermore, while disturbance is often cited as a prerequisite for invasion to occur, this is not always true. Certain types of disturbance (i.e. cyclic fires) may, in fact, prevent invasions.

Box 1.1. Traits of an 'ideal' weed (based on Baker, 1956, 1974).

1. Germinates in a wide range of environmental conditions
2. Long-lived seeds that are internally controlled so that germination is discontinuous
3. Rapid growth from vegetative through to flowering stage
4. Self-compatible, but not completely autogamous or apomictic
5. Cross-pollination (when present) by wind or generalist insects
6. Seeds produced continuously throughout the growth period
7. Seed production occurs under a wide range of environmental conditions
8. High seed output when environmental conditions are favourable
9. Propagules (seeds) adapted to short- and long-distance dispersal
10. If perennial, has a high rate of vegetative reproduction or regeneration from fragments
11. If perennial, ramet attachments fragment easily, so it is difficult to pull from the ground
12. Strong potential to compete interspecifically via allelopathy, rosettes, rapid growth and other means

Impact of weeds

Negative effects

The harmful impacts of weeds can be classified as land-use effects or as ecosystem effects. Land-use effects are easier to quantify because they can be measured in terms of decreased crop yield or increased control costs. Costs to the ecosystem may be just as great, but are less well understood and the impact is harder to quantify in numerical terms.

In managed (agricultural) systems, weeds can decrease the growth of a crop, often in a very predictable and quantifiable way. Zimdahl (1999a) divided the harmful effects of agricultural weeds into nine categories according to the target and type of damage done (Table 1.2). The most commonly known effects are those that either directly affect the crop through competition, increased production costs or reduce the quality of the crop. Less direct effects are those to animal or human health, by increased production or management costs or by decreasing land value. A weed may have one or many of these effects. Attempts to quantify the damage by weeds in agricultural systems have been done (Pimentel *et al.*, 2000); however, these can only be taken as estimates (Zimdahl, 1999a). These have been calculated as a proportion of the potential annual crop yield lost to weeds and as the amount of money spent on weed management.

Quantifying the damage done by weeds to a natural system can be difficult because they cannot be quantified in terms of dollars or time. We can express damage as the cost to control the weed; however, this does not address the actual ecological impact. A weed may effect the survival or growth of other species or change ecosystem processes like nutrient cycling. For example, the fire tree (*Myrica faya*), which was introduced to Hawaiian islands in the 1700s to control erosion in pasture, invaded large tracts of land and replaced the native forest because it increased the nitrogen level of the soil (Vitousek *et al.*, 1987; Vitousek and Walker, 1989). As a legume, it fixes nitrogen causing the nitrogen level of the volcanic soils to increase. This has increased the invasion of other weeds which require higher nitrogen. While the effect of fire tree on the ecosystem is clear, how does one quantify the damage?

Benefits

The benefits of weeds are less well understood than the negative effects, and more difficult to quantify because they occur over a longer time scale. Altieri (1988) and Holzner (1982) reviewed the benefits of weeds in

Table 1.2. Potential harmful effects of agricultural weeds on human land use (based on Zimdahl, 1999a)

Harmful effect	Explanation
Compete with crop	Compete with crop plants for nutrients, water, light and space
Increased protection costs	Weed may harbour crop pests or diseases
Reduced quality of crop	Weed seeds become mixed with crop seed during harvest and will therefore affect the quality of seed crop
Reduced quality of animals	Weeds in rangeland may poison or kill animals, can affect animal products (meat, milk), or affect reproduction
	Weed plants and seed may physically damage animals or their products (wool)
Increased production and processing costs	Cost of weed control (tillage, herbicides)
	Cost of cleaning seeds
Water management	Weeds may impede flow of water through irrigation ditches
Human health	Cause respiratory, digestive or skin ailments, or other health effects
Decreased land value	Cost of restoring land (esp. perennial weeds)
Reduced crop choice	Restrict possible crops that can be grown
Aesthetic value	Recreational land or traffic intersections/thoroughfares

agricultural situations. Weeds may increase crop growth under certain circumstances. For example, in some dry areas of India, three 'weeds' (Arabian primrose, *Arnebia hispidissima*; buttonweed, *Borreria articularis;* and cockscomb, *Celosia argentea*) increase the growth of millet (bajra, *Pennisetum typhoideum*); however, this is not true for sesame (til, *Sesamum indicum*) (Bhandari and Sen, 1979). A fourth weed, indigo (*Indigofera cordiflora*), was beneficial to both crops. Thus, the specific site conditions and species involved must be considered before drawing conclusions about the value of a particular plant.

In some traditional agroecosystems, the importance of certain weeds is recognized even if they are known also to reduce crop yield. These weeds have other functions that compensate for loss of crop yield. For example, in Tabasco, Mexico, some weeds are left because they are recognized for their food, medicinal, ceremonial or soil-improving uses (Chacon and Gliessman, 1982). These weeds are termed 'buen monte' (good weeds) while others are 'mal monte' (bad weeds). In other situations, weeds may be harvested for food, animal fodder or fertilizer. In Australia, *Echium plantagineum* is considered a noxious weed in grazing land, but it also serves as an emergency feed under some conditions (Trenbath, 1985). Its dual names, 'Paterson's Curse' and 'Salvation Jane' reflect this. Weeds are now being recognized for the potential role they may play in mediating crop–predator interactions. Weeds may provide a habitat for some beneficial insects, which could result in higher yields due to a decreased pest load on the crop.

Non-native weeds can be beneficial in non-agricultural situations, especially when the environment has been degraded (Williams, 1997). Non-native species have been useful in a number of restoration projects. For example, natural regeneration of woody plants in subantarctic forests of Argentina is limited due to overexploitation and overgrazing by cattle. However, the introduced European mosqueta rose (*Rosa rubiginosa*) is able to establish in degraded sites, resists grazing and provides shelter for the regeneration of native woody species (De Pietri, 1992).

Finally, weeds may also have beneficial properties such as erosion control (Williams, 1997). However, the properties that make some species excellent at controlling erosion may also make them excellent weeds as well. In the southeastern USA, farmers were encouraged to plant kudzu (*Pueraria montana* var. *lobata*) to control soil erosion; however, after 1953 it was considered a noxious weed by the United States Department of Agriculture (USDA) and was no longer on the list of permissible cover plants. It is now a troublesome weed in the southeastern USA.

A weed is not always a weed

A plant may be both a 'weed' and 'not a weed' depending on where and under what circumstances it is growing. The decision of what is a weed can be quite complex. A plant species may be both a weed and a desired species, depending on its location and on the desired land use. Following are three examples of plants that could be considered weeds or not.

- Proso millet (*Panicum miliaceum*) is a crop grown in Canada and other parts of the world. In the last 30 years, however, weedy biotypes of proso millet have developed and it is now an important agricultural weed in Canada and the USA. The crop and the various weed biotypes differ in seed characteristics, seedling vigour, germination patterns, inflorescence structure and dispersal mechanisms (Cavers and Bough, 1985).
- In Western Australia, where farmers alternate between wheat cropping and sheep pasture, annual grasses (such as annual ryegrass) are either the weed or the crop, depending on the rotation. During the pasture phase, grasses provide early forage and protection from erosion, but they also decrease the growth of nitrogen-fixing clover (*Trifolium*), which can decrease subsequent wheat yields (Trenbath, 1985).
- Monterey pine (*Pinus radiata*) is a native tree species in parts of California, a plan-

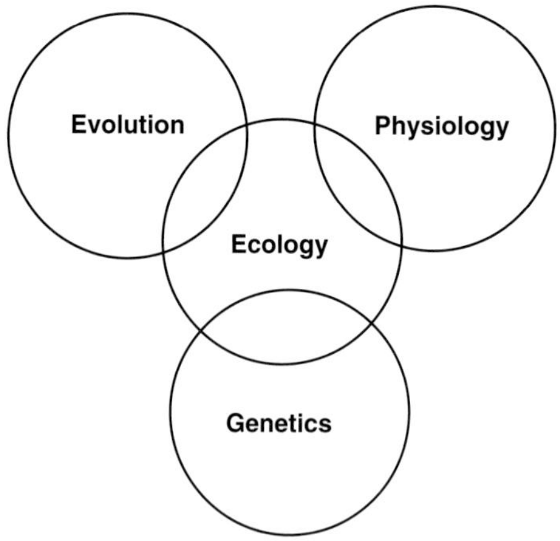

Fig. 1.3. Relationship between ecology and genetics, evolution and physiology.

tation tree in parts of Australia, New Zealand, South Africa and Chile, but is also a weed in natural areas adjacent to plantations.

What is Ecology?

The word ecology was derived from the German word (*oekologie*), which was derived from the Greek words *oikos* meaning 'house' and *logos* meaning 'the study of'. Thus, ecology is the study of organisms and their environment. We can divide the environment into biotic (living) and abiotic (nonliving) factors. Examples of biotic factors are competition and herbivory. Abiotic factors can be physical (e.g. temperature, light quality and quantity) or chemical (e.g. soil nutrient status).

Ecology is closely related to other fields of biology such as physiology, evolution and genetics (Fig. 1.3). There are no distinct boundaries between these fields and ecology, and indeed there is enough overlap that subdisciplines have arisen. The types of questions that these scientists ask are often the same. For example, an ecologist and a physiologist may both ask how a plant's photosynthesis is affected by the surrounding vegetation. To the ecologists, the focus is on the plant growth and survival; to the physiologist, the focus is on the process of photosynthesis.

Levels of ecological study

The field of ecology is vast. It is concerned with areas as diverse as the dispersal of seeds, competition within and between species, and nutrient cycling through ecosystems. Each of these operates on a different temporal (time) and spatial (space) scale, and each has a different focus. Thus, they address different types of questions, and require a different protocol to answer such questions. For convenience, ecological questions can be categorized into subdisciplines (Fig. 1.4). For example, individual organisms can be studied to examine how abiotic factors affect their physiology. Groups of individuals of the same species can be studied to look at population-level processes. Groups of co-occurring populations can be studied to ask community-level questions. Furthermore, interactions between a community and its abiotic factors can be studied to answer ecosystem questions. Each of these categories blends into the next. They are not discrete units of study, rather they are useful, practical and somewhat arbitrary divisions which help to simplify the field of study. In this book we

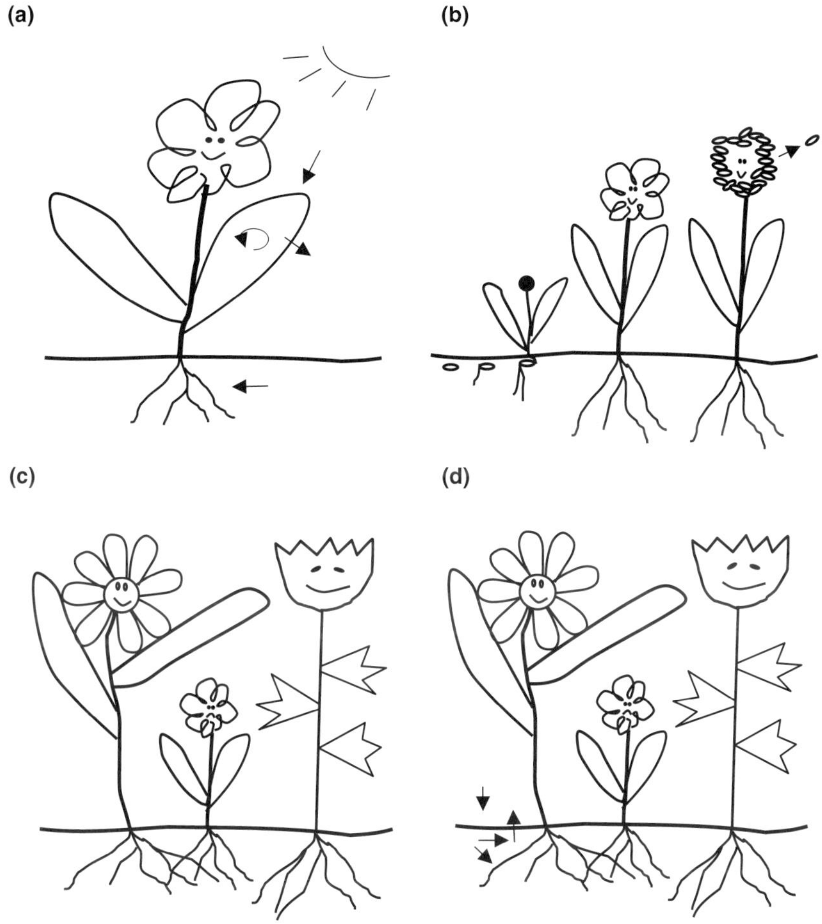

Fig. 1.4. Illustration of (a) ecophysiology, (b) population ecology, (c) community ecology and (d) ecosystem ecology.

are primarily interested in population ecology (Chapters 2–7), interactions between populations (Chapters 8 and 9) and community-level ecology (Chapters 11–14).

Population ecology

A population is a group of potentially interbreeding individuals of the same species found in the same place at the same time. Determining whether individuals are in the 'same space' may pose difficulties. In some cases, the population's distribution will be quite clumped and thus boundaries are easily imposed around these clumps of interacting individuals. Other times, boundaries may be determined by natural or anthropogenic features such as roads, rivers or mountains. Finally, we may impose arbitrary boundaries around our target population. While there is no one 'correct' way to do this, it is important to base one's decision on our knowledge of the organism's biology and on the goals of the study. We should be clear about the reasons for imposing these boundaries and keep in mind their effect when interpreting the results.

Populations can be studied in a number of ways (Table 1.3). A population's density and distribution quantify how it is dispersed over space. Age and sex structure quantifies the demographic characteristics of the population at one time. Population dynamics are quantified by measuring the change in natality (births), mortality (deaths), immigration and emigration over time. Note that each of these measurements is derived from data collected on groups of individuals and could not be a characteristic of any single individual (there is no such thing as the 'density' or 'age structure' of a single organism). Population ecologists ask questions such as:

- What determines a species' distribution and/or density?
- How do physiological, morphological and phenological traits influence the distribution and abundance of a species or population?
- How do biotic or abiotic factors affect a population's growth and reproductive rate?
- What is the age structure of the population?
- Is population size increasing or decreasing?

Community ecology

A community is a group of populations that co-occur in the same space and at the same time (Begon *et al.*, 1990). Definitions of communities are generally vague on where the community boundaries are. Again, we can define boundaries based on the needs of our study. A further difficulty with defining a community is deciding what organisms to include. This is another rather arbitrary decision. Do we look at just plants, animals, fungi or all three? Clearly, we *should* include all organisms within the boundaries of our community because any one may have an important function. However, because of the practical limitations placed on researchers, this is rarely done. Decisions on what constitutes a community can be done at any scale: from the community of fungi colonizing a piece of stale bread, to the community of maize and weeds in a field, to the entire flora and fauna of a boreal forest.

We can describe communities in terms of their structure and function (Table 1.3). Community structure refers to the external appearance of the community. Species composition (species lists, diversity), species traits (life span, morphology) and strata characteristics (canopy, shrubs, vines, herbs) are used to describe community structure. Function refers to how the community 'works'. Nutrient allocation and cycling, biomass production and allocation, and plant productivity are ways to describe community function. Community ecologists ask questions such as:

- How does community structure change over time?
- Can we predict community changes over time?
- Why are there so many (or so few) species in this community?

Table 1.3. Measurements used to characterize populations and communities.

Populations	Communities
Population structure	*Community structure*
Distribution and density of a species (spatial structure)	Species composition and richness
	Physiognomy
Age structure	Species traits
Population dynamics	*Community dynamics*
Natality, mortality, immigration and emigration	Succession
	Disturbance
Population interactions	*Community function*
Competition, herbivory, amensalism, commensalism and mutualism	Nutrient allocation and cycling
	Productivity and biomass allocation

- How does community composition change along spatial gradients?
- How does the addition (or loss) of one species affect the distribution or abundance of other?

What is Weed Ecology?

If ecology is the study of interactions between individuals and their environment, then the only thing that distinguishes weed ecology is that the organisms being studied are weeds. Therefore, weed ecologists ask questions such as:

- Are there specific characteristics or traits of weed populations?
- Do weeds function in a certain way within communities?
- Does the invasion by a weed change the community structure or function in a predictable way?
- What types of communities are easier to invade?

Why are ecology and weed science separate?

Ecology and weed science have developed as separate fields of study. Why is this? The way in which we study a topic is directly related to its historical development. Like familial lineages, there are academic lineages. There is ecology dogma and weed science dogma. There are accepted ways of asking questions, accepted experimental methodology and accepted statistical analyses. Breaking down these barriers is difficult.

To a certain extent, the types of people attracted to these two fields (ecology and weed science) will be different. Some people prefer asking 'applied research' questions while others prefer to ask 'pure science' questions. Ecologists are often biased towards working in natural environments, while weed scientists are often biased towards asking question that have applied 'real' answers. 'Were it not for the general predilections of ecologists to study only systems untouched by human hands, farming-systems research would clearly be called a branch of ecology' (Busch and Lacy, 1983).

The increasing interest in plant invasions into natural communities has expanded the middle ground between these fields. Such workers may ask ecologically based questions, but look for applied answers. For example, they may study the basic population ecology of a weed with an eye to eventually managing it with biological control, and thus both the ecology and weed science literature will be of interest to them. Scientists interested in agricultural and natural habitats may both be concerned with the same species. For example, garlic mustard (*Alliaria petiolata*) and dodder (*Cuscuta* spp.) invade natural and agricultural habitats. There is a renewed call to incorporate ecological thinking into applied fields of study such as weed science (Zimdahl, 1999b). We hope that this exchange of information will increase.

Summary

In this chapter, we have introduced the basic concepts of weed ecology. The term 'weed' is defined many ways; we prefer to use a loose definition that includes all plants that have a negative ecological or economic effect on natural or managed systems. Thus our view of 'weed ecology' is the study of how problematic plants interact with their biotic and abiotic environment. In this book, our goal is to understand why weeds occur where they do. We do not address how to get rid of them. In the next chapter, we begin by looking at plant populations. The first step towards investigating populations is to determine their distribution and abundance.

Questions

At the end of each chapter, you will be asked a series of questions related to a species of your choice. At this point, you should select a species that you wish to focus on. This may take some thought. Are you more interested in natural or managed systems? Are you interested in wide-ranging common weeds, locally problematic weeds or new species weeds? For some species, there will be a lot of literature available, while for others there may be large gaps in our knowledge. In the first case, you will have more information to read and synthesize. In the second case, you will be asked to suggest what information is needed and how this should be obtained. To get started, you may want to refer to a book on weeds in your region. It is a good idea to create a bibliography of references and resources you may need.

1. Name a plant that you would consider to be a weed but that someone else would not. Name a plant that you would not consider to be a weed but that someone else would. Explain how this is possible.

2. Describe why each characteristic listed by Baker (1956, 1974; Box 1.1) might be advantageous for an agricultural weed. Would each characteristic be equally advantageous for a weed in a natural habitat?

3. Why is it possible to define 'weed' in so many ways?

General References

Barbour, M.G., Burks, J.H., Pitts, W.D., Gilliam, F.S. and Schwartz, M.W. (1999) *Terrestrial Plant Ecology*, 3d edn. Benjamin, Cummings, California.

Crawley, M.J. (1997) *Plant Ecology*, 2nd edn. Blackwell Scientific, Oxford.

Zimdahl, R.L. (1999) *Fundamentals of Weed Science*, 2nd edn. Academic Press, San Diego, California.

Literature Cited

Alterieri, M.A. (1988) The impact, uses, and ecological role of weeds in agroecosystems. In: Altieri, M.A. and Liebman, M.Z. (eds) *Weed Management in Agroecosystems: Ecological Approaches.* CRC Press, Boca Raton, Florida, pp. 1–6.

Baker, H.G. (1965) The characteristics and modes of origin of weeds. In: Baker, H.G. and Stebbins, G.L. (eds) *The Genetics of Colonizing Species.* Academic Press, New York, pp. 147–172.

Baker, H.G. (1974) The evolution of weeds. *Annual Review of Ecology and Systematics* 5, 1–24.

Barbour, M.G., Burks, J.H., Pitts, W.D., Gilliam, F.S. and Schwartz, M.W. (1999) *Terrestrial Plant Ecology,* 3rd edn, Benjamin, Cummings, California.

Begon, M., Harper, J.L. and Townsend, C.R. (1990) *Ecology: Individuals, Populations, Communities*, 2nd edn. Blackwell Scientific, Boston, Massachusetts.

Bhandari, D.C. and Sen, D.N. (1979) Agroecosystem analysis of the Indian arid zone. I. *Indigo fera cordiflora* as a weed. *Agro-Ecosystems* 5, 257.

Binggeli, P. (1994) The misuse of terminology and anthropometric concepts in the description of introduced species. *Bulletin of the British Ecological Society* 25, 10–13.

Busch, L. and Lacy, W.B. (1983) *Science, Agriculture and the Politics of Research.* Westview Press, Boulder, Colorado.

Cavers, P.B. and Bough, M.A. (1985) Proso millet (*Panicum miliaceum* L.): a crop and a weed. In: White, J. (ed.) *Studies on Plant Demography: a Festschrift for John L. Harper.* Academic Press, London, pp. 143–155.

Chacon, J.C. and Gliessman, S.R. (1982) Use of the "non-weed" concept in traditional tropical agroecosystems of southeastern Mexico. *Agro-Ecosystems* 8, 1.

Crawley, M.J. (ed.) (1997a) *Plant Ecology*, 2nd edn. Blackwell Scientific, Oxford.

Crawley, M.J. (1997b) Biodiversity. In: Crawley, M.J. (ed.) *Plant Ecology*, 2nd edn. Blackwell Scientific, Oxford, pp. 595–632.

De Pietri, D.E. (1992) Alien shrubs in a national park: can they help in the recovery of natural degraded forest? *Biological Conservation* 62, 127–130.

Gouyon, P.H. (1990) Invaders and disequilibrium. In: di Castri, F., Hansen, A.J. and Debussche, M. (eds) *Biological Invasions in Europe and Mediterranean Basin.* Kluwer, Dordrecht, pp. 365–369.

Harper, J.L. (1957) Ecological aspects of weed control. *Outlook on Agriculture* 1, 197–205
Harper, J.L. (1977) *The Population Biology of Plants.* Academic Press, London.
Harper, J.L. (ed.) (1960) *The Biology of Weeds.* Blackwell Scientific, Oxford.
Hobbs, R.J. and Humphries, S.E. (1995) An integrated approach to the ecology and management of plant invasions. *Biological Conservation* 9, 761–770.
Holzner, W. (1982) Concepts, categories and characteristics of weeds. In: Holzner, W. and Numata, N. (eds) *Biology and Ecology of Weeds.* Dr W. Junk Publishers, The Hague, pp. 3–20.
Joenje, W. (1987) Remarks on biological invasions. In: Joenje, W., Bakker, K. and Vlijm, L. (eds) *The Ecology of Biological Invasions.* Proceedings of the Royal Dutch Academy of Sciences, Series C 90, pp. 15–18.
Kowarik, I. (1995) Time lags in biological invasions with regard the the success and failure of alien species. In: Pyšek, P., Prach, K., Rejmánek, M. and Wade, M. (eds) *Plant Invasions – General Aspects and Specific Problems.* SPB Academic Publishing, Amsterdam, pp. 15–38.
Le Floch, E., Le Houerou, H.N. and Mathez, J. (1990) History and patterns of plant invasion in northern Africa. In: di Castri, F., Hansen, A.J. and Debussche, M. (eds) *Biological Invasions in Europe and Mediterranean Basin.* Kluwer, Dordrecht, pp. 105–133.
Lodge, D.M. (1993) Biological invasions: lessons for ecology. *Trends in Ecology and Evolution*, 8, 133–137.
Luken, J.O. and Thieret, J.W. (1997) *Assessment and Management of Plant Invasions.* Springer-Verlag, New York.
Macdonald, I.A.W., Loope, L.L., Usher, M.B. and Hamann, O. (1989) Wildlife conservation and the invasion of nature reserves by introduced species: a global perspective. In: Drake, J.A., Mooney, H.A., di Castri, F., Groves, R.H., Kruger, F.J., Rejmánek, M. and Williamson, M. (eds) *Biological Invasions. A Global Perspective.* John Wiley & Sons, Chichester, UK, pp. 215–255.
Mack, R.M. (1985) Invading plants: their potential contribution to population biology. In: White, J. (ed.) *Studies on Plant Demography.* Academic Press, London, pp. 127–142.
Mooney, H.A. and Drake, J.A. (1989) Biological invasions: a SCOPE program overview. In: Drake, J.A., Mooney, H.A., di Castri, F., Groves, R.H., Kruger, F.J., Rejmánek, M. and Williamson, M. (eds) *Biological Invasions. A Global Perspective.* John Wiley & Sons, Chichester, UK, pp. 491–508.
Pimentel, D.A., Lach, L., Zuniga, R. and Morrison, D. (2000) Environmental and economic costs of nonindigenous species in the United States. *BioScience* 50, 53–65.
Prach, K. and Wade, P.M. (1992) Population characteristics of expansive perennial herbs. *Preslia* 64, 45–51.
Pyšek, P. (1995) On the terminology used in plant invasion studies. In: Pyšek, P., Prach, K., Rejmánek, M. and Wade, M. (eds) *Plant Invasions – General Aspects and Specific Problems.* SPB Academic Publishing, Amsterdam, pp. 71–81.
Radosevich, S.R., Holt, J.S. and Ghersa, C. (1997) *Weed Ecology: Implications for Management.* John Wiley & Sons, New York.
Rejmánek, M. (1995) What makes a species invasive? In: Pyšek, P., Prach, K., Rejmánek, M. and Wade, M. (eds) *Plant Invasions – General Aspects and Specific Problems.* SPB Academic Publishing, Amsterdam, pp. 3–13.
Salisbury, E.J. (1961) *Weeds and Aliens.* Collins, London.
Schwartz, M.W. (1997) Defining indigenous species: an introduction. In: Luken, J.O and Thieret, J.W. (eds) *Assessment and Management of Plant Invasions.* Springer-Verlag, New York, pp. 7–17.
Stirton, C.H. (1979) Taxonomic problems associated with invasive alien trees and shrubs in South Africa. In: *Proceedings of the 9th Plenary Meeting AETFAT.* pp. 218–219.
Trenbath, B.R. (1985) Weeds and agriculture: a question of balance. In: White, J. (ed.) *Studies on Plant Demography: A Festschrift for John L. Harper.* Academic Press, London, pp. 171–183.
Vitousek, P.M. and Walker, L.R. (1989) Biological invasion by *Myrica faya* in Hawaii: plant demography, nitrogen fixation, ecosystem effects. *Ecological Monographs* 59, 247–265.
Vitousek, P.M., Walker, L.R., Whiteaker, L.D., Mueller-Dombois, D. and Matson, P.A. (1987) Biological invasion by *Myrica faya* alters ecosystem development in Hawaii. *Science* 238, 802–804.
Williams, C.E. (1997) Potential valuable ecological functions of nonindigenous plants. In: Luken, J.O. and Thieret, J.W. (eds) *Assessment and Management of Plant Invasions.* Springer-Verlag, New York, pp. 26–34.
Zimdahl, R.L. (1999a) *Fundamentals of Weed Science*, 2nd edn. Academic Press, San Diego, California.
Zimdahl, R.L. (1999b) My view. *Weed Science* 47, 1.

Part I

Population Ecology

2
Describing the Distribution and Abundance of Populations

Concepts

- A population is a group of individuals of the same species found in the same place at the same time.
- Populations are characterized in terms of their distribution and abundance.
- The distribution of a species can be mapped using historical data, field observations and remote sensing.
- Individuals within a population will not be evenly distributed throughout their range.
- Abundance can be measured as frequency, density, cover or biomass.
- Abundance and distribution do not necessarily reflect a species' ecological impact.

Introduction

A population is a group of individuals of the same species found in the same place at the same time. Like many ecological terms, this definition is flexible, because it can be used to describe populations at many scales. For example, a population may be the number of individuals contained within a small area (e.g. a field) or it may refer to the local or regional distribution of the species. The first step in understanding any species is to document its distribution and abundance. This gives the researcher an idea of the scope of the potential problem (i.e. weediness). Note that we say *potential* problem: while distribution and abundance are useful information, more data must be obtained before a decision is made on a species' weediness.

In this chapter, we discuss how to describe a population's distribution and abundance. Distribution is a measure of the geographical range of a species, and is used to answer questions such as: 'Where does the species occur?', 'Where is it likely to occur?' and 'Where is it able to occur?'. Abundance is a measure of the number or frequency of individuals. It is used to answer questions such as: 'Is the number or frequency of individuals increasing or decreasing?'.

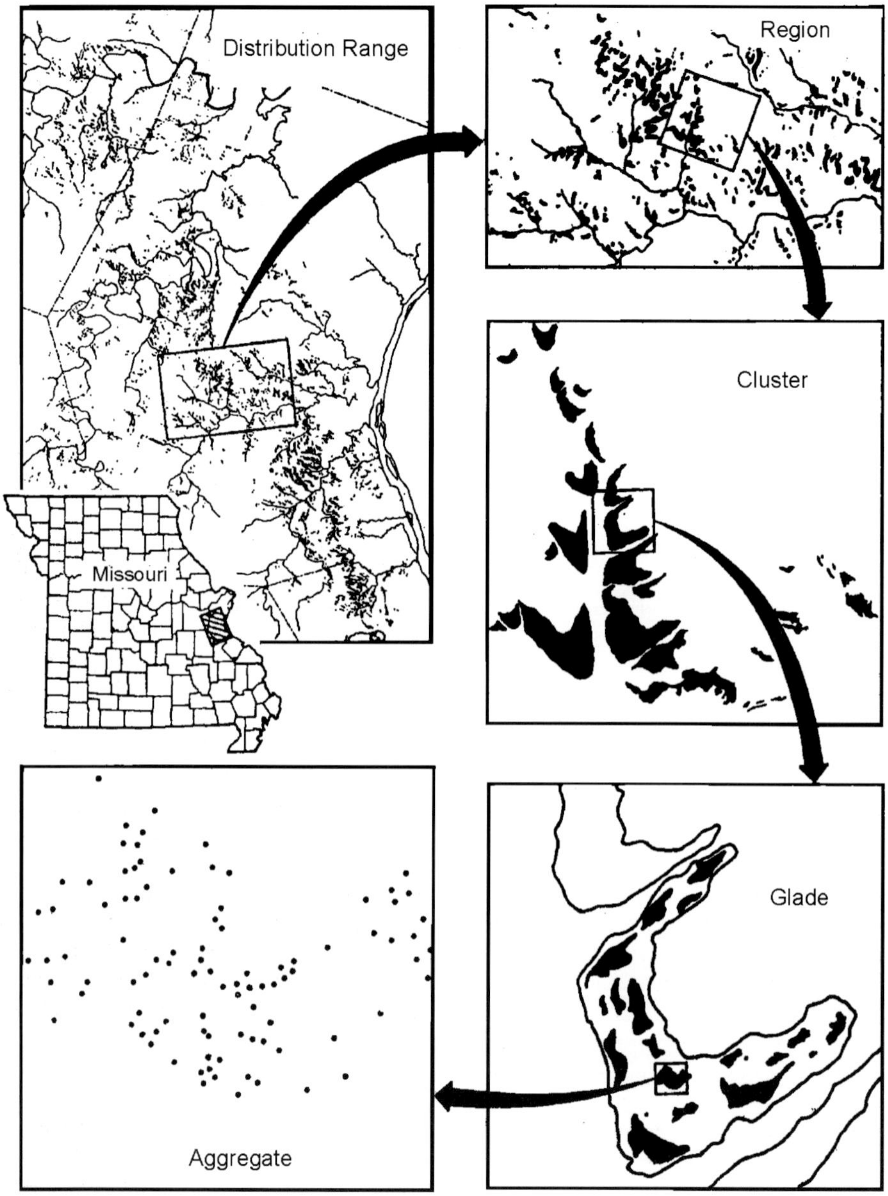

Fig. 2.1. Distribution of Fremont's leather flower (*Clematis fremontii* var *riehlii*) in the Ozarks of Missouri. Shown are distributions at the scale of range, region, cluster, glade and aggregate (Erickson, 1945; with permission of the Missouri Botanical Garden).

Population Distribution

A population's distribution (or range) describes where it occurs. In practical terms, it is a description of where the species has been recorded (Gaston, 1991). Mapping a species' distribution can be done on a number of scales depending on how the information is to be used. For example, Erickson (1945) mapped the distribution of the flowering shrub Fremont's leather flower (*Clematis fremontii* var. *riehlii*) at several scales. This species was restricted to approximately 1100 km^2 in the Missouri Ozarks (Fig. 2.1). Individuals, however, were not distributed throughout the species' range because they live only in sites where the abiotic and biotic conditions are suitable for them. For example, the range of Fremont's leather flower was subdivided into four watershed regions. Within these regions, there were groups of glades (rocky outcrops on south and west facing slopes), and clusters of Fremont's leather flower tended to be located at bases of these glades. Finally, within these clusters, there are loose aggregates of up to 100 individual plants.

Distribution maps have different uses depending on their scale. A researcher wanting to study the pollination of Fremont's leather flower would require a fine-scale distribution map showing the locations of individuals or colonies. Conversely, such a map would not be useful to a researcher interested in the broad-scale environmental controls of the species. They would require a large-scale map of the entire species' distribution.

Distribution change over time

A population's distribution will change over time either naturally or through human influence. Following the retreat of the last North American ice sheet (approximately 10,000 years ago) trees migrated northward, each species at a different rate and following a different route (Davis, 1981). At a smaller scale, a species distribution will change during the process of succession over decades (Chapter 13). Human disturbances, such as changing land-use patterns, will alter the environment such that different species are favoured and therefore population distributions will change. Also, human actions introduce exotic species and this increases their distribution. Thus, a species' distribution is not static; its boundaries are dynamic.

Asking what controls a population's distribution, and whether and why a species' distribution changes over time are fundamental questions of ecology. To better understand a weed species we might want to ask the following questions about its distribution:

- Is the weed at its current limit of distribution?
- Will the weed continue to expand into new locations?
- Is the weed found on specific soil types or land forms?
- Are there likely dispersal routes for this weed?

Distribution boundaries are limited by biotic (living, e.g. interactions with other species) and abiotic (non-living, e.g. temperature) factors. The same factor will not necessarily limit all boundaries of the range equally. For example, abiotic factors are more likely to limit distribution at higher latitudes, while biotic factors are more likely to limit distribution at lower latitudes (Brown *et al.*, 1996). Boundaries are rarely sharp, unless the population abuts against a geographic (e.g. river) or human-made feature (e.g. highway). Typically, individuals within the population become less and less frequent toward the limits of their range.

By following changes in a species' distribution over time, it is possible to tell whether a population is expanding or contracting. In the case of weeds, this may warn us where problems are likely to occur, or alternatively where control measures have been effective. We can also gain information on species' characteristics such as dispersal mechanisms or habitat preferences. Forcella and Harvey (1988) analysed how the distribution patterns of 85 agricultural weeds introduced into the northwestern USA changed between 1881 and 1980. They found that species' migration patterns were dependent on the species' point of entry and on the types of agriculture (e.g. grain, cattle) with which the weed was associated. Furthermore, migration patterns tended to follow land transportation routes. Similarly, Thompson *et al.* (1987) mapped the expansion of purple loosestrife (*Lythrum salicaria*) from 1880 to 1985 along canals, waterways and later along roads (Fig. 2.2). These examples give insight into how future introductions of new plant species might spread depending on their point of origin.

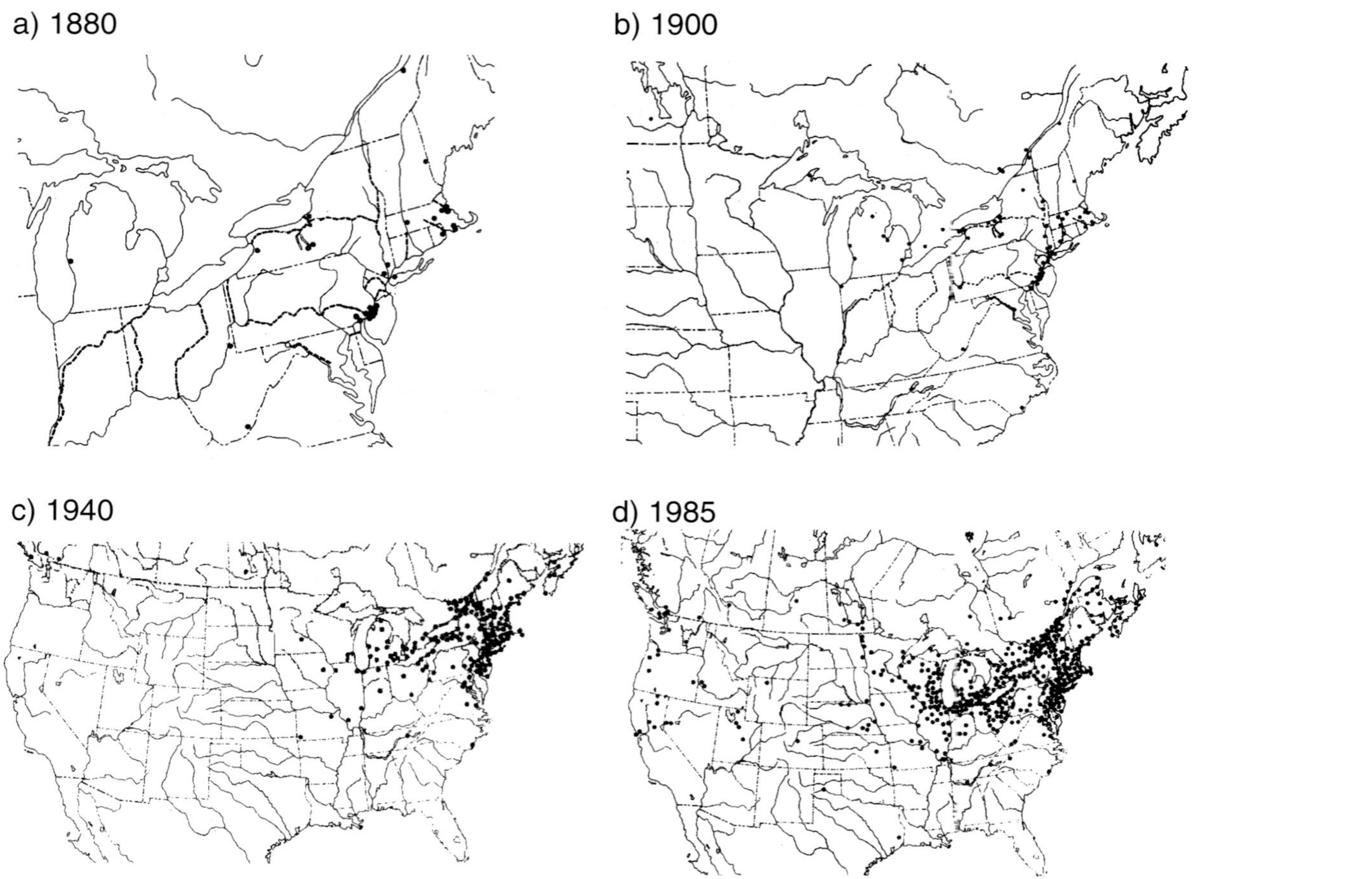

Fig. 2.2. Distribution of purple loosestrife (*Lythrum salicaria*) in North America in 1880, 1900, 1940 and 1985 (from Thompson *et al.*, 1987).

Estimating and mapping distribution

The traditional method for collecting data on the actual distribution of a species is to consult public records such as government documents, herbaria, field notes or academic journals. This type of data allows for the construction of historical distributions as was done by Thompson *et al.* (1987) and Forcella and Harvey (1988) (Fig. 2.2). These give a clear view of a species' regional distribution and change with time. Such records, however, are dependent on the accuracy and precision of the data collected, and this may be difficult to judge. Also, all sites and species will not be sampled equally and therefore, areas with less-intense sampling will be under-represented on maps (Schwartz, 1997). There will also be a sampling bias towards large or more obvious species. For example, purple loosestrife has large purple inflorescences and is more likely to be observed and recorded than a co-occurring weed, Japanese knotweed (*Polygonum cuspidatum*).

Field sampling and herbaria records give us information about the current or recent past distribution of species because

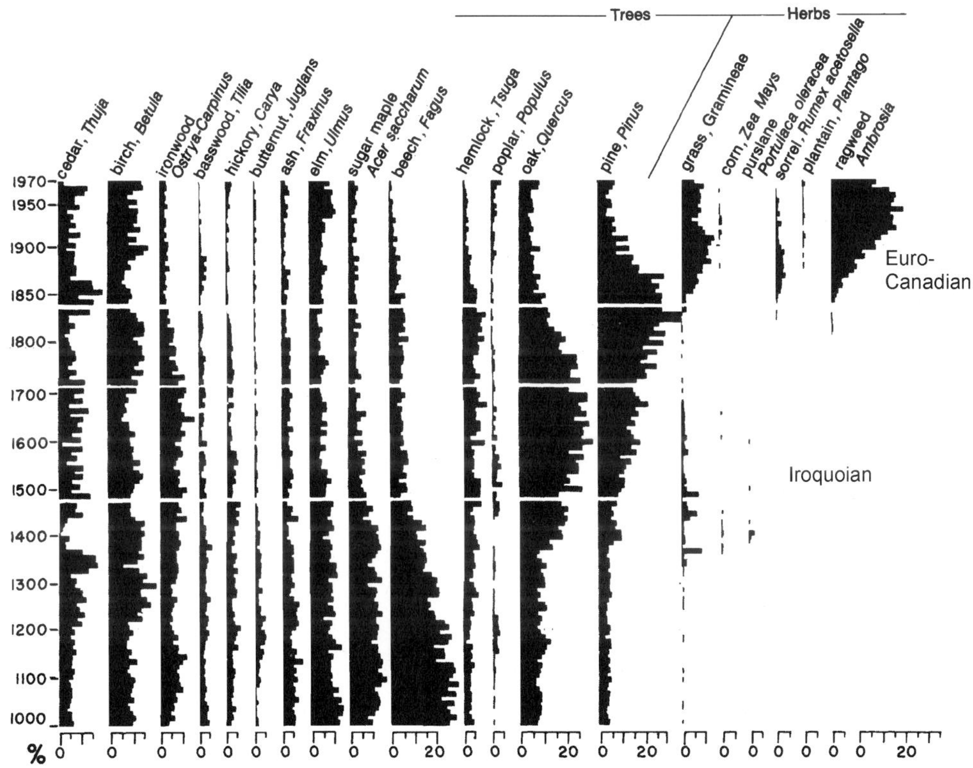

Fig. 2.3. Pollen diagrams of sediment taken from Crawford Lake, Canada. Shown is the per cent of pollen for each species. Note the increase in maize (*Zea mays*), purslane (*Portulaca oleraceae*) and grass (Gramineae) pollen during the Iroquoian period from 1360 to 1660, and the increase in ragweed (*Ambrosia*), dock (*Rumex*), and plantain (*Plantago*) pollen following land clearing by European settlers in 1820. (Adapted from McAndrews and Boyko-Diakonow 1989; with permission of the authors and the Minister of Public Work and Government Services Canada, 2002 and Courtesy of Natural Resources Canada, Geological Survey of Canada.)

these records may only go back a few hundred years. Thus, the initial invasions of some species cannot be tracked in this way. One possible method for tracing early introductions of species and their distribution changes is to use palaeoecological records. Microfossils, such as pollen grains and other plant parts, are preserved occasionally in peat or lakebed sediments. These can be retrieved and then identified (often to species level) to obtain a record of past vegetation. These records can be dated because the sediment is laid down in yearly layers, which can be radiocarbon dated. Changes in species composition over time can be traced by identifying pollen grains in successive layers of the sediment and constructing diagrams that show changes over time (Fig. 2.3). Using this method, extended time series can be constructed. Interestingly, this method has proven that some species previously thought to have been introduced to North America are actually indigenous. The pollen diagrams of Crawford Lake, Ontario, Canada, for example, show that purslane (*Portulaca oleracea*) was not an invasive weed from Europe as previously thought; in fact, it existed in the area from at least *c*. AD 1350 when the Iroquois began cultivating maize (Jackson, 1997).

Collecting field data can be a long, expensive process and therefore new methods to map the distribution of weeds are being developed. Such methods use remote sensing with either aircraft or satellite imagery. Photos or videos are taken to record the spectral reflectance of plants and ground terrain. To detect and map a species using this method, it must be possible to distinguish a species' reflectance pattern from the background of surrounding vegetation, ground, roads and other features. To date there has been some success mapping weeds of rangeland and pasture. Lass *et al.* (1996) were able to map the spatial distribution of common St John's wort (*Hypericum perforatum*) and yellow star thistle (*Centaurea solstitalis*) in rangeland using multispectral digital images taken from aircraft. Everitt *et al.* (1992) were also able to obtain area estimates of falsebroom (*Ericameria austrotexana*), spiny aster (*Aster spinosus*) and Chinese tamarisk (*Tamarix chinensis*) in rangelands and wild land of the southwestern USA. Remote sensing also has the benefit of covering large patches of land, so it can be used to follow the invasion of a species and monitor whether management practices are working. However, before it can be employed, we must have biological information about the species in order to be able to remote sense it properly and interpret the images. For example, a species' spectral reflectance pattern may change over its life cycle: we must know this in order to remote sense at the appropriate time. With advances in the technology, remote sensing may become applicable to more situations in the future.

Potential distribution

A species' ability to grow and reproduce can extend well past the boundaries of its natural or native distribution, because species distributions are not always limited by abiotic conditions. A species' distribution may be limited either by its inability to disperse to other sites or its inability to compete with other species. However, many species thrive after being artificially transplanted into a new habitat for agriculture or forestry. Many weeds were introduced purposefully and, once introduced, were able to rapidly expand their distribution. For example, kudzu (*Pueraria montana* var. *lobata*), which was introduced into the USA in 1876 as an ornamental vine and later used as a forage crop and for erosion control, is now considered to be a serious threat in the southeastern USA.

The area in which a species can (in theory) survive is its potential distribution (i.e. physiological distribution or climatic range). The potential distribution is based on the abiotic environment only and does not take into consideration how the species might survive in 'real situations' where, for example, it competes with other species. The potential distribution of a species may be far greater than its native distribution. For example, the natural distribution of Monterey pine (*Pinus radiata)* is limited to approximately 6500 ha in the coastal fogbelt of California and there have been attempts to place it on the 'threatened species' list in California. While the trees in the native range may be threatened, Monterey pine is also found all over the world and is a weed in some places. How can it be threatened in California, yet be found almost everywhere in the world and even be considered a weed? The answer is that in its native range, Monterey pine has been threatened by development, logging, changing weather patterns and diseases. However, humans have made the tree the most widely planted plantation tree in the world such that it covers over 4,000,000 ha (Clapp, 1995; Lavery and Mead, 1998). It is planted extensively in countries with habitats similar to California, e.g. New Zealand, Australia and Chile, where it is a fast-growing tree that can be harvested in 25-year rotations. Since it does not face the diseases that exist in its native California and is drought tolerant, Monterey pine is a weed in places with a Mediterranean climate and has invaded grasslands and native eucalypt forests (Richardson and Bond, 1991).

By comparing the native and potential distributions of a species, it may be possible to predict where it is likely to spread. The potential distribution of a species can be estimated in several ways. Patterson *et al.* (1996, 1997) estimated the potential distribution of a number of agricultural weeds using laboratory-based studies to determine the temperature and light conditions required by each species. From these data, they can create a mathematical model to predict where the right combinations of

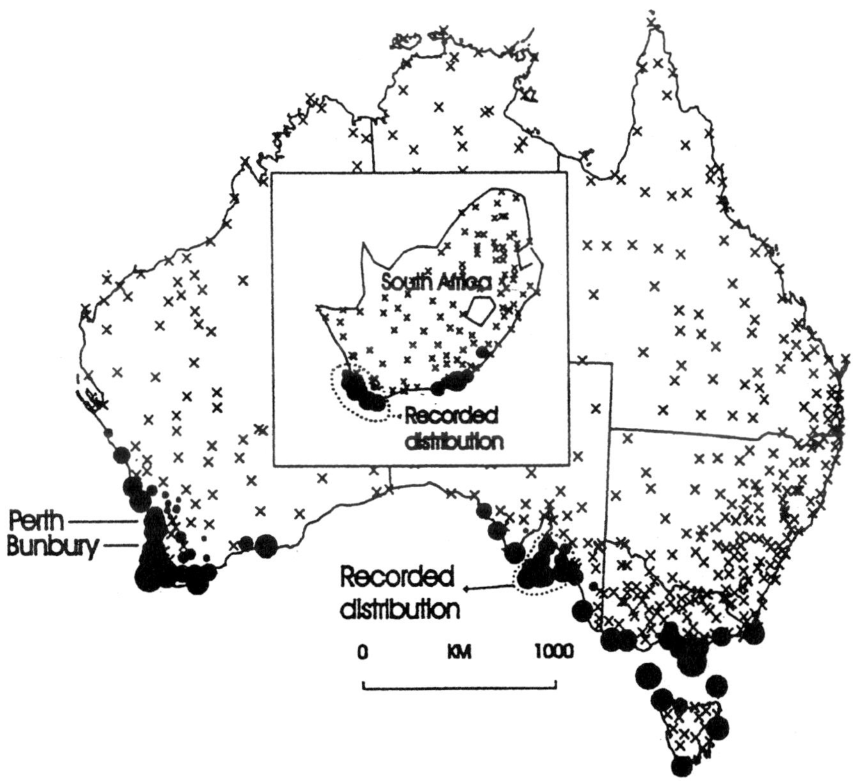

Fig. 2.4. Observed and predicted distributions of bridal veil (*Asparagus declinat*) in Australia. Solid dots indicate the predicted distribution while crosses indicate sites unsuitable to this species. Regions of known infestations are around Adelaide, Perth and Bunbery. The inset shows observed and predicted distributions of bridal veil in South Africa. (Pheloung and Scott 1996; with permission of R.G. and F.J. Richardson and P. Pheloung.)

conditions exist for the species to survive and reproduce. For example, after growing tropical soda apple (*Solanum viarum*) in growth chambers under a variety of day and night temperatures and photoperiods, Patterson *et al.* (1997) compared their results with climatic conditions in 13 southern states of the USA. They concluded that temperature and photoperiod were not likely to limit the expansion of this species and suggested that measures should be taken immediately to control the expansion of soda apple beyond its current distribution in Florida. This type of approach uses only abiotic factors that can be experimentally controlled, and it does not take into account seasonal temperature extremes or precipitation patterns (Patterson *et al.*, 1997).

An alternative way to predict a species' potential distribution is to compare the environmental conditions of the species' native habitat with those of a potential habitat. CLIMEX is one computer model suitable for this (Sutherst and Maywald, 1985). CLIMEX considers measures of growth such as temperature, moisture and daylength, and then adjusts this based on stress indicators such as excessive dry, wet, cold and heat, to give an ecoclimate index. Pheloung and Scott (1996) used CLIMEX to compare the distribution of bridal creeper (*Asparagus asparagoides*) and bridal veil (*Asparagus declinat*) (Fig. 2.4) in their native South Africa to potential habitats in Australia. They concluded that both species had the potential to continue spreading and that measures should be taken to control or eradicate them. Similarly, Holt and Boose (2000) were able to map the potential distribution of velvetleaf (*Abutilon theophrasti*) in California. They concluded that the distribution of velvetleaf was not likely to increase, because its range was limited by water stress.

Thus, potential distribution gives us an idea of the climatic regions where a species is able to survive the physical environment. This does not mean that the species will live there, because a species' distribution is controlled by non-climatic factors such as lack of dispersal or by interactions with other species.

Population Abundance

While distribution describes the geographical extent of a population, abundance describes a population's success in terms of numbers. Individuals will not be equally dispersed throughout their entire range; there will be areas of high and low density. Abundance can be measured in a number of ways. The type of measure selected will depend on the species in question, the habitat type (e.g. forest, field), the goal of the study and the economic resources.

Measures of abundance

Frequency and density

Frequency is the proportion of sampling units (e.g. quadrats) that contains the target species. It is easy to measure because only a species' presence or absence is noted for each quadrat. Frequency is a fast, non-destructive method and is less prone to incorrect estimates by the researcher. Density measures the number of individuals in a given area (e.g. square metre or hectare). It too is non-destructive, and while it is more complicated to measure, it provides more information than frequency.

While frequency and density are probably the most commonly used measures of abundance, there are some difficulties associated with using them. Density assumes that you are able to separate individuals. This is not a problem in higher animals because they are distinct individuals. In plants, however, many species are capable of reproducing vegetatively and therefore, it is often difficult to distinguish one genetic individual from another (see Chapter 5). Frequency does not have this problem.

A further difficulty in identifying individuals is that individuals of the same species may appear morphologically different depending on their age, stage of growth or environment. Many plants differ in appearance from one life stage to another (i.e. they are phenologically plastic). For example, a tree seedling will look very different from a mature adult. In addition, plants may be morphologically plastic: their appearance may differ depending on their environment. Leaves of aquatic plants often appear different depending on whether they are above or below the water, or leaves of terrestrial plants may differ depending on whether the leaf is produced in the sun or the shade. The variable appearance of a species may make it difficult to count. Therefore, measures of frequency and density might exclude individuals that are morphologically different and result in an underestimation of their abundance.

A final problem in using frequency and density as a measure of a population is that they do not distinguish between the sizes of individuals. Therefore, larger individuals are scored the same as smaller ones, even though they will have different influences on the community. Larger plants will probably have more effect on the physical environment (e.g. through shading) and they tend to produce more seed than smaller ones, thereby having a greater influence on subsequent generations. Therefore, frequency and density are better used when vegetation is of uniform size. Other measures of abundance such as cover and biomass can be used when an indication of size is desired.

Cover and biomass

Cover and biomass are sometimes used in place of frequency and density when an indication of individual size is important. Cover is the proportion of ground covered by a given species when viewed from above. Cover is useful when a non-destructive sampling method is required; however, it is sometimes difficult to quantify. It may be difficult to get an accurate value of cover because it is typically done as a visual estimate, so percentage cover estimation is often broadly categorized (e.g. 0%, 1–5%, 5–10%, 10–25%, 25–50%, 50–75% and 75–100%). Measuring cover is subjective and therefore not precise; however, this method is widely used and considered valuable because it provides useful information with relatively low effort by the researcher.

Biomass is the weight of vegetation per area. Biomass is useful when an accurate indication of plant size is needed. It is sampled usually by collecting the shoots and roots from a given area. When collecting, the plant can also be divided into roots, stems, leaves and reproductive structures to observe how plants allocate biomass to different structures. Collecting actual plant samples to determine biomass is not practical for larger organisms such as trees. Therefore, some mathematical equations have been developed to calculate biomass based on size. For example, we may harvest several plants of varying height to establish if there is a correlation between height and biomass. If there is, then height can be measured instead of harvesting the plant. For trees, stem diameter at breast height (dbh) is often taken as a measure of tree size.

Spatial Distribution of Individuals Within a Population

Within a population, individuals are not distributed evenly throughout their range. Individuals can be arranged at random, in clumps or in a regular pattern. These distribution patterns are the result of the abiotic environment, seed dispersal patterns, the species' biology, interactions among species or management practices. When we measure population abundance, it is an estimate of the average value over the entire area. It is important to consider spatial arrangement within a population, especially when determining effects of weeds on crops or on natural communities. Early studies on the effect of weeds on crop yield loss assumed that weeds were randomly distributed; however, it is now clear that this may not be so (Hughes, 1990; Cardina *et al.*, 1997). Crop yield loss due to weeds will be overestimated if weed distribution is not taken into account (Auld and Tisdel, 1988). If weeds are clumped in a few areas of the field, then crop loss estimates for the entire field will be lower than if they were randomly distributed. Another field with the same overall density, but a more random distribution of weeds, will probably have more yield loss.

Problems of Predicting Weediness Based on Distribution and Abundance

Purple loosestrife has been characterized frequently as an invasive species and certainly the distribution and abundance of purple loosestrife has increased dramatically since it was first introduced into the New England states in the mid-1800s (Fig. 2.2) (Thompson *et al.*, 1987). What has not been documented, however, is the effect that this species has on the native vegetation. Just because a species is increasing in distribution and 'appears' to be a dominant species does not mean that it is having a pronounced effect on plant communities. In reality, there is surprisingly little evidence to indicate that purple loosestrife is, in fact, an aggressive weed that has negative effects on other plant populations (Anderson, 1995; Hager and McCoy, 1998). The conspicuous appearance of this plant acts against it, because subjective observation will overestimate its abundance and underestimate the abundance of less conspicuous species. Other species that may disrupt the shoreline component of ecosystems may be more pernicious and problematic but less attention has been given to these species, in favour of the more obvious purple loosestrife (e.g. Japanese knotweed; see Chapter 1).

Summary

The first questions to ask when considering a potential weed problem are: 'Where is it?' 'How abundant is it?' and 'How is it spatially distributed?'. The answers to these questions allow us to characterize the distribution and abundance of the weed. These are important first steps towards understanding the ecology of species, but they are not necessarily good indicators of the species' influence on other populations or on the community as a whole. While we gain some information about whether a species is increasing or decreasing from abundance and distribution data, we need to go further to understand fully the dynamics of a weed and whether it will affect other populations. Although the concepts in this chapter are simple, they are important. If incorrectly applied they could lead to the conclusion that a weed is a problem when in fact, it is not. In the next chapter we begin to 'go further' and look at population structure and dynamics. Individuals within populations are not all identical: they differ in age, size, sex and developmental stage. We look at the repercussions of population structure.

Questions

1. Using the species you selected in Chapter 1, research its distribution. Map the distribution of your species using the appropriate scale (e.g. field, regional, continental) of map. What resources other than maps are available? Consider the following questions:

- At what scale do we know the species' distribution?
- Can we follow changes in its distribution status over time?
- What types of data were used to construct this map?

2. For each of the following environments, which method of estimating abundance (density, cover, biomass or frequency) would be best and why? (i) A natural forest, (ii) planted woodlot, (iii) a maize field, and (iv) a pasture.

3. Why is it important to consider spatial distribution of a weed within: (i) a field of maize, (ii) a natural forest?

4. By understanding abundance and distribution, how would you determine the ecological impact of a weed?

General References

Brown, J.H., Stevens, G.C. and Kaufman, D.W. (1996) The geographic range: size, shape, boundaries, and internal structure. *Annual Review of Ecology and Systematics* 27, 597–623.

Gaston, K.J. (1991) How large is a species range? *Oikos* 61, 434–437.

Weber, E. (2001) Current and potential ranges of three exotic goldenrods (*Solidago*) in Europe. *Conservation Biology* 15, 122–128.

Literature Cited

Anderson, M.G. (1995) Interactions between *Lythrum salicaria* and native organisms: a critical review. *Environmental Management* 19, 225–231.

Auld, B.A. and Tisdel, C.A. (1988) Influence of spatial distribution of weeds on crop yield loss. *Plant Protection Quarterly* 3, 81.

Brown, J.H., Stevens, G.C. and Kaufman, D.W. (1996) The geographic range: size, shape, boundaries, and internal structure. *Annual Review of Ecology and Systematics* 27, 597–623.

Cardina, J., Johnson, G.A. and Sparrow, D.H. (1997) The nature and consequences of weed spatial distribution. *Weed Science* 45, 364–373.

Clapp, R.A. (1995) The unnatural history of the Monterey pine. *Geographical Review* 85, 1–19.

Davis, M.B. (1981) Quarternary history and stability of forest communities. In: West, D.C., Shugart, H.H. and Botkin, D.B. (eds) *Forest Succession: Concepts and Application.* Springer-Verlag, New York, pp. 131–153.

Erickson, R.O. (1945) The *Clematis freemonlii* var. *riehlii* populations in the Ozarks. *Annals of the Missouri Botanical Garden* 32, 413–459.

Everitt, J.H., Escobar, D.E., Alaniz, D.E., Villarreal, R. and Davis, M.D. (1992) Distinguishing brush and weeds on rangelands using video remote sensing. *Weed Technology* 6, 913–921.

Forcella, F. and Harvey, S.J. (1988) Patterns of weed migration in Northwestern U.S.A. *Weed Science* 36, 194–201

Gaston, K.J. (1991) How large is a species range? *Oikos* 61, 434–437.

Hager, H.A. and McCoy, K.D. (1998) The implications of accepting untested hypotheses: a review of the effects of purple loosestrife (*Lythrum salicaria*) in North America. *Biodiversity and Conservation* 7, 1069–1079.

Holt, J.S. and Boose, A.B. (2000) Potential for spread of velvetleaf (*Abutilon theophrasti*) in California. *Weed Science* 48, 43–52.

Hughes, G. (1990) The problem of weed patchiness. *Weed Research* 30, 223–224.

Jackson, S.T. (1997) Documenting natural and human caused plant invasions with paleoecological methods. In: Luken, J.O. and Thieret, J.W. (eds) *Assessment and Management of Plant Invasions.* Springer-Verlag, New York, pp. 37–55.

Lass, L.W., Carson, H.W. and Callihan, R.H. (1996) Detection of yellow thistle (*Centaurea solstitalis*) and common St John's wort (*Hypericum perforatum*) with multispectral digital imagery. *Weed Science* 10, 466–474.

Lavery, P.B. and Mead, D.J. (1998) *Pinus radiata*: a narrow endemic from North America takes on the world. In: Richardson, D.M. (ed.) *Ecology and Biogeography of Pinus.* Cambridge University Press, Cambridge, pp. 432–449.

McAndrews, J.H. and Boyko-Diakonow, M. (1989) Pollen analysis of varved sediment at Crawford Lake, Ontario: evidence of Indian and European farming. In: *Quaternary Geology of Canada and Greenland.* Geology of Canada, No. 1. Geological Survey of Canada, Ottawa, pp. 528–530

Patterson, D.T. (1996) Temperature and photoperiod effects on onionweed (*Asphodelus fistulosus*) and its potential range in the United States. *Weed Technology* 10, 684–689.

Patterson, D.T., McGowan, M., Mullahey, J.J. and Westbrooks, R.G. (1997) Effects of temperature and photoperiod on tropical soda apple (*Solanum viarum* Dunal) and its potential range in the U.S. *Weed Science* 45, 404–418.

Pheloung, P.C. and Scott, J.K. (1996) Climate-based prediction of *Asparagus asparagoides* and *A. declinatus* distribution in Western Australia. *Plant Protection Quarterly* 11, 51–53.

Richardson, D.M. and Bond, W.J. (1991) Determinants of plant distribution: evidence from pine invasions. *American Naturalist* 137, 639–668.

Schwartz, M.W. (1997) Defining indigenous species: an introduction. In: Luken, J.O. and Thieret, J.W. (eds) *Assessment and Management of Plant Invasions*. Springer-Verlag, New York, pp. 7–17.

Sutherst, R.W. and Maywald, G.F. (1985) A computerized system of matching climates in ecology. *Agriculture, Ecosystems and Environment* 13, 281–299.

Thompson, D.Q., Stuckey, R.L. and Thompson, E.B. (1987) Spread, impact, and control of purple loosestrife (*Lythrum salicaria*) in North American wetlands. Fish and Wildlife Research Report No. 2, US Department of Interior, Fish and Wildlife Service, Washington, DC.

3
The Structure and Dynamics of Populations

Concepts

- Populations are dynamic – they change over time, space and with the environment.
- Population change over time is related to rates of birth, death, immigration and emigration.
- Populations interact across space. A group of spatially isolated, conspecific populations that occasionally interact through migration of seeds or pollination is called a metapopulation.
- Individuals within a population are unique; they vary in their age, size, stage of development, and other physical and genetic features. This variation gives a population structure.
- Life history strategies are a way of understanding a population.

Introduction

In Chapter 2, we discussed ways of describing populations in terms of their distribution and abundance. Populations were treated as whole entities. We then discussed the spatial distribution of individuals within a population, and how this would influence estimates of distribution and abundance. For the most part, we treated individuals as identical entities. Populations, however, are made up of individuals that vary in age, size, genetic structure (genotype) and appearance (phenotype). As a result, populations are structured by this variation. Population structure refers to the organization of individuals within a population, based on specific characteristics. For example, in a human population we could compare the age structure of men and women.

Demography is the study of population size and structure, and how they change over time. Populations are also dynamic: their size and structure change over time. Population size refers to the total number of individuals or the density of individuals within a specific population. A change in population structure will affect population dynamics; as population size increases or decreases, the structure will be affected. In this chapter, we will first look at how population size changes over time. We then look

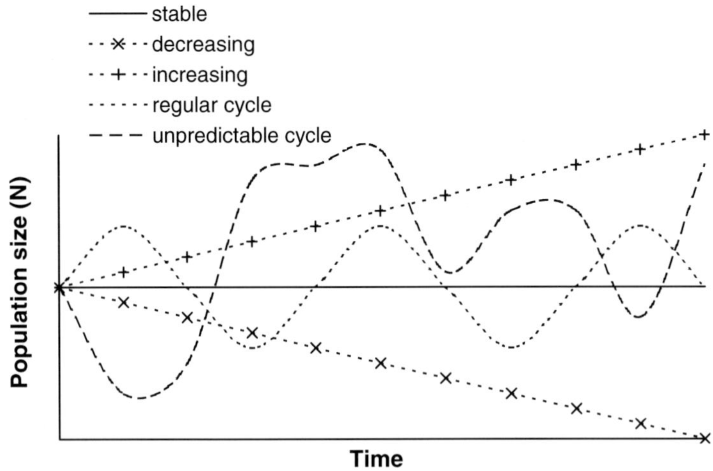

Fig. 3.1. Population size changes over time.

at how immigration and emigration can influence a population's demography. Third, we examine the different ways that populations can be structured. Finally, we look at life history strategies.

Population Dynamics: Size Changes over Time

In nature, a population's size will rarely remain constant. Within a short time frame, population size may remain stable, steadily increase or decrease, or it may cycle regularly, or in an unpredictable fashion (Fig. 3.1). The rate of population change is dependent on the ratio of individuals entering the population through births (*B*) or immigration (*I*) to individuals leaving through deaths (*D*) or emigration (*E*). Thus, the change in a population's size (*N*) from one time period (*t*) to the next (*t*+1) can be represented by the equation:

$$N_{(t+1)} = N_t + B - D + I - E$$

Birth (or natality) is the addition of individuals to the population. For plants, births may refer either to the number of seeds pro-

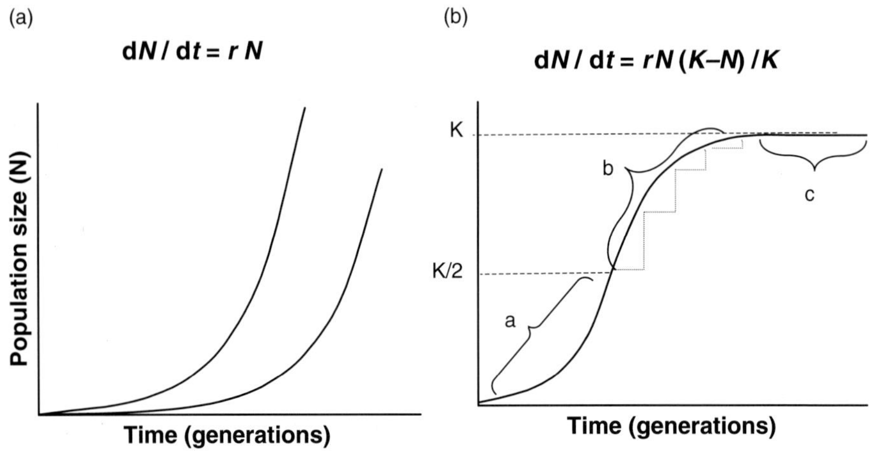

Fig. 3.2. The (a) exponential and (b) logistic growth curves.

duced or seeds germinating (Chapter 6), or to individuals produced via vegetative reproduction (Chapter 5). Mortality is the loss of individuals from the population through death. Mortality rates and causes will change over time. In the following sections we look at population growth curves, first using the exponential and logistic models of growth and then by looking at real populations.

Exponential and logistic growth curves

As long as births outnumber deaths (ignoring immigration and emigration), population growth will be positive. Over generations, a population with a constant positive growth rate will exhibit exponential growth (Fig. 3.2a). The greater the difference between birth rate and death rate, the more rapid the increase. The difference between birth rate and death rate is the instantaneous rate of population increase (r). Therefore the exponential population growth can be shown as:

$$dN/dt = rN \text{ or } N_{t+1} = Nt\, e^{rt}$$

where dN/dt is the change in N during time(t).

Many plants have the potential to produce a huge number of offspring. This is especially true for some weeds where a single individual may produce more than 1,000,000 seeds per season (Table 3.1). Given that plants produce so many seeds, why then do their populations not continue to increase exponentially? Many seeds will not be viable, while others will not germinate because environmental conditions are not appropriate, or because the seed dies due to predation or disease. In spite of this, there can still be many viable seedlings produced per adult plant. During the early stages of population growth, density may increase exponentially (Fig. 3.3), but at some point, the growth will slow and density may even begin to decrease. Why is this so? Exponential growth cannot be maintained because populations are limited by a lack of resources. At some point there will not be enough resources (e.g. nutrients, light or space) to satisfy the needs of every new individual and so population density will level off.

The logistic curve is a model of population growth under limiting resources. Once a seed germinates, there are many biotic and abiotic forces that cause mortality and reduce population growth rate. For example, each seedling requires resources (space, nutrients, water, light) to survive, and individuals that fail to acquire adequate resources will fail to reproduce or may die.

The lack of adequate resources will cause the population growth curve to level off. The growth of all populations will eventually level off. The carrying capacity (K) is the maximum number of individuals the environment can support. To incorporate K into the population growth equation, the exponential equation can be modified by including an additional term that causes the growth rate to level off. It looks like this:

Table 3.1. Plant size and seed production of various weed species (from Holm *et al.*, 1977).

Species	Common name	Plant height (cm)	Seeds per plant (number)
Amaranthus spinosa	Spiny amaranthus	to 120	235,000
Anagallis arvensis	Scarlet pimpernel	10–40	900–250,000
Chenopodium album	Common lambsquarters	to 300	13,000–500,000
Digitaria sanguinalis	Large crabgrass	to 300	2000–150,000
Echinochloa crus-galli	Barnyardgrass	to 150	2000–40,000
Eleusine indica	Goosegrass	5–60	50,000–135,000
Euphorbia hirta	Garden spurge	15–30	3000
Polygonum convolvulus	Wild buckwheat	20–250	30,000
Solanum nigrum	Black nightshade	30–90	178,000
Striga lutea	Witchweed	7–30	50,000–500,000
Xanthium spinosum	Spiny cocklebur	30–120	150

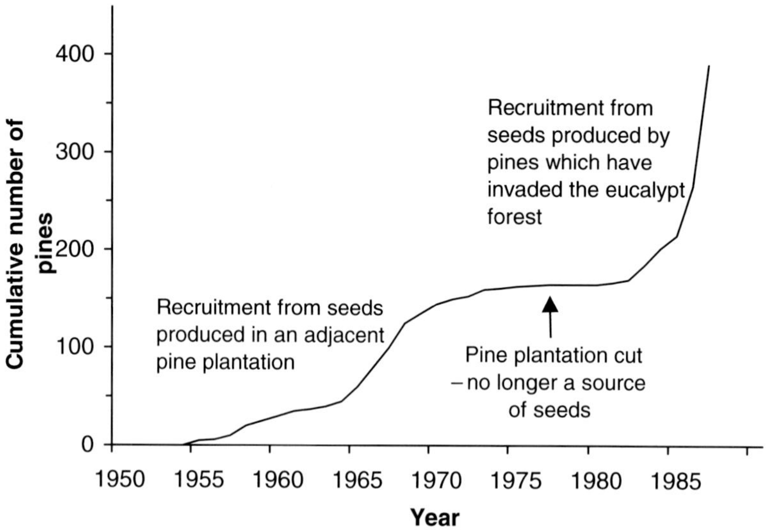

Fig. 3.3. Increase in Monterey pine (*P. radiata*) in a eucalypt dry sclerophyll forest in Australia. Initially pine recruitment occurred from seed imported from an adjacent pine plantation. After 1980, recruitment rate increased even though the pine plantation was cut because pines established in the eucalypt forest were becoming mature and producing seed (redrawn from Burdon and Chilvers, 1994).

$$dN/dt = rN\,(K\text{-}N)/N$$

This is the logistic growth-curve equation which incorporates limits to population growth over time. When population density (N) is less than K, the term $(K\text{-}N)/N$ will be positive and population growth will be positive. As the value of N approaches K, the rate of growth decreases until N=K when the rate of population growth (dN/dt) becomes zero. The population size is stable because births equals deaths at this time.

There are three parts to the logistic growth curve (Fig. 3.2b). Initially, population size increases at an exponential rate. The maximum rate of growth occurs at half the value of K. Beyond this, the rate of population increase slows down but is still positive. This occurs because not all individuals will be affected by limiting resources at the same time because of differences in size, age, health and reproductive status. Over time, the proportion of individuals affected by limiting resources will increase and this causes the curve to level off at K.

Real population growth curves

The exponential and logistic growth curves are idealized mathematical descriptions of how population size will change over time. They provide a conceptual framework on which to base more complex approaches to population growth. In real situations, population growth is more variable over time (Fig. 3.4). There are a number of reasons why population size fluctuates over time. We will address a few here and you will see other examples in the rest of this text.

- The logistic growth model assumes that the environment is stable over time and therefore K remains stable. This is unrealistic because the abiotic environment is naturally variable: temperature, nutrients, water and light change over time. Even small changes in one factor can affect the number of individuals the environment can support.
- There is random variation in birth and death rates. This is termed demographic stochasticity. An occasional low birth rate or high death rate can cause the population to become extinct.

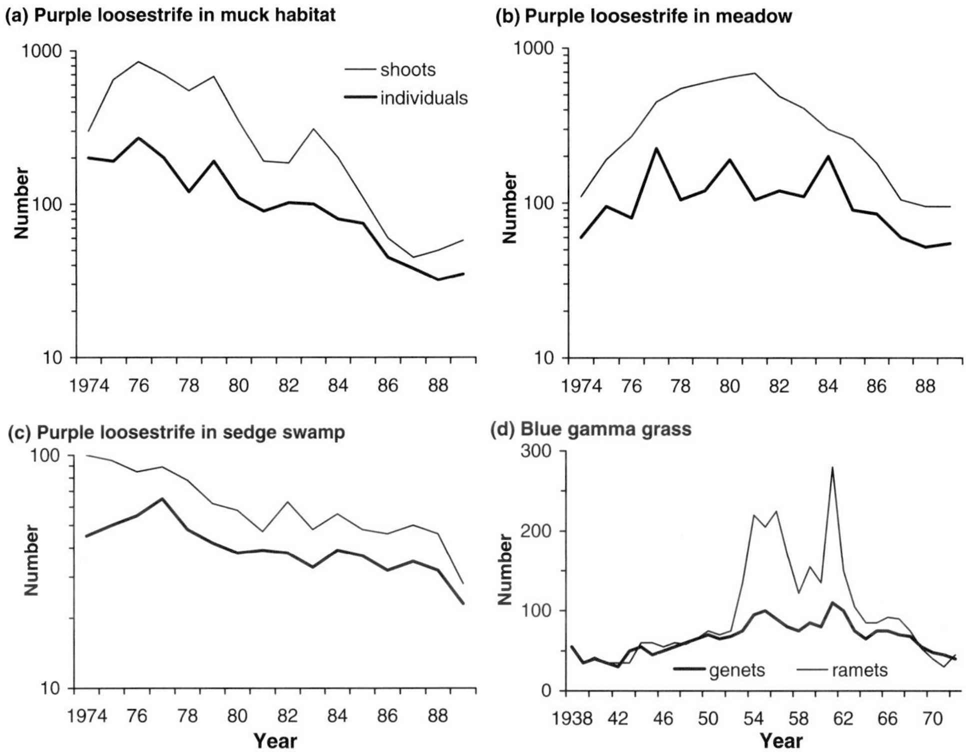

Fig. 3.4. Examples of plant population changes over time showing: (a), (b), (c) the number of individuals and shoots of purple loosestrife (*L. salicaria*) in three habitat types (Falińska, 1991), and (d) the mean number (number per/5 m^2) of ramets and genets of blue grama grass (*Bouteloua gracilis*) (Fair *et al.*, 1999).

- The logistic and exponential growth curves assume that populations are independent of other populations. Populations, however, interact (through competition, herbivory) and this causes population size to fluctuate. Population interactions are addressed in Chapters 8 and 9.

Effects of Migration (Immigration and Emigration) on Population Size

Sometimes it may be possible to ignore the effects of immigration and emigration (migration) by assuming that they are equal, or that their effect on population size is negligible. However, many will be dependent on the immigration of individuals from other populations. A population with fewer births than deaths will remain viable only when supported by seeds imported from other populations.

Migration demographically links populations. Determining whether migration is an important demographic process has two problems. First, the concept of migration assumes that there are specific boundaries over which individuals (seeds) move. In human populations we have political boundaries, so we can keep track of the movement of (most) individuals. As discussed in Chapter 2, plant population boundaries are rarely discrete. Second, even if 'real' boundaries do exist, tracking the movement of individuals can be challenging. Therefore, it is difficult to establish if migration is occurring.

Migration among populations: creating metapopulations

Traditionally, populations have been described as a collection of individuals that are capable of interbreeding. In reality, most populations are scattered and clustered into smaller subgroups. This clustering may be a random process but it usually reflects the heterogeneity of the landscape, i.e. there are a limited number of areas where individuals of various species can live and these individuals cluster in amenable habitats. When populations become divided into clusters, we can say that each cluster becomes spatially isolated from each other. If spatially isolated populations interact through migration (e.g. of seed) or distant pollination, then the aggregate of interacting populations is called a 'metapopulation'. The implication of using the term 'metapopulation' is that interactions among populations are not always common but they do occur.

Each population within a metapopulation will likely be genetically distinct because each is adapted to local environmental conditions. Although individuals within a population will mostly mate with individuals from their own population, metapopulation dynamics will introduce some genetic material from surrounding populations. Since the continued existence of a population is determined mainly by whether there is enough local genetic variation to withstand environmental change (including diseases, herbivory, drought) and ensure births exceed deaths, metapopulation dynamics may prevent the extinction of local populations. For example, immigrants (or at least their genetic material via pollen) from other populations can help maintain a population that otherwise would become extirpated (locally extinct) because it is not genetically suited to changes in its environment (e.g. decreasing light levels). Populations that are maintained only

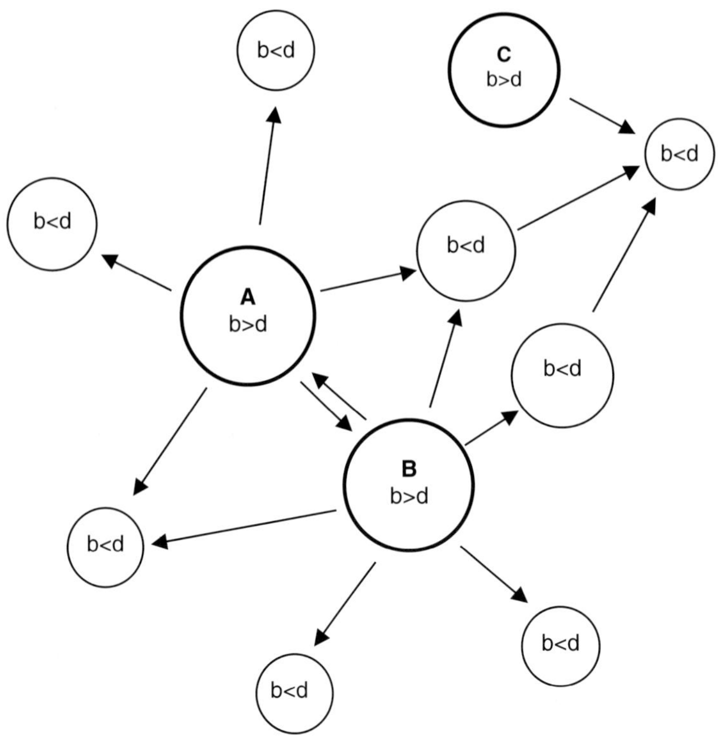

Fig. 3.5. Metapopulation dynamics: population patches may be a source (bold) or a sink for seeds (or other propagules).

through immigration from other (source) populations are called 'sink' populations (Pulliam, 1988) (Fig. 3.5). In weedy white campion (*Silene alba*), for example, isolated populations survive only because new genetic material arrives via immigration from surrounding populations – in this case, the immigrant genetic material is delivered via pollen (Richards, 2000; see Chapter 4 on pollination).

Perhaps the most important aspect of the metapopulation concept is the implication for conservation. Because populations may contain relatively few individuals, be restricted to a small area or have low genetic variation, they are subject to local extinction. However, the metapopulation is usually persistent because local adaptations in populations increase the total amount of genetic variation. If the landscape-scale environment changes suddenly, the chances are good that at least one population has the genes needed to allow for recolonization of habitats vacated by local extinctions. This means that should a disease or a drought strike, then some of the populations will survive. Over time, this means that the local habitats that populations occupy often are 'emptied' and recolonized many times. Therefore while local populations may go extinct and the habitats emptied, the metapopulation of a species will continue. This has become important in understanding how to conserve species. It is possible

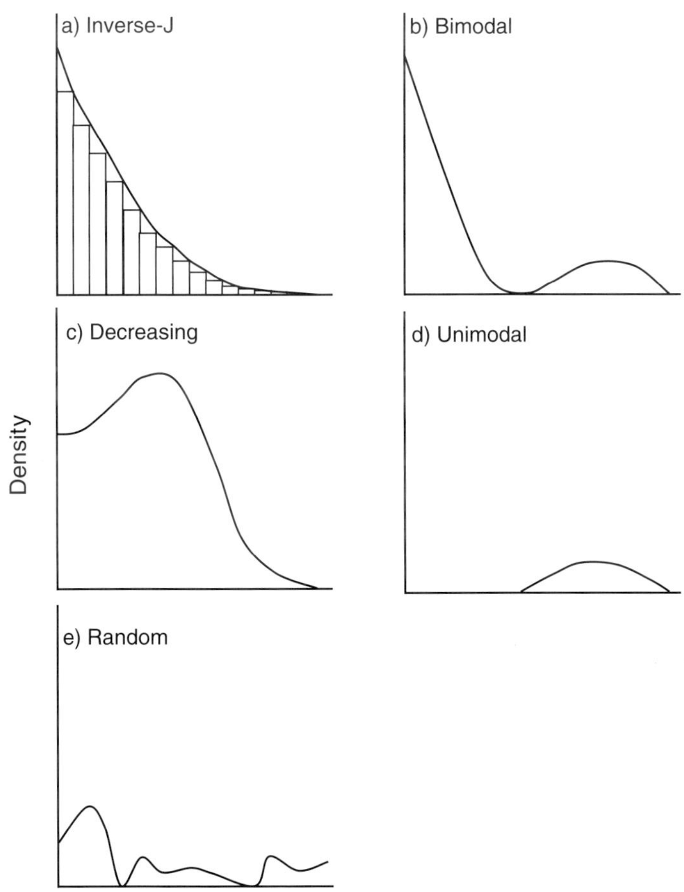

Fig. 3.6. Theoretical age structure distribution used to assess population trends. The *x*-axis is the age class and the *y*-axis is the tree density (redrawn from Whipple and Dix, 1979).

that a large contiguous reserve that does not allow for spatial isolation, local adaptation, and development of a metapopulation can actually hasten extinction of a species, as it is vulnerable to sudden environmental change (Beeby, 1994; Hanski and Gilpin, 1997; Schwartz, 1997; Honnay *et al.*, 1999; Etienne and Heesterbeek, 2000).

Population Structure

Populations are characterized based on the age, size, appearance or genetic structure of individuals. In fact, population structure could be based on any characteristic that is variable within a population. Population structure is not a static feature of a population because individuals age, grow, reproduce and die at different rates depending on their individual characteristics and their environment. In this chapter, we focus on age, size and developmental stage structure of populations.

Age structure

The distribution of ages within a population can be characteristic of the species itself, or it can reflect the 'health' of the population, or the environment inhabited by the population. In a 'healthy' population, younger individuals will outnumber older individuals because a proportion of young individuals will die before they reach maturity. Whipple and Dix (1979) proposed five age-class distributions to explain population trends of trees (Fig. 3.6). The 'inverse-J' curve shows a population with many more juveniles than adults; this population is likely to be relatively constant or increasing. The 'bimodal' distribution is a result of pulse recruitment (addition of new individuals) where periods of lower recruitment are followed by periods of higher recruitment. This population will likely be stable or increase as long as recruitment pulses are frequent enough to replace dying individuals. A 'decreasing' population distribution means the population is not replacing itself because recruitment is not high enough to replace those that are dying. If recruitment is zero the distribution will become 'unimodal' as the population ages and no young individuals are added. Although individuals are present, the population will become extinct unless increased reproduction occurs. Finally, a random distribution is typical of a population in a marginal habitat, or one that is responding to disturbance. Populations that have recently invaded a site are also likely to exhibit this distribution (Luken, 1990).

Age structures can be difficult to interpret because they do not always fit the theoretical distributions described above, nor are they consistent over time. Montana populations of spotted knapweed (*Centaurea maculosa*) tended to have inverse-J distributions in 1984, but in 1985 the distribution decreased (Fig. 3.7). This occurred following a severe drought in 1984, when young individuals experienced higher mortality than older individuals (Boggs and Story, 1987). While overall population density decreased by 40% between 1984 and 1985, the density of younger individuals (years 2 and 3) was reduced by 83%. This resulted in a change of age structure from one year to the next. The observed structure of a population is the result of abiotic and biotic forces encountered by previous generations of a population. It is important for scientists tracking changes in population density to be aware of age structure, because future changes in abundance depend very much on the current age distribution. As seen in spotted knapweed, harsh conditions may differentially affect age groups causing demographic changes.

There are complications, however, associated with characterizing populations based solely on age structure data. First, seeds that are persistent in the soil (seed bank) are seldom accounted for when assessing age structure of a population. The seed bank represents potential individuals that replenish the population when no new seeds are produced. Therefore, a population with no apparent seed production ('unimodal') may increase again via the seed bank rather than through renewed seed production. Second, not all plant species can be aged accurately

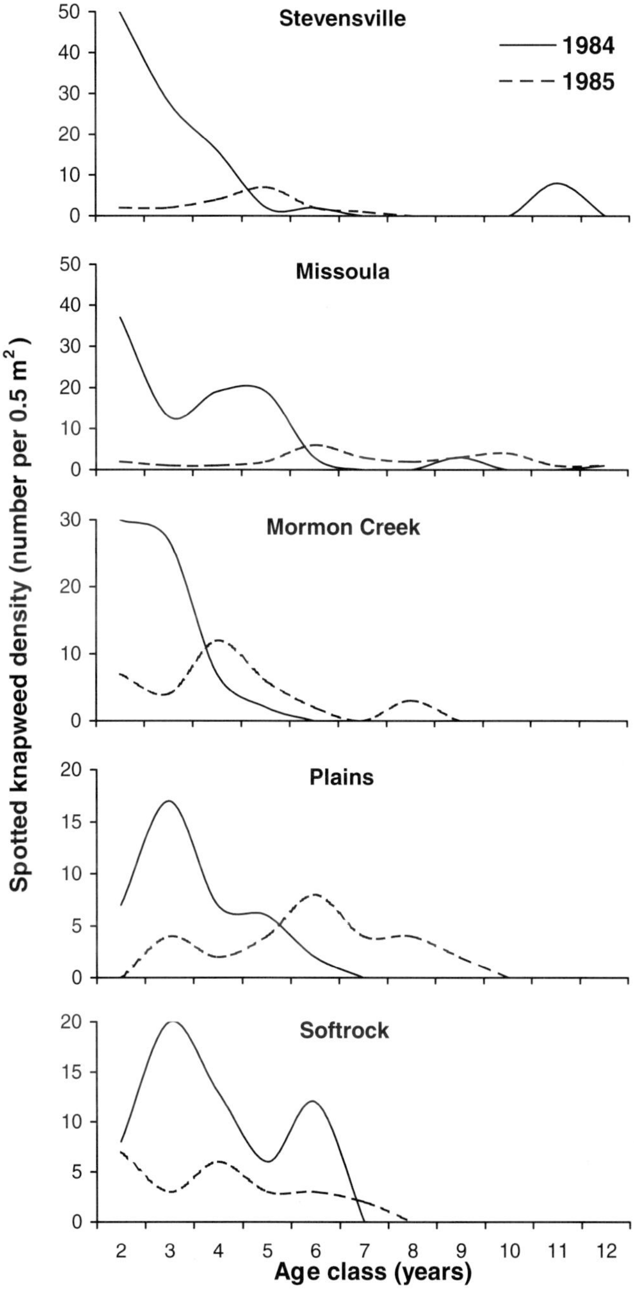

Fig. 3.7. Population age structure of spotted knapweeed at five sites in Montana (with individuals <1 year removed) (adapted and redrawn from data in Boggs and Story, 1987).

and so age structure data may be suspect. Woody species (most trees and some shrubs) are easier to age than herbaceous species because they often produce annual growth rings which can be counted; however, not all woody plants produce annual rings, and some produce more than one ring in a year. Species producing multiple main stems will also be difficult to age. Some woody species can be aged by counting morphological features such as bud scars. Annual rings in the roots of some herbaceous perennials can also be used (Boggs and Story, 1987; Dietz and Ullman, 1998).

A third problem with using age structure data to characterize populations is that age may not be biologically relevant to population processes such as reproduction, growth or death (Werner, 1975). Two genetically identical individuals of the same age may differ physically depending on their environment and this will influence when they reproduce, the number of offspring they produce and when they die.

Size structure

Most populations will tend to have fewer large individuals and many smaller ones. However, larger individuals can have a disproportionate effect on the rest of the population because they tend to live longer and produce more offspring than smaller individuals of the same age (Leverich and Levin, 1979). Larger individuals can also directly affect smaller individuals through shading. Plant size is a measure of the success of an individual because the larger individuals have acquired more resources than smaller individuals. For this reason, it is often more useful to structure populations by size rather than age. Furthermore, size may be a better predictor of an event (e.g. reproduction or death) than age (Werner, 1975; Werner and Caswell, 1977; Gross, 1981). Werner (1975) found that rosette size of teasel (*Dipsacus fullonum*) was a better predictor than age of whether a plant remained a vegetative rosette, flowered or died. For example,

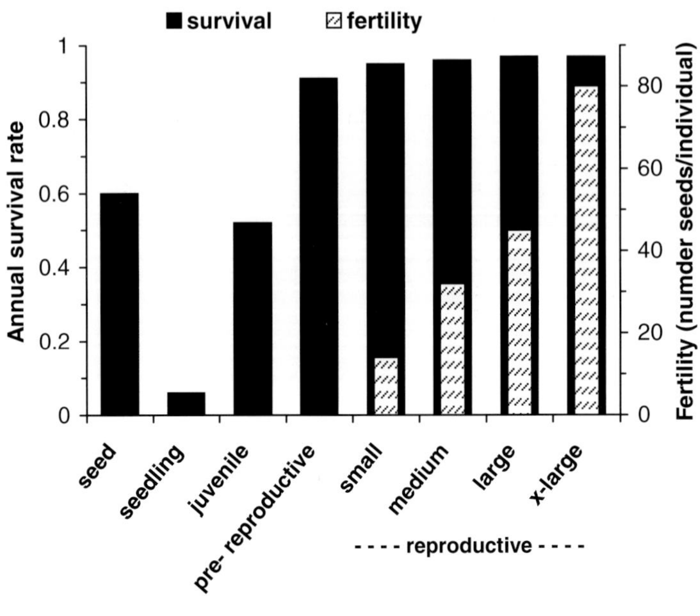

Fig. 3.8. Annual survival and fertility (number of seeds per individual) of prayer plant (*Calathea ovandensis*). Individuals were classified into five stage classes (seed, seedling, juvenile, pre-reproductive and reproductive) with four size classes of reproductives (small, medium, large and extra large) (redrawn from data in Horvitz and Schemske, 1995).

rosettes attaining 30 cm in diameter had an 80% chance of flowering. Still, size is not a perfect predictor of life cycle events. An example of this is when small, repressed agricultural weeds flower even when they are tiny compared with their neighbours.

The simplest way to measure 'size' is to measure some visible aspect of growth such as plant height, diameter (e.g. of the stem), or number or size of leaves. Biomass is a more exact measure of size because it is a more direct measure of acquired resources, but biomass measurements require harvesting, drying and weighing the plant, and is a destructive sampling method.

A strong linear correlation between size and age rarely exists for many reasons. Some species of trees (e.g. sugar maple, *Acer saccharum*) remain as slow-growing or suppressed individuals for decades until a canopy gap appears, after which they grow rapidly (Canham, 1985). Alternatively, plants may grow rapidly during the early life stages until they reach a maximum size and then divert resources to reproduction and maintenance rather than growth. Size structure also develops in shorter-lived species. In jewelweed (*Impatiens capensis*), size structure developed because larger individuals grew faster and had a lower risk of death than smaller ones (Schmitt *et al.*, 1987). One should never assume that age and size are correlated until the relationship has been tested.

Phenology

A plant's phenology (stage of development) can be used in conjunction with or instead of plant age and size to examine population structure (Sharitz and McCormick, 1973; Werner and Caswell, 1977; Gatsuk *et al.*, 1980; Horvitz and Schemske, 1995; Deen *et al.*, 2001). This measure may be more biologically meaningful than age or size alone because an individual's phenological stage may be more linked to its likelihood of survival or reproduction. Horvitz and Schemske (1995) showed that the annual

Table 3.2. Life table of Drummond phlox (*P. drummondii*) (adapted from Leverich and Levin, 1979).

Age at start of interval (days) x	Length interval (days)	No. surviving on day x n_x	Survivorship l_x	No. dying during interval d_x	Mean mortality rate/day m_x
0	63	996	1.00	328	0.0052
63	61	668	0.67	373	0.0092
124	60	295	0.30	105	0.0059
184	31	190	0.19	14	0.0024
215	16	176	0.18	2	0.0007
231	16	174	0.17	1	0.0004
247	17	173	0.17	1	0.0003
264	7	172	0.17	2	0.0017
271	7	170	0.17	3	0.0025
278	7	167	0.17	2	0.0017
285	7	165	0.17	6	0.0052
292	7	159	0.16	1	0.0009
299	7	158	0.16	4	0.0036
306	7	154	0.15	3	0.0028
313	7	151	0.15	4	0.0038
320	7	147	0.15	11	0.0107
327	7	136	0.14	31	0.0325
334	7	105	0.11	31	0.0422
341	7	74	0.07	52	0.1004
348	7	22	0.02	22	0.1428
355	7	0	0	–	–

survival and fertility of the prayer plant (*Calathea ovandensis*), varied depending on the individual's phenological stage (Fig. 3.8). Seedlings had less than 10% survival, seeds and juveniles had moderate survival while other stage classes had over 90% survival. Reproductive individuals produced different numbers of seeds per plant depending on their size.

Illustrating population structured data

Data on age structure can be tabulated into life tables (Table 3.2). These tables summarize age-specific survival, mortality and reproductive rates. Survival data is used to construct survivorship curves that display the proportion of individuals surviving to the beginning of each age class (Fig. 3.9). Survivorship curves are easier to interpret when presented on a log scale because they show constant mortality rate as a straight line. A steeper slope indicates a higher mortality rate.

Pearl and Miner (1935) presented three general survivorship curves (Fig. 3.10). These model curves are often referred to as Deevey curves after Deevey (1947). Type I is typical of species, such as some human populations, with low early mortality, and high mortality later in the life span. A Type II curve shows a constant mortality rate throughout the life span. Some birds have this type of survivorship curve. A Type III curve has high early mortality that decreases later in the life span. This is typical of many plant species where seedling mortality is very high (e.g. agricultural weeds). When presented on an arithmetic scale, the curves appear different. The methodology for collecting and calculating survivorship data will be explained in Chapter 10.

Why does population structure matter?

Interpreting population structure can be difficult and time consuming. Why, then, do we do it? Why not simply calculate population means (e.g. mean age or height) and use these simple numbers to describe a popula-

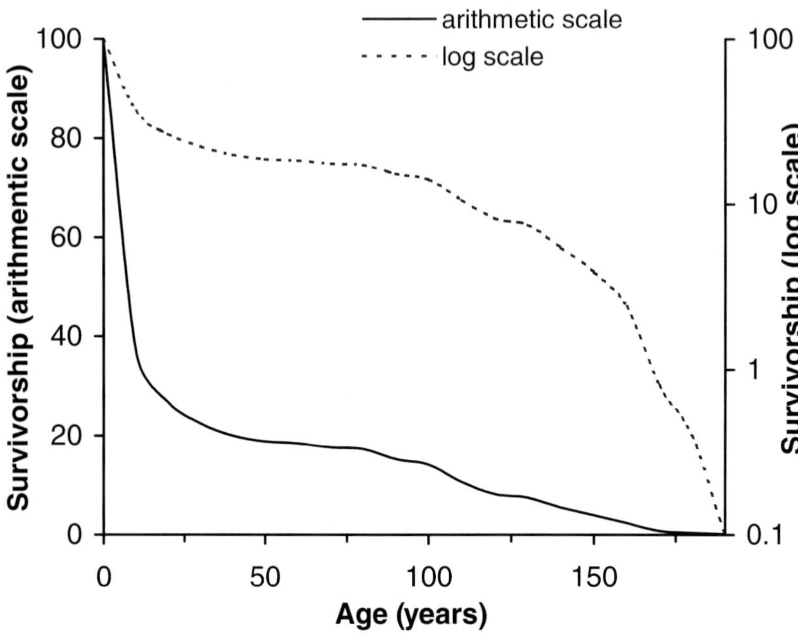

Fig. 3.9. Survivorship curve of Drummond phlox (*Phlox drummondii*) shown on arithmetic and logorithmic scales (data from Leverich and Levin, 1979).

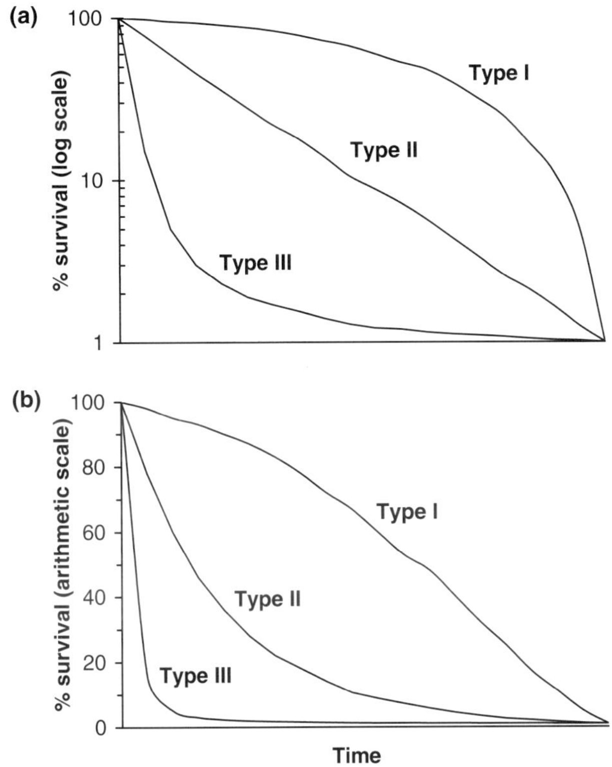

Fig. 3.10. Idealized survivorship curves shown on: (a) log scale, and (b) arithmetic scale. Type I shows low early mortality and high late mortality. Type II shows constant mortality rate over time. Type III shows high early mortality and low mortality late in life.

tion? The answer is that there is a lot of valuable information in the variability of a population and that by reducing this to a mean value we lose information (Hutchings, 1997). Populations may have the same mean values but different structure; for example, the four populations shown in Fig. 3.11 have the same mean stem diameter (x=20 cm) but the proportion of individuals in each size class differs. The fate of these populations will most likely differ.

Another reason we may be interested in the structure of a population is that it is a way to identify specific individuals of interest. For example, we may only be only interested in plants of a certain size (age or stage). If we know, for example, that only individuals above a specific size will impact crop yield, then we can focus our research on the larger age classes. Recognizing population structure helps us to focus on specific individuals within a population, and it gives us a glimpse of possible population dynamics to come.

Life History Strategies in Plants: Population Structure and Life Cycles

A strategy is 'a grouping of similar or analogous genetic characteristics which recurs widely among species or populations and causes them to exhibit similarities in ecology' (Grime *et al.*, 1990). The term 'strategy' is sometimes criticized because it is anthropomorphic (has human attributes) and teleological (has a purpose) (Grime *et al.*, 1990); however, few suitable alternative terms exist

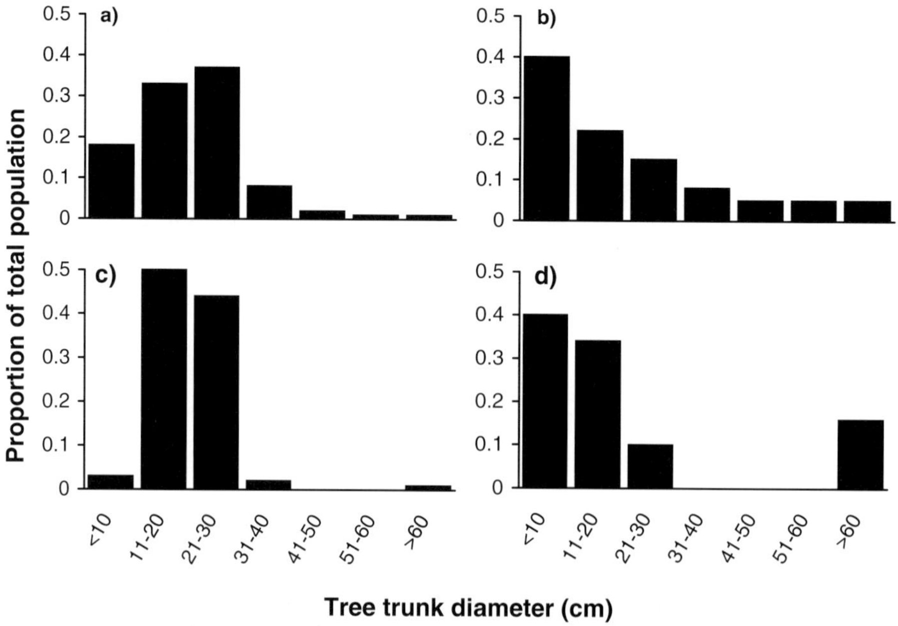

Fig. 3.11. Age structure of four imaginary populations, each with the same mean age ($\bar{x}$ = 20), but differing age distributions.

(e.g. 'set of traits', 'syndrome'). Life history comprises both the general description of a life cycle of a plant (annual, biennial, perennial) and all of the more specific aspects of life cycles within population (age, stage, size).

Many individuals, populations and species have adapted or been genetically constrained to adopt similar life histories to survive. Such common patterns of life histories suggest that there are general 'life history strategies'. There is no one optimum plant strategy that maximizes survival in all situations. If there were, there would be only one plant species. Environmental conditions vary drastically over time and space, and therefore different traits will be favoured in different situations and at different times. Based on life history strategies, we can make general predictions about what traits are likely to allow individuals, populations or species to exist under different environmental conditions.

r- and K- selection

One way to classify plants by life history strategy is to refer to them as being '*r*' or 'K' selected (e.g. Beeby, 1994). Following disturbances, the species that will recolonize most rapidly are generally small annuals that have a rapid growth rate, reproduce early and produce many small seeds. This set of traits allows the species to arrive, germinate, establish and reproduce quickly. Therefore, if further disturbance occurs, there will be seed available to re-establish. This specific set of traits is called an '*r*-strategy'; the '*r*' refers to the high intrinsic rate of population growth displayed by species with this strategy. In situations where disturbance is infrequent, and environmental conditions are relatively stable, traits such as large size, longevity, delayed reproduction are favoured. Plants with this set of traits are 'K-strategists' because the populations are theoretically maintained at or near the carrying capacity (K). Table 3.3 summarizes characteristics found in *r*- and K-strategists.

Table 3.3. Features of *r*- and K-selected species (adapted from Pianka, 1970).

	r-selected species	K-selected species
Climate	Unpredictable and/or variable; uncertain	Predictable or constant; more certain
Mortality	Occasional catastrophic mortality, density dependent	Mortality rate lower and more constant, density independent
Survivorship	Usually Deevy Type III	Usually Deevy Type I or II
Population size	Variable over time, often below carrying capacity	Constant over time, often at or near the carrying capacity
Life span	Short, usually <1 year	Long, usually >1 year
Body size	Small	Large
Competition	Often low	Often intense
Rate of development	Rapid	Slow
Reproduction	Usually early, monocarpic	Usually late, polycarpic
Offspring	Produce many, small offspring	Produce few, large offspring
Leads to ...	Productivity	Efficiency

Many plants cannot be divided neatly into *r*- or K-strategies because they represent ends on a continuum; most plants actually use both strategies in the appropriate environmental conditions. Although valuable as a tool, it is naïve to use *r*- and K-selection as the sole criteria in predicting the potential colonization ability or the weediness of a plant.

The contrast between *r*- and K-selection is clearly illustrated by two different varieties of barnyardgrass (*Echinochloa crus-galli*) in California (Barrett and Wilson, 1983). *E. crus-galli* var. *crus-galli* has numerous, small dormant seeds. This allows it to survive in unpredictable, heterogeneous habitats and hence it is more cosmopolitan. *E. crus-galli* var. *oryzicola* does not exhibit dormancy; it has large seeds that germinate with the rice crop (*Oryza sativa*), and large, vigorous seedlings. It is, therefore, more K-selected as it is adapted to homogeneous, predictable environments (rice paddies) and it is the more noxious variety of weed in rice paddies. However, it is restricted to this habitat and is less of a problem worldwide than *E. crus-galli* var. *crus galli*.

Agricultural weeds are commonly characterized as being *r*-selected. These weeds are adapted to frequent disturbance through tillage, herbicides or other agronomic practices. Their life span is short, reproduction is early, fecundity is high and seeds are small (Pianka, 1970). Nevertheless, it would be wrong to state that all agricultural weeds are *r*-selected. There is a degree of stability in the regularity of disturbance, and so some K-selected species also persist. Such species may be perennial weeds with polycarpic reproduction, and few seeds with abundant nutrient reserves. With the increase of no-till farming, K-selected weeds may increase in agriculture systems (Swanton *et al.*, 1993; Buhler *et al.*, 1994).

A weed may be anywhere on the spectrum between *r*- and K-selected. For example, johnsongrass (*Sorghum halpense*) and cocklebur (*Xanthium strumarium*) are two of the world's 'worst' agricultural weeds; however, johnsongrass is K-selected and cocklebur is *r*-selected (Holm *et al.*, 1977; Radosevich and Holt, 1984). Additionally, despite being *r*-selected (in general), cocklebur is an effective competitor (for water) and undergoes both early and late germination, characteristics not traditionally associated with *r*-selected species (Pianka, 1970; Scott and Geddes, 1979).

C-S-R selection

Because many plants may exhibit a 'compromise' of *r*/K-selected attributes, a modified theory of plant strategy and selection was developed (Grime, 1977, 1979). Grime used characteristics of the established phase of the life cycle to characterize plants based

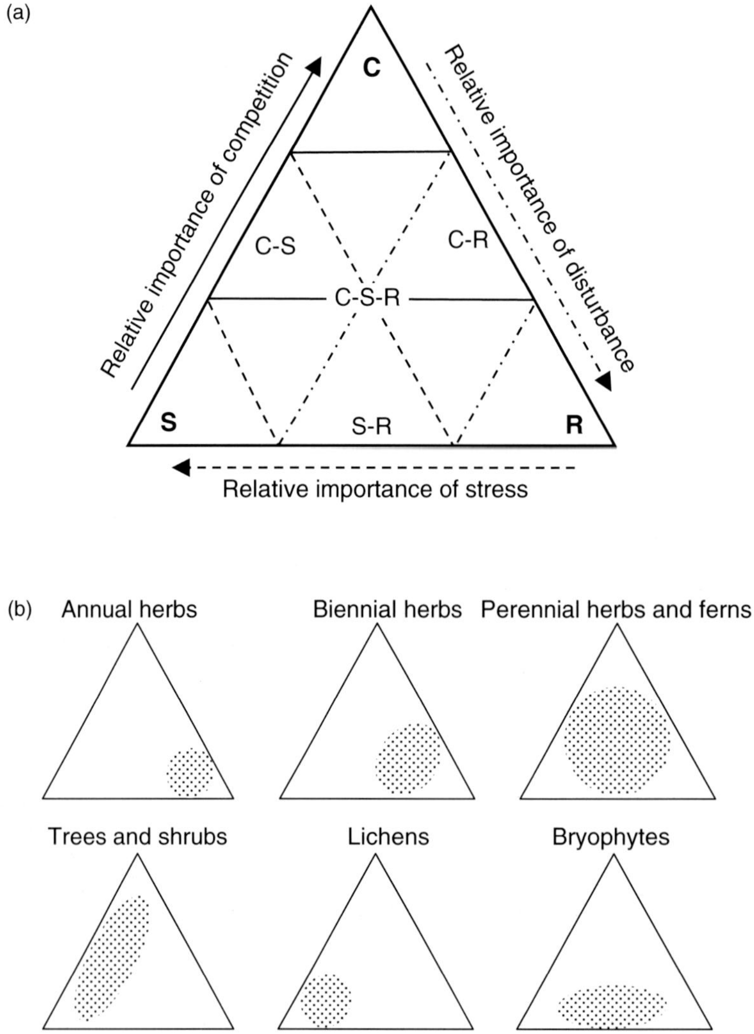

Fig. 3.12. The C-S-R model showing: (a) the location of the three main strategy types (C= competitors, S= stress tolerators, R= disturbance tolerant ruderals) and secondary strategies, and (b) the placement of various types of vascular and non-vascular plants along the three axes (redrawn from Grime, 1977).

on their ability to withstand competitors, disturbance and stress. In his triangular conceptual model, the corners represent ruderals (disturbance tolerators) (R), competitors (C) or stress-tolerators (S) (Fig. 3.12a). C-strategists maximize resource capture in undisturbed but productive habitats by increasing vegetative production and reducing allocation to reproduction. R-strategists maximize reproduction and growth, and are adapted to disturbed but potentially productive environments. These two strategies are somewhat analogous to K- and *r*-selection, respectively. The S-strategists are adapted to stressful, harsh environments where disturbance is rare and competition is unimportant. By reducing vegetative growth and reproduction they maximize their survival.

Characterization of C, S and R species is

Table 3.4. Characteristics of competitive, stress-tolerant and ruderal plants (adapted from Grime, 1977).

	Competitive C	Stress-tolerant S	Ruderal R
Morphology			
Life forms	Herbs, trees, shrubs	Lichens, herbs, trees, shrubs	Herbs
Morphology	Leaves form high, dense canopy, extensive lateral spread of roots and shoots	Variable	Small stature, little lateral spread
Leaf form	Robust	Often small, leathery or needle-like	Various
Life history			
Longevity of established phase	Variable	Long	Short
Longevity of leaves and roots	Relatively short	Long	Short
Frequency of flowering	Usually every year	Variable	Produced early in life history
Annual production allocated to seeds	Small	Small	Large
Structures persisting in unfavourable conditions	Dormant buds and seeds	Stress-tolerant leaves and roots	Dormant seeds
Regeneration strategies	Vegetative growth, small seeds, persistent seed bank	Vegetative growth, persistent seedling bank	Small seeds, persistent seed bank
Physiology			
Maximum potential relative growth rate	Rapid	Slow	Rapid
Response to stress	Rapid response to maximize vegetative growth	Slow, limited response	Rapid response to divert from vegetative growth to flowering
Storage of mineral nutrients from photosynthesis	Into vegetative structures, some stored for new growth in following season	Storage in leaves, stems, and/or roots	Seeds
Other			
Litter	Copious, often persistent	Sparse, sometimes persistent	Sparse, not usually persistent
Palatability to unspecialized herbivores	Variable	Low	Variable, often high

based on a plant's morphology, physiology, life history and other traits (Table 3.4). Intermediate species are shown in the central region of the triangular model (Fig. 3.12b). Weeds are usually classified as ruderals (R), or competitive ruderals (CR). Both strategies are adapted to productive habitats, but CR-strategists would be found in less frequently disturbed habitats than R-strategists who have short life spans which allow species to re-establish after disturbance. While Grime's strategies have been discussed widely in reference to weed species, some have pointed out its limitations (Tilman, 1987). Grime's model relies on a narrow definition of competition (Grace, 1991). This will be dealt with in the next chapter on competition.

Summary

Describing population dynamics, population structures, life cycles and life history strategies is difficult because of genetic and environmental variation and the complex interactions and combinations that can occur. This complexity is the reason why our convenient measures and descriptions of populations are often not adequate even if they do a reasonable job of approximating the real world. This complexity explains why:

- simple logistic and exponential equations do not adequately describe populations;
- spatial isolation within metapopulations influences survival and conservation decisions;
- classifying plant population structure by age, growth stage, size and life cycle can be difficult; and
- life history strategies are good rules of thumb but not all that accurate in predicting the population dynamics and impact of plants, especially weeds.

Population dynamics and structure are good concepts to understand, but they need to be developed and studied in the context of ecological interactions and genetic variation. This means it is not enough to understand the general patterns of populations. We should also understand how populations change with genetic diversity, variation in reproduction, and with the presence of competitors, herbivores and disease. In short, population dynamics and structure influence and are influenced by many other factors that we will be discussing in future chapters.

Questions

1. What is known about the population structure and dynamics of your selected species of weed? Suggest ways that that your species can be structured, i.e. by age, size, phenology. Describe the life history strategy of your species. Is it an *r*- or K-selected species – or somewhere in between? Place your species on Grime's C-S-R model and explain why you placed it there.
2. Describe the size distributions of the four populations shown in Fig. 3.11. Assuming that age is correlated with size, what is the likely fate of each of these populations? Explain why. Would your answer change if age were not correlated with size? Explain why.
3. How might metapopulation dynamics be considered in controlling a recently introduced invasive weed?
4. Explain what it means to have a Type I, II or III survivorship curve.
5. How might the carrying capacity (K) of a weed be modified by changes in management practices?
6. Explain why a plant's population size does not increase indefinitely.

General References

Beeby, A. (1994) *Applying Ecology*. Chapman and Hall, New York.

Cousens, R. and Mortimer, M. (1995) *Dynamics of Weed Populations*. Cambridge University Press, Cambridge.

Hanski, I. and Gilpin, M. (1991) Metapopulation dynamics: brief history and conceptual domain. *Biological Journal of the Linnaean Society* 42, 3–16.

Literature Cited

Barrett, S.C.H. and Wilson, B.F. (1983) Colonizing ability in the *Echinochloa crus-galli* complex (barnyardgrass). II Seed biology. *Canadian Journal of Botany* 61, 556–562.

Beeby, A. (1994) *Applying Ecology*. Chapman and Hall, New York.

Boggs, K.W. and Story, J.M. (1987) The population age structure of spotted knapweed (*Centaurea maculosa*) in Montana. *Weed Science* 35, 194–198.
Buhler, D.D., Stoltenber, D.E., Becker, R.L. and Gunsolus, J.L. (1994) Perennial weed populations after 14 years of variable tillage and cropping practices. *Weed Science* 42, 205–209.
Burdon, J.J. and Chilvers, G.A. (1994) Demographic changes and the development of competition in a native Australian eucalypt forest invaded by exotic pines. *Oecologia* 97, 419–423.
Canham, C.D. (1985) Suppression and release during canopy recruitment of *Acer saccharum*. *Bulletin of the Torrey Botanical Club* 112, 134–145.
Deen, W., Swanton, C.J. and Hunt, L.A. (2001) A mechanistic growth and development model of common ragweed. *Weed Science* 49, 723–731.
Deevey, E.S. (1947) Life tables for natural populations of animals. *Quarterly Review of Biology* 22, 283–314.
Dietz, H. and Ullman, I. (1998) Ecological applications of 'herbchronology': comparative stand age structure analyses of the invasive plant *Bunias orientalis* L. *Annals of Botany* 82, 471–480.
Etienne, R.S., and Heesterbeek J.A.P. (2000) On optimal size and number of reserves for metapopulation persistence. *Journal of Theoretical Biology* 203, 33–50
Fair, J., Lauenroth, W.K. and Coffin, D.P. (1999) Demography of *Bouteloua gracilis* in a mixed prairie: analysis of genets and individuals. *Journal of Ecology* 87, 233–243.
Falińska, K. (1991) *Plant Demography in Vegetation Succession*. Kluwer, Dordrecht.
Gatsuk, E., Smirnova, O.V., Vorontzova, L.I., Zaugolnova, L.B. and Zhukova, L.A. (1980) Age-states of plants of various growth forms: a review. *Journal of Ecology* 68, 675–696.
Grace, J.B. (1991) A clarification of the debate between Grime and Tilman. *Functional Ecology* 5, 583–587.
Grime, J.P. (1977) Evidence for the existence of three primary strategies in plants and its relevance to ecological and evolutionary theory. *American Naturalist* 111, 1169–1194.
Grime, J.P. (1979) *Plant Strategies and Vegetation Processes*. John Wiley & Sons, Toronto.
Grime, J.P., Hodgson, J.G. and Hunt, R. (1990) *The Abridged Comparative Plant Ecology*. Unwin Hyman Inc., Boston.
Gross, K.L. (1981) Predictions of fate from rosette size in four 'biennial' plant species: *Verbascum thapsus*, *Oenothera biennis*, *Daucus carota*, and *Tragopogon dubius*. *Oecologia* 48, 209–213.
Hanski, I. and Gilpin, M.E. (eds) (1997) *Metapopulation Biology: Ecology, Genetics and Evolution*. Academic Press, New York.
Holm, L., Plunknett, D.L., Pancho, J.V. and Herberger, J.P. (1977) *The Worlds's Worst Weeds: Distribution and Biology*. University Press of Hawaii, Honolulu.
Honnay, O., Hermy, M. and Coppin, P. (1999) Effects of area, age and diversity of forest patches in Belgium on plant species richness, and implications for conservation and reforestation. *Biological Conservation* 87, 73–84.
Horvitz, C.C. and Schemske, D.W. (1995) Spatiotemporal variation in demographic transistions for a neotropical understorey herb: projection matrix analysis. *Ecological Monographs* 65, 155–192.
Hutchings, M.J. (1997) The structure of plant populations. In: Crawley, M.J. (ed.) *Plant Ecology*, 2nd edn. Blackwood Scientific, Oxford, pp. 325–358.
Leverich W.J. and Levin, D.A. (1979) Age-specific survivorship and reproduction in *Phlox drummondii*. *American Naturalist* 113, 881–903.
Luken, J.O. (1990) *Directing Ecological Succession*. Chapman and Hall, London, pp. 251.
Pearl, R. and Minor, J.R. (1935) Experimental studies on the duration of life. XIV. The comparative mortality of certain lower organisms. *Quarterly Review of Biology* 10, 60–79.
Pianka, E.R. (1970) On r- and K- selection. *American Naturalist* 104, 592–597.
Pulliam, H.R. (1988) Sources, sinks, and population regulation. *American Naturalist* 132, 652–661.
Radosevitch, S.R. and Holt, J.S. (1984) *Weed Ecology: Implications for Management*. John Wiley & Sons, New York.
Richards, C.M. (2000) Inbreeding depression and genetic rescue in a plant metapopulation. *American Naturalist* 155, 383–394.
Schmitt, J., Eccleston, J. and Ehrhardt, D.W. (1987) Dominance and suppression, size dependent growth and self-thinning in a natural *Impatiens capensis* population. *Journal of Ecology* 75, 651–665.
Schwartz, M.W. (1997) *Conservation in Highly Fragmented Landscapes*. Chapman and Hall, New York.
Scott, H.D. and Geddes, R.D. (1979) Plant water stress of soybean (*Glycine max*) and common cocklebur (*Xanthium pensylvanicum*): a comparison under field conditions. *Weed Science* 27, 285–289).

Sharitz, R.R. and McCormick, J.F. (1973) Population dynamics of two competing annual species. *Ecology* 54, 723–740.

Swanton, C.J., Clements, D.R. and Derksen, D.A. (1993) Weed succession under conservation tillage: a hierarchical framework for research and management. *Weed Technology* 7, 286–297.

Tilman, D. (1987) On the meaning of competition and the mechanisms of competitive superiority. *Functional Ecology* 1, 304–315.

Werner, P.A. (1975) Predictions of fate from rosette size in teasel (*Dipsacus fullonum* L.). *Oecologia* 20, 197–201.

Werner, P.A. and Caswell, H. (1977) Population growth rates and age versus stage distribution models for teasel (*Dipsacus sylvestris* Huds.). *Ecology* 58, 1103–1111.

Whipple, S.A. and Dix, R.L. (1979) Age structure and successional dynamics of a Colorado subalpine forest. *American Midland Naturalist* 101, 142–158.

4
Sexual Reproduction

Concepts

- Benefits of sexual reproduction relate to increased fitness – specifically, this includes the production of genetically variable offspring (often expressed as hybrid vigour) that allows at least some offspring to survive in a heterogeneous environment.
- Costs of sexual reproduction include disrupting well-adapted genotypes and the cost of producing reproductive structures.
- Plants can have complex combinations of gender expression; for example, they can be exclusively male or female, or both male and female at the same time.
- Pollination by animals is more accurate, but more energetically expensive because the floral structures are elaborate; wind pollination requires less initial resource commitment because the floral structures are small; since it is less accurate, a lot of resources used to produce pollen and ovules can be wasted.
- Self-compatibility guarantees some degree of mating success by ensuring pollination occurs even when only one individual is present.
- Self-incompatibility prevents inbreeding depression by ensuring that ovules are fertilized by pollen from genetically different individuals.

Introduction

Plants have two general means by which they reproduce: asexually and sexually. Since humans only reproduce sexually, asexual reproduction (= vegetative reproduction) is not as familiar to us, but it is rather common in plants. Asexual reproduction involves the replication of chromosomes without the production of gametes or the need for sex. Asexual reproduction produces offspring that are genetically identical to their parents. Typical examples of this form of reproduction are the stolons ('runners') produced by strawberries (*Fragaria* species), and root sprouting ('suckering') by aspens (*Populus* species). We will discuss asexual reproduction in the next chapter; in this chapter, we focus on sexual reproduction.

As in any organism, plant sexual reproduction requires the fusion of two gametes (a sperm and ovum) to form a zygote. Each gamete normally contains one set of chromosomes and the zygote will normally have two sets of chromosomes: one from each parent. Therefore, sexually produced offspring possess a unique recombination of their parents' genes and are genetically different from parents. In flowering plants, sexual reproduction is facilitated by pollination. Pollination occurs when pollen is transported to the stigma on a flower. Once on a compatible stigma, pollen produces a pollen tube that delivers the sperm to the female gametes (ova) and, ultimately, a seed will be produced. Through sexual reproduction, there are many ways to successfully transmit at least some portion of an individual's genotype (= all the genes of an individual) to its offspring, for example:

- Plants may express different combinations of gender, e.g. individuals can be either genetically programmed to be only male or female ('diclinous') or be both male and female at some time during their lifespan ('monclinous') (Table 4.1).
- Because they cannot move, plants package sperm inside specialized protective

Table 4.1. Gender expression in plants.

Term	Description
Monocliny	Each *individual* is genetically capable of expressing both genders. Whether both genders are actually expressed can be influenced by genetic and environmental factors:
Sequential monocliny	Each *individual* expresses only one gender at a given time. Gender changes over a growing season or from year to year, e.g. saltbushes (*Atriplex*) (Freeman and McArthur, 1984)
Simultaneous monocliny	An *individual* expresses both genders at a given time, but not all *flowers* necessarily express both genders, at the same time or ever
Sexual monomorphism	Male and female gender will be expressed in the same *flower*, though not necessarily at the same time:
Protandry	Male gender expressed before female gender, e.g. wild carrot, *(Daucus carota*) (Dale, 1974)
Protogyny	Female gender expressed before male gender, e.g. common mullein (*Verbascum thapsus*) (Gross and Werner, 1977)
True monomorphism	Both genders expressed at same time in same flower, e.g. sow thistles (*Sonchus*) (Hutchinson *et al.*, 1984)
Monoecy	In at least some *flowers*, only one gender is ever expressed:
Gynomonoecy	Female flowers and perfect flowers exist, e.g. plantains (*Plantago*) (de Haan *et al.*, 1997)
Andromonoecy	Male flowers and perfect flowers exist, e.g. horsenettle *(Solanum carolinense*) (Steven *et al.*, 1999)
True monoecy	All flowers are either male or female; no perfect flowers exist, e.g. nettles (*Urtica*) (Bassett *et al.*, 1977)
Dicliny	Each *individual* is genetically capable of expressing only one gender during its existence. However, species or populations are not always totally diclinous:
Dioecy (true dicliny)	All individuals are either entirely male or entirely female, e.g. poison ivy *(Rhus radicans*) (Mulligan and Junkins, 1977)
Gynodioecy	Some individuals are entirely female; others are monoclinous, e.g. viper's bugloss (*Echium vulgare*) (Klinkhamer *et al.*, 1994)
Androdioecy	Some individuals are entirely male; others are monoclinous, e.g. annual mercury (*Mercurialis annua*) (Pannell, 1997)

Note that some terms apply to the gender expression of the *individual* as a whole, while others refer to gender expression within the *flowers*. Cruden and Lloyd (1995) give alternative terminology for gender expression.

tissue ('pollen') that can be sent across short and long distances via gravity, water, wind and animals.
- Plants can mate with themselves or with another individual.

We will discuss these in detail, but first we will examine the benefits and costs of reproducing sexually.

The Benefits and Costs of Sex

Sex costs resources – this encompasses everything a plant uses and produces, e.g. fats, proteins, carbohydrates and water. Resources used in sexual reproduction will not be available for anything else, like making leaves that will increase photosynthesis and sugar production to feed the plant. Therefore, in order for sex to exist, its benefits must outweigh its costs. Benefits include producing genetically variable genotypes that are both adaptable and less likely to suffer from genetically related 'medical' problems that are contained in a mobile unit (a seed) that can escape the parental environment. Costs include disrupting well-adapted genotypes and having to use resources to produce floral structures to facilitate sexual reproduction. In general, this 'trade-off' between benefits and costs of sex relate to the concept of fitness.

Fitness

Fitness provides a relative measure of how well an individual succeeds at continuing its lineage. Individuals that are 'fit' to their environment are ones that can survive and reproduce successfully. In any population, the genotype with the highest relative fitness is the one that produces the most offspring that will survive and reproduce themselves. Relative fitness is often measured by testing for significant effects of any phenomenon (e.g. low nitrogen concentrations in the soil) on specific fitness components. These components are usually tangible traits of plants that can be measured empirically, for example:

- number and mass of seeds produced;
- success of seed dispersal;
- seed survival in the soil (the 'seed bank');
- amount and rate of seed germination;
- mass, height and growth rate of seedlings;
- resource allocation to roots, shoots and flowers (measured by examining their relative masses);
- pollination and fertilization success.

The effective measure of relative fitness itself can be thought of as how much of the original parental genotype survives from generation to generation, both in terms of the genetic composition of direct descendants and how much of the population eventually contains some portion of a parental genotype. If the environment generally remains constant, then individuals continue to produce offspring that are very close copies of themselves, i.e. the most fit offspring genotype will be those that are most similar to the parental genotype. This is what happens with agricultural weeds when farming practices do not change over time – weeds adapt to these specific practices and produce many similar offspring because these offspring encounter an equally favourable and unchanged environment. When environmental conditions change, however, the fitness of these weeds may decrease.

This same principle applies when plants are introduced to new environments. Effectively, the environment for the plant has 'changed' since it encounters a new location and habitat. If the physical environment of the originating habitat is similar, but the biological environment differs, then the plant often has greater fitness than existing plants. One example in North America is the introduction of garlic mustard (*Alliaria petiolata*). Garlic mustard originates from Eurasia, where the physical environment is reasonably similar to northeastern North America. Unlike Eurasia, garlic mustard in North America does not appear to have any effective natural enemies (e.g. pathogens, herbivores). Also, garlic mustard prefers forest edge habitats; many species native to the forest understories of North America are not adapted to this type of environment. With forest fragmentation increasing rapidly, garlic mustard can outcompete native species because of its high growth rate and fecundity in its new environment of forest edges without natural enemies (Anderson *et al.*, 1996; Nuzzo, 1999).

Benefits of sexual reproduction

New, better fit and better adaptable genotypes

The main benefits of sexual reproduction are the potential for genetic combinations that may be better fit to the current or new environments and, concomitantly, producing genetically variable offspring that can adapt to changing or new environments. Each seed is a unique genotype containing different alleles ('versions') of different genes, hence whatever type of environment the offspring encounter, there is a high probability that some of them will survive to reproduce. Formally, we sometimes refer to genetic variation as 'hybrid vigour' (= 'heterosis'). Intraspecific ('within species') hybrids are very common in sexual organisms (including diverse organisms like garlic mustard and humans). Intraspecific hybrids are less likely to express deleterious combinations of alleles, i.e. 'bad' products from 'bad' versions of genes that reduce the ability of an individual to germinate, grow and survive to reproduce. The reason that sex avoids this problem is because when two individuals of the same species ('conspecifics') that have dissimilar genotypes mate, genetic recombination occurs so their offspring are less likely to receive a copy of the same deleterious allele from each parent.

In plants, there is another reasonably common form of hybridization: interspecific ('between species'). Again, this is less familiar to humans because it does not happen with us. However, in plants, interspecific hybrids are formed when individuals from two different species mate. Not all species can mate to produce viable hybrids and, in fact, hybrids usually form from mating between closely related species because their genomes must be similar enough to successfully produce offspring capable of reproducing themselves. Interspecific hybrids may have higher fitness because of new genetic combinations. They usually have characteristics that are intermediate to their parents (Table 4.2) (Bailey *et al.*, 1995; Clements *et al.*, 1999). New species with weedy characteristics can arise through hybridization either when two weeds hybridize or when a weed and native species hybridize (Briggs and Walters, 1984). Examples of some hybrids (indicated by the '×' below) between weed species are:

- bitter yellow dock (*Rumex* × *crispo-obtusifolius* = *R. crispus* × *R. obtusifolius*);
- false leafy spurge (*Euphorbia* × *pseudo-esula* = *E. cyparissias* × *E. esula*);

Table 4.2. Characteristics of giant knotweed (*Fallopia sachalinesis*) and Japanese knotweed (*Fallopia japonica* var. *japonica*) and their hybrid *Fallopia* x *bohemica*. (Adapted from Bailey *et al.*, 1995).

Characteristic	Giant knotweed	Hybrid	Japanese knotweed
2*n*	44	44 or 66	88
Height	Gigantic plant, up to 4 m tall	Intermediate in size, 2.5–4 m tall	Large plant 2–3 m tall
Leaf size	Up to 40 cm long by 22 cm wide	Up to 23 cm long by 19 cm wide	10–15 cm long
Leaf length:width ratio	Approx 1.5	1.1–1.8	1–1.5
Leaf underside	Scattered, long flexible hairs	Larger leaves have many short, stout hairs	Smooth
Floral sex expression	Male-fertile flowers and male-sterile flowers borne on separate plants	Male-fertile flowers and male-sterile flowers borne on separate plants	Flowers usually male-sterile

- tall cat-tail (*Typha* × *glauca* = *T. angustifolia* × *T. latifolia*);
- goat's-bladder (*Tragopogon* × *mirus* = *T. dubius* × *T. porrifolius*);
- hybrid goat's-beard (*Tragopogon* × *miscellus* = *T. dubius* × *T. pratensis*).

In some cases a hybrid can be found beyond the distribution of the parent species. For example, the range of hybrid goat's-beard (*T.* × *miscellus*) has increased substantially beyond the range of at least one of the parent species in Washington state (Novak *et al.*, 1991). Thus, this species may have a greater ecological amplitude than its parents.

Hybridization between a native species and a related weed species can be more serious than hybridization between two weed species, because it can cause extinction of the native species if the hybrid species has greater fitness. For example, Freas and Murphy (1988) determined that the widespread Australian saltbush (*Atriplex serenana*) appeared to be hybridizing with the one remaining population of Bakersfield saltbush (*Atriplex tularensis*). Several native sunflowers (*Helianthus* species) in the southern USA are vulnerable to extirpation or extinction because of hybridization with the introduced annual sunflower (*Helianthus annuus*) (Rhymer and Simberloff, 1996).

Getting away from your parents: the mobility of offspring

Sexual reproduction generally has an ancillary benefit of producing mobile offspring, i.e. seeds or seeds inside fruits. Sexually produced offspring are usually dispersed away from the maternal parent so there is less chance of competing with their parents, siblings or other relatives. When the environment is not favourable to the parent (and hence the offspring are also likely to suffer), dispersal away from the parent is important. The benefits of dispersal are discussed in more detail in Chapter 6 but it is useful to keep in mind that dispersal is an indirect benefit of sex.

Costs of sexual reproduction

Sex disrupts well-adapted genotypes

Plants that reproduce sexually risk breaking up well-adapted genotypes, because it results in genetic recombination. In a relatively unchanging environment, offspring that are similar to the maternal genotype are usually better adapted than ones with recombined genotypes. This fitness disadvantage of a recombined genotype is called 'outbreeding depression' (Waser and Price, 1989, 1993; Parker, 1992). The offspring lose the complex genetic structure that made their parents so successful in a local environment.

Cost of producing reproductive structures

To reproduce sexually, plants must allocate resources to produce sexual organs, and floral structures that increase the chances of pollen dispersal. These structures can be quite resource expensive. Milkweed, for example, allocates 37% of its photosynthate to nectar production (Southwich, 1984). Plants that reproduce only once in their life span ('monocarpic') must maximize reproductive output per unit of resource expended. Even in plants that have repeated reproductive events in their life span (polycarpic), the costs of sexual reproduction are important because it may result in resources being directed away from growth and maintenance. Sexually reproducing plants often have to commit resources to reproduction early in the growing season. This increases the risk associated with sexual reproduction because if the weather prevents pollination or if seeds are destroyed, the plant may not have enough resources left to survive.

Ecology of Flowers and Flowering

Gender expression

In most animals, an individual is either male or female. Defining gender in a plant, however, is complicated. In plants, gender can apply to individual flowers or to the individual as a whole (Table 4.1). The reason for this complexity is related to sexual selection, i.e. the factors that influence the relative ability of individuals to obtain mates and reproduce offspring (Willson, 1994). Generally, the more options plants have in expressing gender, the more likely they are to reproduce (sexually) successfully, no matter what environment they encounter. We will also show, however, that there can be risks for an individual to express many combinations of gender and, consequently, there can be benefits of expressing only one or few combinations of gender.

Gender based on flowers of individuals

Most people recognize that flowers can be male and female at the same time because we are taught to recognize the basic structures of a typical flower, i.e. petals, sepals, stamens and carpels (Fig. 4.1). However, some or all flowers on an individual may express only one gender. The sexual expression of a flower also can be separated in time with male structures (anthers and pollen) maturing first and then the female structures (e.g. stigma) becoming receptive, or vice versa. There are many complex variations of this with equally complex terminology (Table 4.1).

Gender based on the individual

The gender of an individual plant can be controlled genetically, environmentally or both. Plants often have the genetic ability to be male and female but the relative expression of male and female traits varies with the short-term environmental conditions and perhaps long-term selection pressures (Barrett, 1998; Campbell, 2000). Plants where individuals are (genetically) one sex are called 'diclinous'; plants that (genetically) can express more than one sex are called monoclinous (Table 4.1). Humans would considered to be 'diclinous', using this terminology. Like its flowers, an entire individual plant can be: male, female, both male and female at the same time, male and then female, female and then male, continually changing from female to male or vice versa.

Allocation strategies for expressing genders in flowers and individuals

Generally, environmental stress tends to increase the expression of male gender in plants (Freeman *et al.*, 1980; Escarre and Thompson, 1991). In a resource-poor environment, it is better to be male than female.

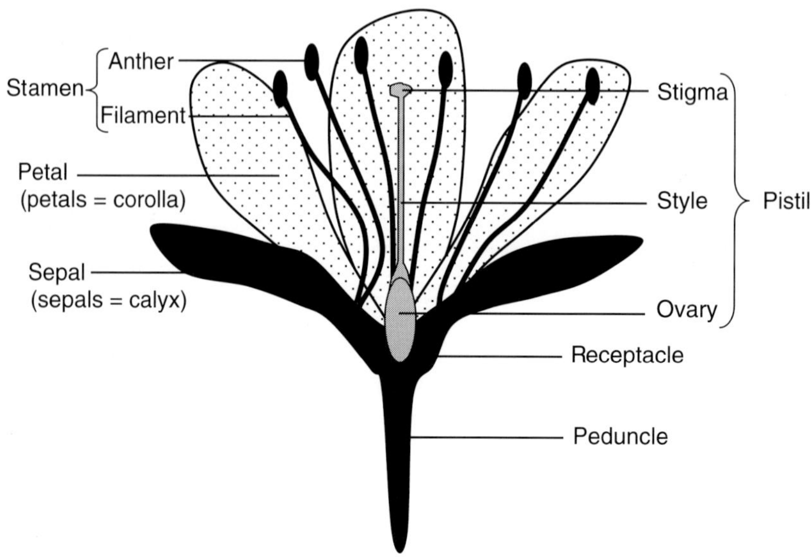

Fig. 4.1. Drawing of archetypal flower.

This is because male structures (like pollen) require fewer resources to develop, whereas female structures (like ovules) are where offspring develop and they require allocation of more resources to be nurtured and dispersed. Conversely, in resource-rich environments, it may be advantageous to express more female gender as there is no question that any offspring produced are, in part, carrying the female parent's genotype (as we shall see, males have little control of their reproductive success as pollen can go astray). Since plants cannot predict their future environment, any allocation and gender expression strategy is risky and generally depends on whatever previous and current selection pressures favour. Some plants try to use 'bet-hedging' by allocating equal amounts of resources to both male and female genders; however, even this may reduce fitness if the environment currently or eventually changes to favour the expression of one gender rather than both. Further complicating gender expression is the fact that all of the resources allocated may be wasted because the process of mating is rather risky in plants as they rely on a 'third party' to facilitate sex. The 'third party' relates to pollination mechanisms, (i.e. what carries the pollen from male to female).

Pollination mechanisms

For plants to reproduce sexually, there must be ample pollen available to carry sperm that will fertilize ova. Because of its micro-

Table 4.3. Suites of floral traits associated with pollination syndromes (adapted from Howe and Westley, 1977).

Pollinating agent	Anthesis	Colour	Odour	Flower shape
Insect pollination				
Beetles	Day and night	Dull	Fruity or aminoid	Flat or bowl-shaped; radial symmetry
Carrion or dung flies	Day and night	Purple–brown or greenish	Decaying protein	Flat or deep; radial symmetry; often traps
Bees	Day and night or diurnal	Variable, but not pure red	Usually sweet	Flat to broad tube; bilateral or radial symmetry; may be closed
Butterflies	Day and night or diurnal	Variable; pink very common	Sweet	Upright; radial symmetry; deep or with spur
Vertebrate pollination				
Bats	Night	Drab, pale, often green	Musty	Flat 'shaving brush' or deep tube; radial symmetry; much pollen; often upright, hanging outside foliage, or borne on trunk or branch
Birds	Day	Vivid, often red	none	Tubular, sometimes curved; radial or bilateral symmetry, robust corolla; often hanging
Abiotic pollination				
Wind	Day or night	Drab, green	None	Small; sepals and petals absent or reduced; large stigmata; much pollen; often catkins
Water	Variable	Variable	None	Minute; sepals and petals absent or reduced; entire male flower may be released

scopic size (micrometers), pollen is usually produced in large quantities in order to increase the chance of reaching non-mobile ova. For a plant to mate successfully, pollen must be transferred from the anther to a genetically compatible stigma, style and ovum. Pollen generally is delivered via three mechanisms: by animals (zoophily), wind (anemophily) or water (hydrophily).

Animal pollination (zoophily)

Animal-pollinated species must allocate resources to create floral morphologies that attract animals and these can be very resource expensive (Harder and Barrett, 1995). Floral morphology varies with the type of animal pollinator, as do the pigments used to colour flowers, the height and breadth of the inflorescence, and the provision of nectar (Table 4.3) (Wyatt, 1983).

Plants vary in their pollination strategy, i.e. whether to use many types of pollinators or very specialized pollinators (Johnson and Steiner, 2000), but some general trends do exist. Of most relevance here, weeds tend not to need elaborate floral morphologies because they are not usually co-adapted with their pollinators or use abiotic vectors for pollination (Baker, 1974). For example, wild carrot (*Daucus carota*) has an open flat inflorescence that enables a variety of insects to access pollen (Dale, 1974); many weeds use a similar strategy. Regardless, there are no guarantees of successful pollination because the inflorescence can be eaten, pathogens or parasites can infest the flowers, or animals can rob nectar without transferring pollen. Plants may increase the likelihood of successful pollination by:

- deceiving pollinators (using chemicals that resemble nectar to lure them);
- trapping pollinators in a flower to ensure they are covered in pollen;
- forcing pollinators to specialize by hiding rewards like nectar or having specialized flowers that require structures like uniquely shaped proboscises;
- flowering only when other species are not flowering.

Wind pollination (anemophily)

Wind-pollinated flowers are more drab, have small or absent petals and no nectar. They are less showy (though not necessarily less elaborate) but often are less energetically expensive (Whitehead, 1983). Wind-pollinated plants must produce vast quantities of pollen to ensure success. Wind pollination presents risks because most of the pollen does not reach the proper stigma, and successful pollination depends on appropriate environmental conditions such as precipitation, temperature, relative humidity and wind direction (Whitehead, 1983; Murphy, 1999). Examples of wind-pollinated weeds include ragweeds (*Ambrosia* species), quackgrass (and other weeds in the grass family (*Poaceae*)), and Monterey pine (*Pinus rigida*). It may be advantageous for weeds and other colonizing species to be wind-pollinated to avoid reliance on other organisms to ensure successful pollination.

Water pollination (hydrophily)

Water pollination is unique to submergent aquatic plants (see Les, 1988, for an extensive review). Submergent weeds that are water pollinated include horned pondweeds (*Zannichellia*) and pondweeds (*Najas*). Generally, water pollination is inefficient since the pollen (or sometimes the entire male parts of a flower) must float on the water or be transported in the water to reach stigmas. We emphasize that many familiar aquatic weeds actually are not water pollinated. Emergent aquatic plants like cattails (*Typha* species) are wind pollinated, while other emergents (pickerelweed, *Pontederia cordata*) and floating plants (water hyacinth, *Eichhornia crassipies*) are animal pollinated.

Pollination problems

Earlier, we discussed the concept that interspecific hybrids can be quite fit. However, we emphasized that not all species can mate with each other. Improper pollen transfer refers to situations where pollen from an individual of a different species ('heterospecific') lands on a stigma and does not produce any hybrid offspring. This is a problem for the pollen donor because a lot of pollen is therefore 'wasted' on individuals where fertilization will not occur. The pollen recipient is also affected if heterospecific pollen contains toxins ('allelochemicals'), pathogens or parasites (Murphy, 1999). For both donors and recipients, the result is lower pollination success and production of fewer viable seeds. Plants also suffer reduced pollination success and seed set if pollen or pollinator limitation exists (e.g. Lalonde and Roitberg, 1994; Collevatti *et al.*, 1997) because:

- few compatible mates are nearby;
- compatible individuals produce low numbers of flowers, pollen or ovules because of genetic defects or poor environmental conditions;
- the weather is poor for wind pollination or animal pollinators;
- there are few appropriate animal pollinators in the community.

Self-compatibility and self-incompatibility

Some individuals can successfully mate with themselves if pollen is transferred from stigma to style because they have: (i) viable male and female flowers open simultaneously, or (ii) their flowers have both male and female reproductive parts that are viable simultaneously. This is called 'self-compatibility'.

The benefits and costs of self-compatibility

Self-compatibility can be important to colonizing species because it means a single individual can invade a site, and be able to self-fertilize and produce seed. With this advantage, it is not surprising that many exotic or native plants considered weedy are self-compatible (Mulligan and Findlay, 1970; Baker, 1974; Barrett, 1992). A second advantage of self-compatibility is that it can be less costly if resource allocation to floral structures is reduced, as pollinators may not be necessary or as important if the individual simply uses gravity to collect pollen from its anthers to land on its stigma.

Though self-compatibility might be advantageous, not all weeds use this, e.g. jimsonweed (*Datura stramonium*) (Motten and Antonovics, 1992). This is because self-compatibility has costs as well as benefits. Consequently, many individuals are 'self-incompatible', i.e. they cannot mate with themselves. Self-incompatible plants can avoid some costs of self-mating. The main costs avoided are the otherwise increased chances of accumulating harmful alleles and decreased adaptability to new or changing environments. You have seen these explanations given in the discussion of the benefits and costs of sex. The only novel aspect here is that the reduced fitness caused by mating with a close relative and accumulating multiple copies of deleterious alleles is formally termed 'inbreeding depression'. Even self-incompatible plants have no guarantees of avoiding costs since their genetically recombined genotypes may not be adapted to the environment. Additionally, a self-incompatible individual still may mate with a close relative since its likely mates (close neighbours) often are close relatives (see Madden, 1995; Lefol *et al.*, 1996; Nunez-Farfan *et al.*, 1996; Guttieri *et al.*, 1998; Sun and Ritland, 1998; see Stanton, 1994; Wilson and Payne, 1994, for discussion of mate selection to avoid this problem).

How self-incompatibility is enforced

If an individual is to avoid inbreeding depression in its most extreme form (self-mating), there must be mechanisms available to block self-pollen from eventually producing offspring. Some individuals use a mechanism described as 'histochemical incompatibility'; this is a bit like a pollen grain causing an allergic reaction in the female so that the tissues change and fertilization cannot occur. The basis for histochemical incompatibility is a class of compounds called 'glycoproteins' expressed in pollen, stigma and style. The glycoproteins are signals that identify incompatible mates, usually relatives and self-pollen. The histochemical incompatibility reaction can occur

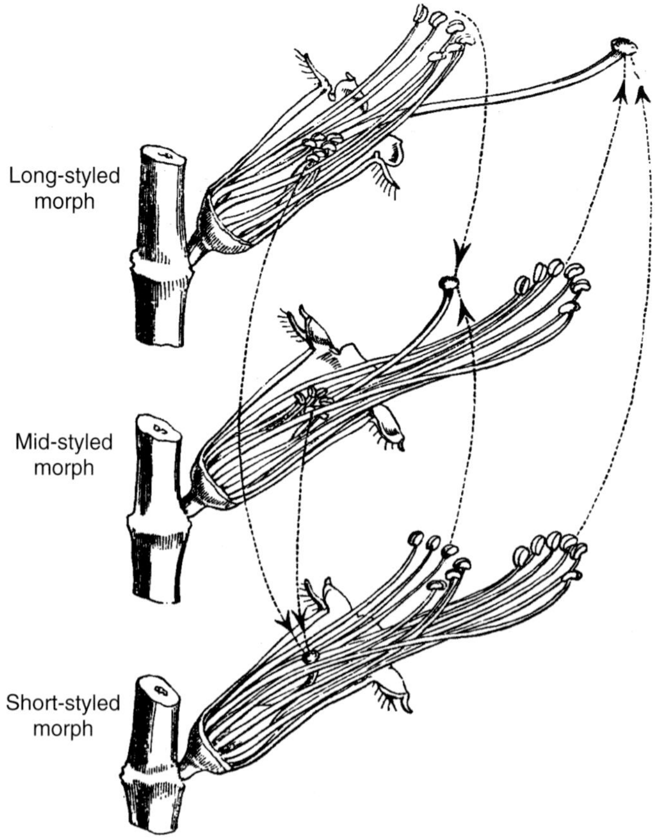

Fig. 4.2. Illustration of tristyly in purple loosestrife (*Lythrum salicaria*). The three floral morphs are: long-styled (with short and mid anthers), mid-styled (with long and short anthers) and short-styled (with long and mid anthers). Petals and calyx on the close side are removed to reveal flower parts. Arrows show direction that pollen must travel from anther to stigma to ensure full fertilization (Darwin 1877).

at different parts of the flower and at different times of its life cycle (Sims, 1993). Sporophytic incompatibility occurs when pollen is on the stigma; the glycoproteins signal the stigma not to exude the water needed for pollen to germinate. Gametophytic incompatibility usually occurs as the pollen tube is trying to grow in the style towards the embryo sac with the ovules.

Structurally, plants may avoid incompatible mates by having pollen that cannot physically adhere to certain stigmas, i.e. it is too big, too small, the wrong shape or the wrong texture. Additionally, the stigmas may be located above the pollen-bearing anthers so pollen cannot fall on top of the stigma. In some species, 'heterostyly' occurs where different types of flowers have stamens and styles of distinct lengths. Figure 4.2 illustrates heterostyly in purple loosestrife (*Lythrum salicaria*). The subsequent physical separation of stigmas from anthers (with self-pollen) enforces outcrossing.

Summary

Sexual reproduction can be energetically expensive and wasteful because of the floral structures needed, but is usually necessary to produce offspring that are better fit because they are genetically capable of adapting to new or changing environments. To maximize the benefits of sex, plants have evolved elaborate ranges of gender expression, floral morphologies and pollination mechanisms. All of these generally facilitate the production of hybrid offspring. However, sometimes self-mating, as an extreme form of inbreeding, can be more reliable as a means of producing offspring and does allow plants to produce new offspring rapidly, even if mate choice is limited. The risk of inbreeding depression may be less than the risk of not producing offspring at all. None the less, inbreeding can pose such risks of reduced genetic variation and relative fitness in offspring that individuals may have self-incompatibility mechanisms to prevent it.

Questions

1. Using the plant you selected, describe its method(s) of pollination, whether it is self-compatible or self-incompatible, and its form(s) of gender expression
2. Why do weeds (and plants in general) have sex?
3. Plants are sessile (they don't move). What are the implications of this in terms of sexual reproduction?
4. If pest management using insecticides accidentally eliminated most of the insect pollinators, do you think weeds or native plants would suffer more? Why?

General References

Ellstrand, N.C. and Schierenbeck, K.A. (2000) Hybridization as a stimulus for the evolution of invasiveness in plants? *Proceedings of the National Academy of Sciences USA* 97, 7043–7050.

Real, L. (1983) *Pollination Biology*. Academic Press, New York.

Wheeler, M.J., Franklin-Tong, V.E., and Franklin, F.C.H. (2001) The molecular and genetic basis of pollen–pistil interactions. *New Phytologist* 151, 565–584.

Literature Cited

Anderson, R.C., Dhillon, S.S. and Kelley, T.M. (1996) Aspects of the ecology of an invasive plant, garlic mustard (*Alliaria petiolata*) in central Illinois. *Restoration Ecology* 4, 181–191.

Bailey, J.P., Child, L.E. and Wade, M. (1995) Assessment of the genetic variation and spread of British populations of *Fallopia japonica* and its hybrid *Fallopia* x *bohemica*. In: Pyšek, P., Prach, K., Rejmanek, M. and Wade, M. (eds) *Plant Invasions: General Aspects and Special Problems*. SPB Academic Publishing, Amsterdam, pp. 141–150.

Baker, H. G. (1974) The evolution of weeds. *Annual Review of Ecology and Systematics* 1, 1–24.

Barrett, S.C.H. (1992) Genetics of weed invasions. In: Jain, S.K. and Botsford, L.W. (eds) *Applied Population Biology*. Kluwer, Dordrecht, pp. 91–119.

Barrett, S.C.H. (1998) The evolution of mating strategies in flowering plants. *Trends in Plant Science* 3, 335–341.

Bassett, I.J., Crompton, C.W. and Woodland, D.W. (1977) The biology of Canadian weeds. 21. *Urtica dioica* L. *Canadian Journal of Plant Science* 57, 491–498.

Briggs, D. and Walters, S.M. (1984) *Plant Variation and Evolution*, 2nd edn. Cambridge University Press, New York.

Campbell, D.R. (2000) Experimental tests of sex-allocation theory in plants. *Trends in Ecology and Evolution* 15, 227–232.

Clements, D.R., Upadhyaya, M.K. and Bos, S.J. (1999) The biology of Canadian weeds. 110. *Tragopogon dubius* Scop., *Tragopogon pratensis* L., and *Tragopogon porrifolius* L. *Canadian Journal of Plant Sciences* 79, 153–163.

Collevatti, R.G., Amaral, M.E.C. and Lopes, F.S. (1997) Role of pollinators in seed set and a test of pollen limitation in the tropical weed *Triumfetta semitriloba* (Tiliaceae). *Revista de Biologia Tropical* 45, 1401–1407.

Cruden, R.W. and Lloyd, R.M. (1995) Embryophytes have equivalent sexual phenotypes and breeding systems: why not a common terminology to describe them? *American Journal of Botany* 82, 816–825.

Dale, H.M. (1974) The biology of Canadian weeds. 5. *Daucus carota. Canadian Journal of Plant Science* 54, 673–685.

Darwin, C. (1877) *The Different Forms of Flowers of the Same Species.* Murray, London.

de Haan, A.A., Luyten, R.M.J.M., Bakx-Schotman, T.J.M.T. and van Damme, J.M.M. (1997) The dynamics of gynodioecy in *Plantago lanceolata* L. I. Frequencies of male-steriles and their cytoplasmic male sterility types. *Heredity* 79, 453–462.

Escarre, J. and Thompson, J.D. (1991) The effects of successional habitat variation and time of flowering on seed production in *Rumex acetosella. Journal of Ecology* 79, 1099–1112.

Freas, K.E. and Murphy, D.D. (1988) Taxonomy and the conservation of the critically endangered Bakersfield saltbush, *Atriplex tularensis. Biological Conservation* 46, 317–324.

Freeman, D.C. and McArthur, E.D. (1984) The relative influences of mortality, nonflowering, and sex change on the sex ratios of six *Atriplex* species. *Botanical Gazette* 145, 385–394.

Freeman, D.C., Harper, K.T. and Charnov, E.L. (1980) Sex change in plants: old and new observations and new hypotheses. *Oecologia* 47, 222–232.

Gross, K.L. and Werner, P.A. (1978) The biology of Canadian weeds. 28. *Verbascum thapsus* L. and *V. blattaria* L. *Canadian Journal of Plant Science* 58, 401–413.

Guttieri, M.J., Eberlein, C.V. and Souza, E.J. (1998) Inbreeding coefficients of field populations of *Kochia scoparia* using chlorsulfuron resistance as a phenotypic marker. *Weed Science* 46, 521–525.

Harder, L.D. and Barrett, S.C.H. (1995) Mating cost of large floral displays in hermaphrodite plants. *Nature* 373, 512–515.

Howe, H.F. and Westley, L.C. (1997) Ecology of pollination and seed dispersal. In: Crawley, M. (ed.) *Plant Ecology*, 2nd edn. Blackwell Scientific, London, pp. 262–281.

Hutchinson, I., Colosi, J. and Lewin, R.A. (1984) The biology of Canadian weeds. 63. *Sonchus asper* (L.) Hill and *S. oleraceus. Canadian Journal of Plant Science* 64, 731–744.

Johnson, S.D. and Steiner, K.E. (2000) Generalization versus specialization in plant pollination systems. *Trends in Ecology and Evolution* 15, 140–143.

Klinkhamer, P.G.L., de Jong, T.J. and Neil, H.W. (1994) Limiting factors for seed production and phenotypic gender in the gynodioecious species *Echium vulgare* (Boraginaceae). *Oikos* 71, 469–478.

Lalonde, R.G. and Roitberg, B.D. (1994) Mating system, life-history, and reproduction in Canada thistle (*Cirsium arvense*; Asteraceae). *American Journal of Botany* 81, 21–28.

Lefol, E., Danielou, V., Fleury, A. and Darmency, H. (1996) Gene flow within a population of the outbreeding *Sinapsis arvensis*: isozyme analysis of half-sib families. *Weed Research* 36, 189–195.

Les, D.H. (1988) Breeding systems, population structure and evolution in hydophilous angiosperms. *Annals of the Missouri Botanical Garden* 75, 819–835.

Madden, A.D. (1995) An assessment, using a modelling approach, of inbreeding as a possible cause of reduced competitiveness in triazine-resistant weeds. *Weed Research* 35, 289–294.

Motten, A.F. and Antonovics, J. (1992) Determinants of outcrossing rate in a predominantly self-fertilizing weed, *Datura stramonium* (Solanaceae). *American Journal of Botany* 79, 419–427.

Mulligan, G.A. and Findlay, J.N. (1970) Reproductive systems and colonization in Canadian weeds. *Canadian Journal of Botany* 48, 859–860.

Mulligan, G.A. and Junkins, B.E. (1977) The biology of Canadian weeds. 23. *Rhus radicans* L. *Canadian Journal of Plant Science* 57, 515–523.

Murphy, S.D. (1999) Pollen allelopathy. In: Inderjit, Dakshini, K.M.M. and Foy, C.L. (eds) *Principles and Practices in Plant Ecology: Allelochemical Interactions.* CRC Press, Boca Raton, Florida, pp. 129–148.

Novak, S.J., Soltis, D.E. and Soltis, P.S. (1991) Ownbey's *Tragopogons*: 40 years later. *American Journal of Botany* 78, 1586–1600.

Nunez-Farfan, J., Cabrales-Vargas, R.A. and Dirzo, R. (1996) Mating system consequences on resistance to herbivory and life history traits in *Datura stramonium. American Journal of Botany* 83, 1041–1049.

Nuzzo, V. (1999) Invasion pattern of the herb garlic mustard (*Alliaria petiolata*) in high quality forests. *Biological Invasions* 1, 169–179.

Pannell, J. (1997) Mixed genetic and environmental sex determination in an androdioecious population of *Mercurialis annua*. *Heredity* 78, 50–56.

Parker, M.A. (1992) Outbreeding depression in a selfing annual. *Evolution* 46, 837–841.

Rhymer, J.M. and Simberloff, D. (1996) Extinction by hybridization and introgression. *Annual Review of Ecology and Systematics* 27, 83–109.

Sims, T.L. (1993) Genetic regulation of self-incompatibility. *CRC Reviews in Plant Sciences* 12, 129–167.

Southwich, E.E. (1984) Photosynthate allocation to floral nectar: a neglected energy investment. *Ecology* 65, 1775–1779.

Stanton, M.L. (1994) Male–male competition during pollination in plant populations. *American Naturalist* 144, S40–S68.

Steven, J.C., Peroni, P.A. and Rowell, E. (1999) The effects of pollen addition on fruit set and sex expression in the andromonoecious herb horsenettle (*Solanum carolinense*). *American Midland Naturalist* 141, 247–252.

Sun, M. and Ritland, K. (1998) Mating system of yellow starthistle (*Centaurea solstitalis*), a successful colonizer in North America. *Heredity* 80, 225–232.

Waser, N.M. and Price, M.V. (1989) Optimal outcrossing in *Ipomopsis aggregata*: seed set and offspring fitness. *Evolution* 43, 1097–1109.

Waser, N.M. and Price, M.V. (1993) Crossing distance effects on prezygotic performance in plants: an argument for female choice. *Oikos* 68, 303–308.

Whitehead, D.R. (1983) Wind pollination: some ecological and evolutionary perspectives. In: Real, L. (ed.) *Pollination Biology*. Academic Press, Toronto, Ontario, pp. 97–108.

Willson, M.F. (1994) Sexual selection in plants: perspective and overview. *American Naturalist* 144, 513–539.

Wilson, H.D. and Payne, J.S. (1994) Crop/weed microgametophyte competition in *Cucurbita pepo* (Cucurbitaceae). *American Journal of Botany* 81, 1531–1537.

Wyatt, R. (1983) Pollinator–plant interactions and the evolution of breeding systems. In: Real, L. (ed.) *Pollination Biology*. Academic Press, Toronto, pp. 51–95.

5

Asexual Reproduction

Concepts

- Asexual reproduction (apomixis) can occur through the production of seeds without fertilization (agamospermy) or clonal reproduction (vegetative growth).
- Agamosperm reproduction has some of the benefits of seed production; however, the lack of genetic recombination means that novel genotypes are not formed and that deleterious mutations may accumulate.
- Agamospermic species vary in their ability to colonize.
- Clonal reproduction allows the individual to bypass the seedling stage of growth lowering the mortality of new individuals, but, again, there is no new genetic recombination.
- Reproducing clonally increases a species' colonizing ability and persistence.
- One species may reproduce via a combination of sexual reproduction, agamospermy and clonal growth, but there is a trade-off of resources among these types of reproduction.

Introduction

Plants do not need to rely on sexual reproduction to pass on their genes to the next generation. Asexual reproduction is the creation of new individuals without involving genetic recombination. As a result, asexual offspring are genetically identical to a single parent plant. The ability to reproduce without sex can be a great advantage to organisms (such as plants) that cannot move to find a mate or avoid an inhospitable environment. However, as we will discuss, reproduction without sex has as many costs and benefits as sexual reproduction.

The two main types of asexual reproduction are clonal growth and agamospermy. Clonal growth (or vegetative reproduction) is the creation of new, potentially independent plants through vegetative growth. Agamospermy is the creation of seed without fertilization. The term 'apomixis' is occasionally used as a synonym for agamospermy (Mogie, 1992); however, we prefer apomixis as a synonym for asexual reproduction because we are concerned more with the ecological consequence of this type of reproduction rather than the genetic or cellular aspects.

Asexual reproduction is of interest to

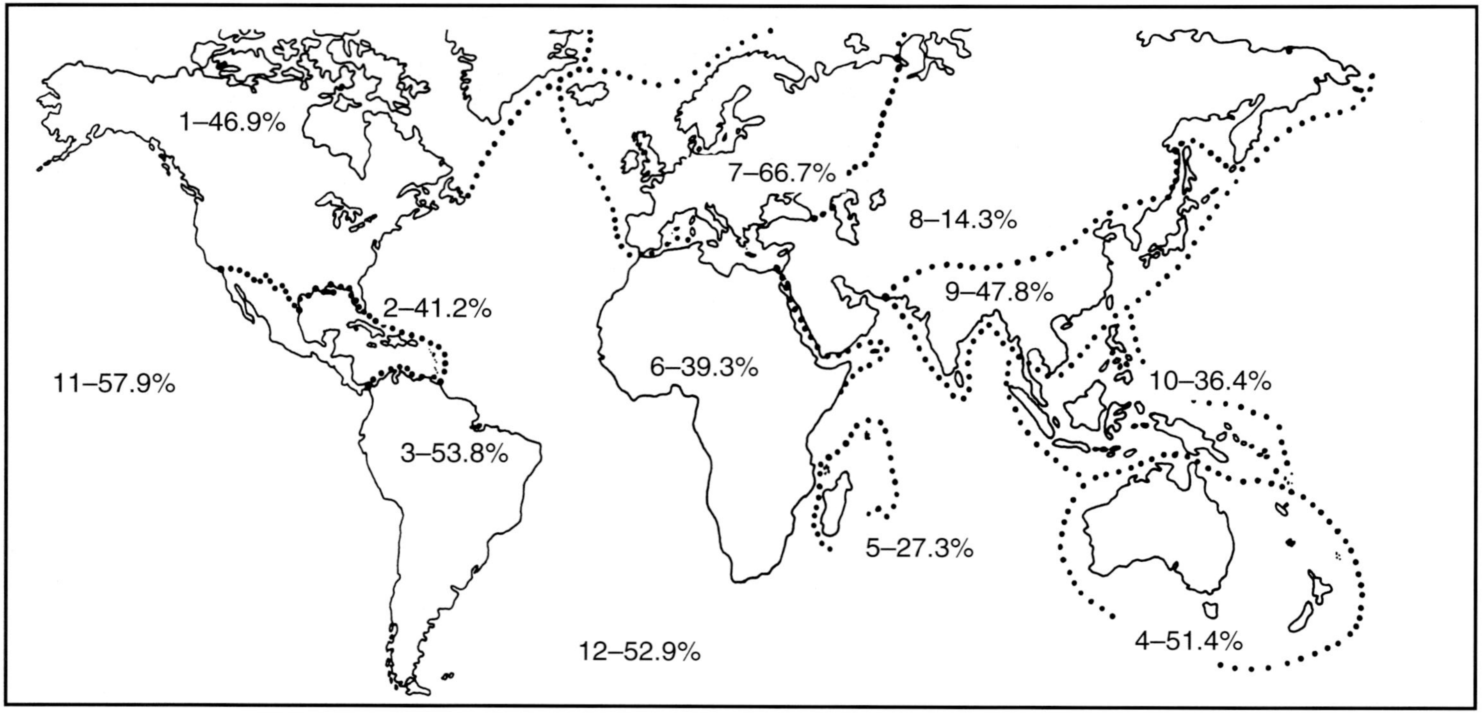

Fig. 5.1. Proportion of the most aggressive non-native species in natural habitats that are capable of clonal growth. Regions are 1) North America *n* (number of species) =36; 2) Central America; 3) South America *n*=134; 4) Australasia *n*=81; 5) Malagassia *n*=23; 6) Africa *n*=59; 7) Europe *n*=24; 8) North Asia *n*=7; 9) South Asia *n*=23; 10) Malesia *n*=12; 11) Pacific *n*=59; 12) Oceanic Islands *n*=17 (Pyšek 1997; with permission of Backhuys Publishers).

weed ecology, because it allows one individual to invade a new habitat and become established as a population without requiring a mate. Many weed species are capable of uniparental reproduction either through self-pollination (see Chapter 4 for examples), agamospermy (e.g. dandelion, *Taraxacum officinale*) or clonal propagation (e.g. quackgrass, *Elytrigia repens*) (Barrett, 1992). The importance of asexual reproduction varies with climate and habitat type. Harsh environmental conditions and lack of mates favours individuals that can reproduce asexually. Therefore, the distribution of species that have asexual capabilities increases towards the North and South Poles (Pyšek, 1997) (Fig. 5.1).

Agamospermy

Agamospermy is the production of seed without fertilization (i.e. the fusion of gametes – sperm and ovum). There are three main types of agamospermy (diplospory, apospory and adventitious embryony) but there are numerous and often complex variations (Fig. 5.2). Normally, meiosis occurs and the gamete (ovum) contains one copy of all chromosomes, i.e. the ovum is 'haploid'. After being fertilized by sperm, the seed will have the normal number of copies of chromosomes (i.e. it will be 'polyploid'). In apospory, meiosis occurs as usual, but the cells that would normally form the ovum degenerate and a polyploid somatic (non-

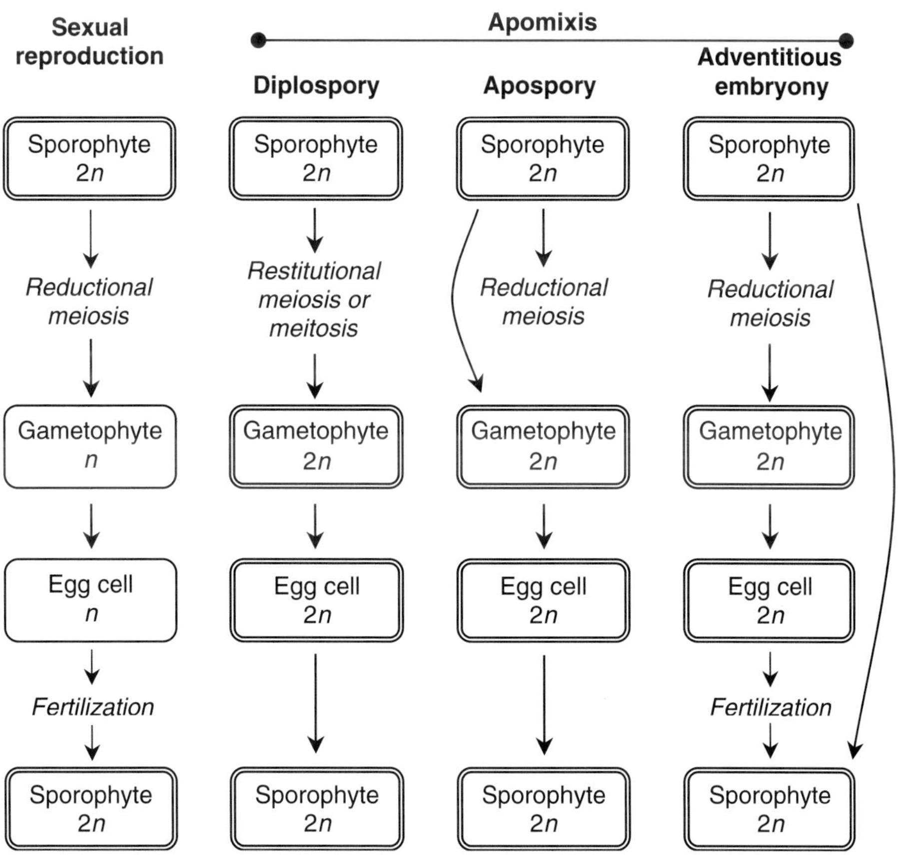

Fig. 5.2. Comparisons of sexual and asexual reproduction. Reduced (*n*) life cycle stage(s) have single border while, unreduced stages (2n) have double borders (redrawn from van Dijk and van Damme, 1999).

sexual) cell replaces the ovum and will form a seed. In diplospory, meiosis does *not* proceed as usual. As the embryo sac containing the ovum is produced, the numbers of chromosome copies are not reduced as is normal. The result is an ovum that has at least two (and usually more) complete copies of all chromosomes and this will form the seed. In adventive embryony, meiosis also does *not* proceed as normal and is altered so much that both embryo sac and ovum are not produced, and somatic cells form the embryo directly. In some cases, an asexual and sexual embryo can occur in the same seed because the normal sexual processes may still occur (van Dijk and van Damme, 2000).

Agamospermy is very common among ferns. It is not present in gymnosperms and it is present in only about 10–15% of angiosperm families (Richards, 1997). In angiosperms, approximately 75% of agamospermic taxa are in the daisy (*Asteraceae*), grass (*Poaceae*) and rose (*Rosaceae*) families. A high proportion of the species in the dandelion (*Taraxacum*), hawkweed (*Hieracium*), and raspberry (*Rubus*) genera are agamospermic.

Facultative agamospermy is the production of asexual seeds if pollination fails. It is present in some cinquefoils (*Potentilla*). This trait is particularly useful to weeds because they can produce seeds both with and without pollen, and this can aid the spread of a species when pollinators are absent in the new habitat. For example, the dioecious species screwpine (*Pandanus tectorius*) was able to invade islands because it could produce agamospermic seeds and therefore male plants were not necessary for it to colonize (Cox, 1985).

Obligate agamosperms are only able to produce seeds asexually; however, agamospermy rarely occurs to the total exclusion of sexual reproduction. Many raspberries and most dandelions are obligate agamosperms. Species with obligate agamospermy are often triploids or pentaploids and therefore cannot reproduce via pollen.

The occurrence of agamospermy is often associated with the following traits or conditions: polyploidy, phenotypic plasticity, perennial habit, hybridization and pollen limitation (Table 5.1). These traits do not necessarily cause agamospermy to develop: they may be either conducive to its development, or occur as a result of it. In some cases, the association of agamospermy with these traits is not fully understood. For example, some perennial weeds are agamospermous whereas others are not, and it is not always clear why a particular species has developed this trait.

Table 5.1. Traits and conditions associated with agamospermy (based on information from Asker and Jerling, 1992 and Richards, 1997).

Trait or character	Description
Hybridization	Hybridization is thought to bring about the conditions necessary for agamospermy. Hybrids may be more vigorous, long-lived and partially sterile.
Polyploidy (multiple sets of base chromosomes)	Polyploidy 'may buffer against the effects of deleterious mutations'. Polyploidy is also associated with other changes such as altered secondary metabolism, increased seed size and seedling vigour, and a switch from annual to perennial habit.
Phenotypic plasticity	As in inbreeding population, agamospermic species tend to have higher phenotypic plasticity. Selection is more likely to encourage phenotypic plasticity in populations with less genetic variability.
Polycarpic perennials, often rosette forming	Very few annuals, biennials and monocarpic species are agamospermic.
Pollen limitation	When seed production is limited by the lack of pollen, seeds produced by individuals carrying an agamospermic mutation are more likely to persist in higher numbers.

Costs and benefits

Richards (1997) proposed several costs and benefits of agamospermy. Agamosperm reproduction has some of the benefits of sexual reproduction (e.g. seed production), often without the costs of pollen production (Table 5.2). While agamospermy may avoid the cost of meiosis, the lack of recombination means that deleterious mutations may accumulate and novel genotypes are not formed. Not every cost and benefit will apply to all agamosperm species. For example, some agamosperms require pollen chemicals to help form the endosperm, although the pollen's gametes are not used in the creation of the new individual (e.g. blackberry, *Rubus fruticosus*).

Ecological aspects

We have said that possessing the ability to reproduce via agamospermy can improve the chances of colonization success and gave screwpine as an example. Not all agamospermic species are equally good colonizers. While agamospermy increases the chance of colonization, other traits are required. For example, two closely related species of agamospermous dandelion (*Taraxacum*), which co-occur in sand dunes of Northumberland, UK, have different life history strategies in spite of their similar morphologies. Rock dandelion (*Taraxacum lacistophyllum*) is more opportunistic than *Taraxacum brachyglossum* because it has a faster growth rate, shorter life span, reproduces earlier, has lighter and more dispersible seeds, and can respond faster to the addition of nutrients (Ford, 1985). Thus, rock dandelion is a more successful

Table 5.2. Costs and benefits associated with agamospermy (based on text in Richards, 1997).

	Description
Benefits	
Assured reproduction	In the absence of pollination, seed production is assured (although some agamosperms still require the 'cue' from pollination to create asexual seeds)
Advantages of seed	Obtain dispersal and dormancy but maintain advantages of vegetative reproduction
Avoid 'cost of meiosis'	No 'unfit' zygotes created through recombination that may disrupt co-adapted genotypes. Offspring have same fitness as maternal parent
Avoid 'cost of males'	Energy does not go towards the creation of pollen (although many agamosperms do produce pollen)
Benefit from 'extremely fit genotype'	Many agamosperms are highly heterozygous and thus have high fitness. Less fit genotypes will decrease through natural selection
Costs	
Accumulate detrimental mutations	Non-lethal detrimental mutations will remain in the population because there is no recombination and selection to remove them from the population
No recombination	Agamospecies lack genetic recombination which can create novel advantageous genotypes that may be more fit, especially in cases of habitat or climatic changes
Narrow niche	Outcrossing creates genetic variation among individuals of a population that will lead to increased likelihood of inhabiting more niches. This is lacking in agamosperm populations, although there is some evidence of high levels of somatic mutations in asexual lineages
Lack 'fine-tuning'	Recombination can create genotypes better adapted to local environments. Agamosperm populations are more likely to be generalists (weedy)

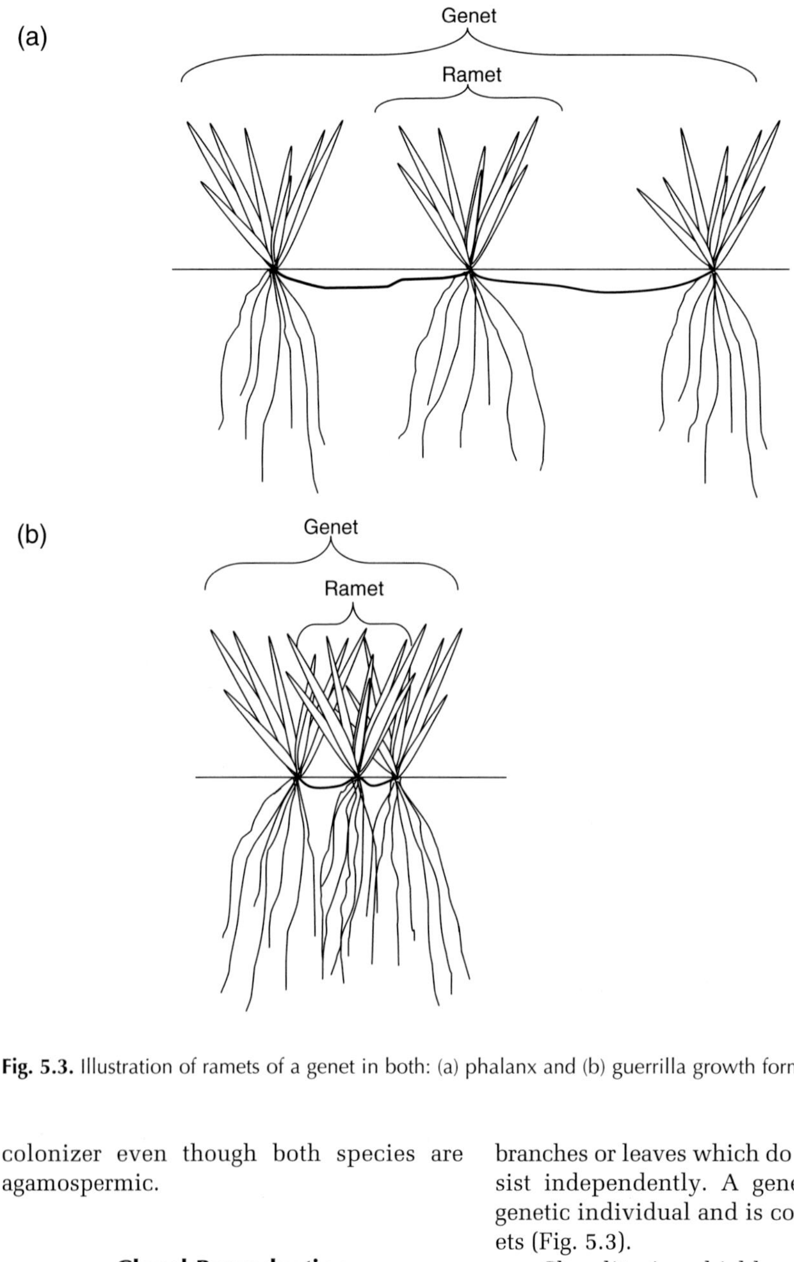

Fig. 5.3. Illustration of ramets of a genet in both: (a) phalanx and (b) guerrilla growth forms.

colonizer even though both species are agamospermic.

Clonal Reproduction

Clonal reproduction (also known as vegetative reproduction, clonal growth and vegetative multiplication) results in the formation of new individuals that are genetically identical to the parent plant and capable of physiologically independent growth (ramets). This differs from the production of branches or leaves which do not usually persist independently. A genet is the entire genetic individual and is composed of ramets (Fig. 5.3).

Clonality is a highly successful plant strategy and has evolved independently many times in individuals of species that are not otherwise similar or closely related, i.e. it is a 'polyphyletic' trait. Clonal plants have a global distribution. About 28% of dicotyledons have some sort of clonal reproduction (Leakey, 1981), and about 40% of these are predominantly clonal. In North America,

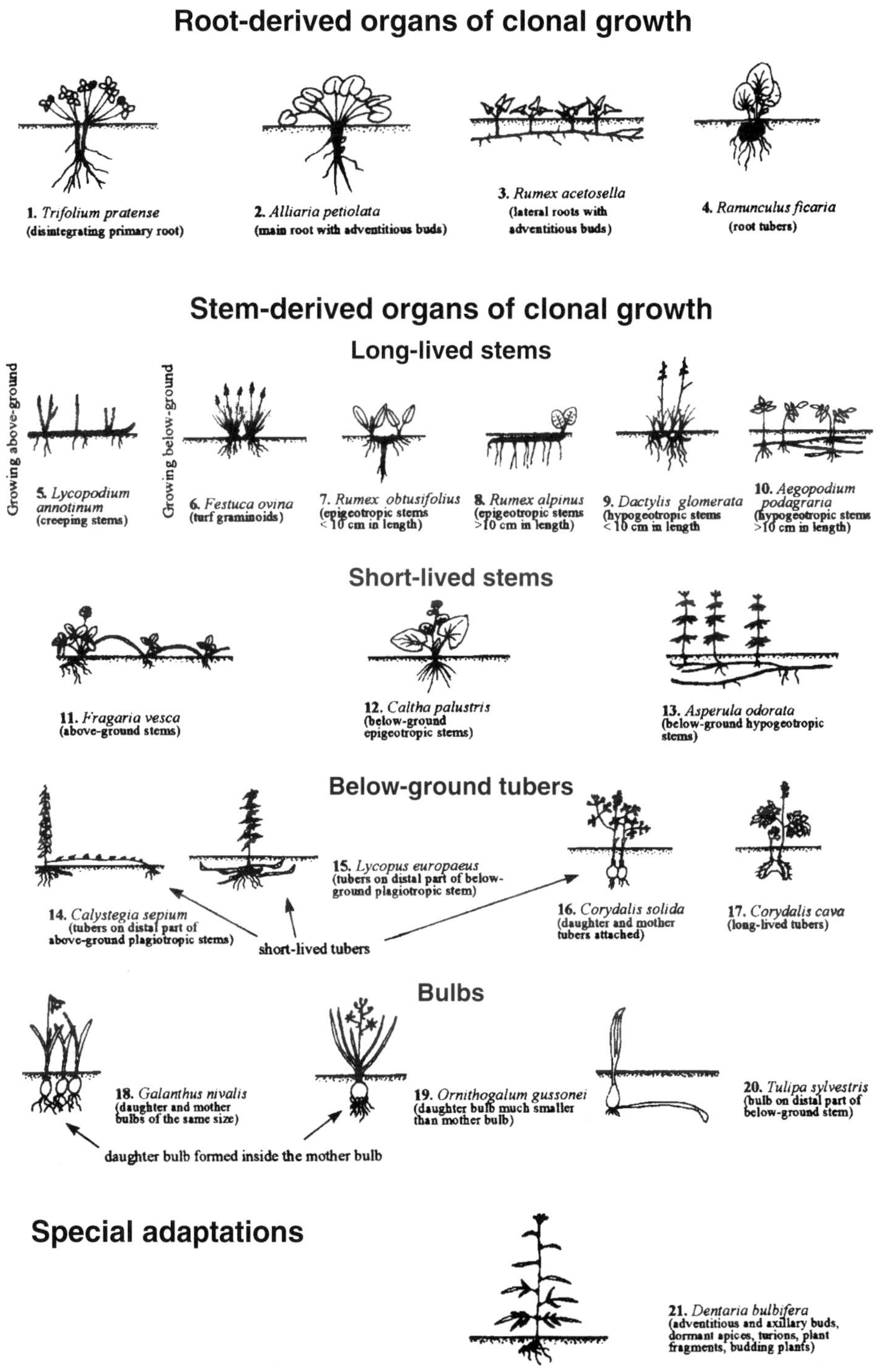

Fig. 5.4. Classification of clonal growth types based on Central European vegetation (Klimeš *et al.*, 1997; with permission of Backhuys Publishers).

Table 5.3. Definitions of structures common in clonal species with weed examples.

Term	Definition	Examples
Creeping stems		
Rhizome	A horizontal, underground structure connecting ramets. It may bear roots and leaves and it may be cordlike or fleshy	Bermudagrass, *Cynodon dactylon* Quackgrass, *Elytrigia repens* Kentucky bluegrass, *Poa pratensis* Field horsetail, *Equisetum arvense*
Stolon and runner	An above-ground, horizontal branch (stolon) or stem (runner) connecting ramets or plantlets. Roots and shoots develop from nodes	Bermudagrass, *Cynodon dactylon* Redtop grass, *Agrostis stolonifera* Strawberry, *Fragaria* spp. Crabgrass, *Dactylis glomerata*
Tuber	An underground storage organ formed from the stem or root and lasting only one year. New tubers are formed each year from different tissue	Yellow nutsedge, *Cyperus esculentus* Purple nutsedge, *Cyprus rotundus* Field horsetail, *Equisetum arvensis*
Shoot bases		
Bulb	A fleshy underground storage organ composed of leaf bases and swollen scale leaves	Wild onion, *Allium vineale* Lily, *Lilium* spp. Wild garlic, *Allium sativum*
Bulbit	A small bulb developing from an above-ground shoot either in place of a flower (vivipory) or on a lateral shoot	Wild onion, *Allium vineale* Wild garlic, *Allium sativum*
Corm	A non-fleshy underground storage organ formed from the swollen base of the stem	Buttercup, *Ranunculus bulbosus* Oat grass, *Arrhenatherum etatius*
Root suckers		
	Above-ground shoots that emerge from creeping roots, tap roots or root tubers	Canada thistle, *Cirsium arvense* Field bindweed, *Convolvulus arvensis*

over 70% of monocot families are predominantly clonal. When considered on an area-covered basis, clonal species are important. For example, the ten most widespread species in Britain are clonal and together they cover 19% of the landmass (Callaghan *et al.*, 1992). While individual ramets may have shorter life spans, genets may survive for thousands of years and cover thousands of square metres (Cook, 1985). Clonality is common in perennials but not annuals or biennials. Many of the worst weeds are clonal.

Clonality is a successful strategy in stable but harsh conditions, such as in the Arctic (van Groenendael *et al.*, 1996; Peterson and Jones, 1997). Historically, clonal families have had greater success during periods of climatic stress. Their current distribution reflects their habitat preferences. For example, in Central Europe, clonal plants tend to be in colder, wetter, nutrient poor habitats (Klimeš *et al.*, 1997).

Mechanisms of clonal growth

Both herbaceous and woody plants can be clonal. For herbaceous plants, the classification of clonal structures is based on: the tissue of origin (stem or root), the position of the growing tip (above or below ground), the structure of storage organs (bulbs or tubers), and the length and longevity of the connections (spacers) between ramets (Klimeš *et al.*, 1997) (Fig. 5.4). Definitions of clonal structures are given in Table 5.3. A clonal plant may possess one or many of these characteristics. Wallaby grass (*Amphibromus scabri-*

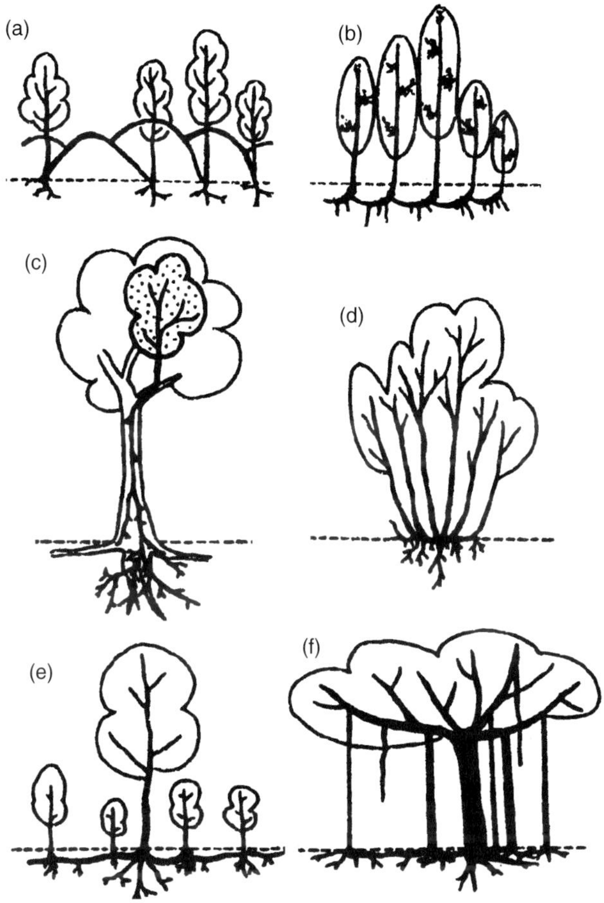

Fig. 5.5. Examples of how clonal growth occurs in woody plants. Shown are (a) layering from drooping branches, (b) sprouting rhizomes, (c) reiteration by aerial shoots or from within roots, (d) basal rooting of coppice shoots, (e) suckering from root buds, and (f) rooting of freehanging roots (Jeník, 1994; with permission of the Institute of Botany of the Czech Academy of Sciences).

valvis), for example, produces rhizomes and corms in addition to flowers (Cheplick, 1995).

Clonal reproduction of woody plants long has been exploited by nursery workers. It allows growers to bypass the stages of seed production, seed germination and seedling establishment, decreasing both the time and mortality rate inherent in these stages of growth. The mechanisms of woody clonality differ somewhat from herbaceous clonal growth. New ramets develop from woody individuals either when shoots (trunk, branch and twigs) bear a root primordia, or when roots bear a shoot bud (Fig. 5.5). The most common types of clonality in woody plants are through sprouting of roots (root suckers) and from layering of branches and stems when they come in contact with soil. Root suckering is exhibited almost exclusively by angiosperms, whereas gymnosperms are more likely to layer.

Costs and benefits of clonal growth

As with agamospermy, there are costs and benefits associated with clonal reproduc-

Table 5.4. Costs and benefits of vegetative reproduction.

	Description
Benefit	
Increased growth rate	New individuals (ramets) bypass the seedling stage and are capable of rapid growth that will increase survivorship and reproductive potential
Movement to better environment	Creation of new ramets allows the genet to move spatially and thus invade new, possibly better environments. This buffers the genet from spatial variability
Sequestering of biological space	The occupation of space and increased potential to capture resources will decrease the chance of invasion by other species
Lower mortality	New ramets have a lower mortality rate than seedlings
Invasion potential	Movement of large genets increases potential to invade and displace competitors. Ramets that remain attached can draw resources from a wide patch supporting the invasion front
Increased resource acquisition	Spreading plants have a high potential to invade nutrient rich environments. This may be of benefit in spatially or temporally heterogeneous habitats
Buffering of temporal variability	Storage organs increase survival during stressful periods and changing environments
Risk aversion to the genet	The risk to a genet is spread among the ramets
No 'cost of sex'	Creation of ramets does not incur the costs associated with sexual reproduction.
Persistence	Some clones are extremely long lived
Costs	
Loss of genetic recombination	Lack of genetic recombination through sexual reproduction means the benefits from novel genotypes are lost
Vulnerable to disturbance	Spatial integrity of clones can make them more vulnerable to large scale disturbances such as floods, fire and frost heave
Mortality of individual ramets	Nutrients are shared among ramets, therefore survivorship of an individual ramet may be decreased in favourable habitat
Transmission of disease	A disease may be able to spread throughout the portions of a genet that remain connected
Decreased sexual reproduction	The creation of new clones decreases the allocation of resources to sexual reproduction

tion (Table 5.4). One of the main benefits to clonal growth is that it allows the individual to bypass the juvenile stage of growth necessary for individuals that reproduce by seeds. The seedling stage is often where the highest mortality occurs for plants. Thus, new clonal individuals have a higher growth rate, lower mortality and can take up 'biological space' that might otherwise become occupied by competitors. A second general benefit to clonal growth is that new ramets can move into other habitats allowing the genet to invade new space or enter into a better environment while maintaining a presence in the 'old' habitat. The main cost to clonal reproduction is that there is no new genetic recombination, which reduces an individual's ability to adapt to new environments. In addition, if ramets remain attached, it more likely that the entire genet may be killed by disturbance, disease or herbicides.

Ecological aspects: the phalanx vs. guerrilla strategies

Two general types of clonal growth are guerrilla and phalanx (Lovett Doust, 1981) (Fig. 5.3). Guerrilla-type growth forms loosely packed, often linear patches. Guerrilla growth is a foraging strategy that maximizes movement of a species into new habitats. Such species are likely to invade new habitats and vacate other ones over the course of a season (Hutchings and Mogie, 1990). For example, at your local golf course or in your own lawn, redtop grass (*Agrostis stolonifera*) will spread by sending ground level stems (stolons) into an area that is occupied by other plants. Once there, the stolons produce new shoots that, in turn, produce more stolons for further colonization. At first, only a few stems of redtop grass appear as if by stealth, but eventually the habitat area is taken over as more stolons and shoots are produced.

Phalanx-type growth is the result of slow growing, branched clones which form dense patches. Phalanx growth exploits space by maximizing the occupation of a site and deterring invasion from other species. Such species form dense monocultures with individuals of approximately equal size. Patches have little movement over the course of a season (Hutchings and Mogie, 1990). Often, peripheral (younger) ramets are dependent on interior (older) ramets for resources while interior ramets flower and set fruit (Waller, 1988). Quackgrass spreads in a phalanx pattern.

Ecological aspects of clonality

Species persistence

Clonal species often have the ability to persist at the edges of their distribution because they are not dependent on sexual reproduction. This pattern is thought to occur because seed production requires higher temperatures than clonal growth, and because fewer appropriate pollinators are found in stressful habitats (Abrahamson, 1980). Japanese knotweed (*Fallopia japonica*), for example, is distributed widely in northern Europe and North America, and yet seed production has never been observed in places like Britain or the USA (Brock *et al.*, 1995). Therefore, preventing seed production may not be enough to eradicate a weed because it may survive and spread through clonal growth.

Clonal species may form remnant populations when conditions become unsuitable for seed production. Arctic dwarf birch (*Betula glandulosa*), for example, forms clonal stands at the northern edge of its distribution in sites where the species was once widely distributed (Hermanutz *et al.*, 1989). Even populations that appear entirely clonal may revert to sexual seed production when conditions improve and isolated clonal stands can act as seed sources for recolonization when conditions improve. Whorled wood aster (*Aster acuminatus*) remains in small clonal populations under forest canopy, but produces seed when a canopy gap opens (Hughes *et al.*, 1988).

Woody plants that are clonal benefit from better physical stability and protection from most risks (e.g. fire, wind and herbivores) (Peterson and Jones, 1997). Peripheral ramets may protect inner ones by buffering them from damage. In addition, if a disturbance removes the above-ground biomass, then sprouts from an existing rootstock will provide a 'sprout bank' (Ohkubo *et al.*, 1996). Species with root sprouts have a better chance of establishing than species reliant on seeds.

Physical and physiological integration of ramets

Ramets may remain physically attached through connectors (e.g. stolons or rhizomes) or they may fragment into independent parts. The degree of ramet integration can vary from highly integrated compact patches to fragmented genets forming only loose associations. Separation of ramets occurs naturally when specialized tissues (called 'plantlets') are abscised or when parts of the plant are separated through decay of the tissue that connects them. Separation also occurs through fragmentation caused by disturbance. In agricultural systems, for example, tillage will fragment quackgrass rhizomes.

The longevity of connections between ramets determines the success of reproduction, exploitation and persistence of the genet (van Groenendael *et al.*, 1996)

Table 5.5. Costs and benefits associated with maintaining physiological integration among ramets (based on text in Jonsdottir and Watson, 1997).

Benefits	Costs
Support new ramets Buffering environmental heterogeneity and stress Resources sharing, division of labour among ramets Regulation of competition among ramets through the control of ramet production Recycling of resources	Cost of maintaining rhizomes, stolons and other connecting tissue Higher risk of genet mortality and extinction Resource dilution

(Table 5.5). When ramets remain integrated, they continue to share resources. For example, older established ramets may support younger ones, during early establishment, by sending resources to them. Integration benefits the entire genet because it increases the longevity of the clone and prolongs the occupation of the site.

The benefits of remaining integrated increase in heterogeneous environments (Wijesinghe and Handel, 1994). Genets with integrated ramets effectively live in two or more places at once, because each shoot section is anchored in a different microhabitat (Alpert and Stuefer, 1997). Having multiple rooting sites reduces the risk to the intact genet, because resources can be shared between ramets and, therefore, ramets in poorer sites are supported by ones in better sites. Sharing resources may result in less biomass accumulation of individual ramets, but total biomass of the genet will increase.

Genets that remain integrated are typical of nutrient poor environments (van Groenendael *et al.*, 1996). Integrated clones tend to interact more intra-clonally than with other species. Ramets of woody species remain integrated for long periods of time and this results in woody clones being long lived and spreading extensively. In fact, the largest plant is argued to be a clonal patch of trembling aspen (*Populus tremuloides*) found in Utah, USA. A single male clone contains approximately 47,000 trees (ramets) and covers 43 ha (Grant, 1993).

Integrated ramets may 'specialize' or create a 'division of labour' (Alpert and Stuefer, 1997). Resources within a genet can be re-allocated quickly because of phenotypic plasticity. This allows genets to exploit nutrient-rich sites by rapidly increasing ramet density in a localized microhabitat (Hutchings and Mogie, 1990) or modifying root and shoot structure to optimize resource use within their environment (de Kroon and Hutchings, 1995; van Groenendael *et al.*, 1996).

Genets are more likely to fragment in nutrient rich environments and this allows them to colonize and monopolize large tracts of land. Examples of fragmenting species include bracken fern, (*Pteridium aquilinum*), Kentucky bluegrass (*Poa pratensis*) quackgrass and white clover (*Trifolium repens*); these all form larger patches in open habitats with adequate moisture (Jonsdottir and Watson, 1997).

In heterogeneous environments with both nutrient-rich and nutrient-poor microhabitats, the genets may fragment but the fragments are much smaller in size because the microhabitat is smaller than in consistently nutrient-rich environments. Tall goldenrod (*Solidago altissima*), Canada goldenrod (*Solidago canadensis*), eastern lined aster (*Aster lanceolatus*) and New York aster (*Aster novi-belgii*) are examples of species that produce small fragments and form small patches in moderately disturbed, shaded environments. While fragmented genets allow for greater colonization, the trade-off is that such species are less likely to spread to new, favourable habitats because they concentrate their resources in one place (Hutchings and Mogie, 1990). The lower colonization ability is ironic because fragmented genets often must be better interspecific competitors than integrated genets.

Table 5.6. Expected differences between asexually produced and sexually produced offspring (adapted from Williams, 1975, and Abrahamson, 1980).

Asexual offspring	Sexual offspring
Mitotically standardized	Meiotically diversified
Produced continuously	Produced seasonally
Develop close to parent	Can be widely dispersed
Develop immediately	Can be dormant
Develop more directly to reproductive stage	Develop more slowly from seedling stage to reproductive stage
Environment and optimum genotype predictable from those of parent since they are genetically the same as parent	Environment and optimum genotype unpredictable because genetic recombination has occurred
Low mortality rate	High mortality rate – especially during seedling stage.

This occurs because when ramets fragment, the genet will encounter other species more often than its own ramets.

Sexual Reproduction in Asexually Reproducing Species

Clonal reproduction rarely occurs to the total exclusion of sexual reproduction, although there are examples of this, such as Japanese knotweed. Clonal growth may occur at the expense of seed production (creeping buttercup, *Ranunculus repens*, and Canada goldenrod). A trade-off occurs between these two methods of reproduction because only a finite amount of resources is available to allocate to reproduction (Abrahamson, 1980). The allocation of resources to sexual vs. asexual reproduction will change over the life of the genet. Sexually and asexually produced offspring will have different genetic and ecological characteristics (Table 5.6). For example, offspring produced through sexual reproduction will differ genetically from their parents and, have the ability to disperse, but suffer a high mortality rate in the seedling stage. Offspring produced asexually will develop immediately and have a low mortality rate, but have less dispersal potential.

In some cases there can be individual plants reproducing sexually and asexually alongside conspecific (same species) individuals that are reproducing only one way. For example, individuals of the introduced wild garlic (*Allium vineale*) produce a stalk which has either only sexually reproducing flowers, only asexual bulbits (bulbs produced on shoots where the flowers normally are located) or both, in addition to producing two types of bulbs at their base (Ronsheim and Bever, 2000). The relative allocation of resources to bulbs, bulbits and flowers is under strong genetic control, as genotypes do not vary allocation patterns in response to nutrient addition.

Seedling recruitment may occur only at some times during the life of a clonal species. For example, seedling recruitment of Canada goldenrod occurred only in the first 3–6 years after colonization, and successful genets were established mainly in the first year (Hartnett and Bazzaz, 1983). This pattern of recruitment is called 'initial seedling recruitment' (ISR). It results in a population with an even age structure because new individuals are recruited at approximately the same time. ISR genets may be long-lived because once established they can be virtually immortal unless a disturbance kills the entire genet.

White clover is an example of the opposite type of recruitment pattern where there is continual recruitment of new genets into the population via seed production (Barrett and Silander, 1992). This type of recruitment is called repeated seedling recruitment (RSR). Such populations have an uneven age structure. Following a disturbance, some genets die making room for new recruitment of genets. RSR genets have

Table 5.7. Expected life-history trends for clonal plants, in relation to their seedling recruitment patterns: initial seedling recruitment (ISR) and repeated seedling recruitment (RSR) (from Eriksson, 1989).

	ISR	RSR
Key features in recruitment phase	Dispersal	Competitive ability
Genetic diversity in local population	Low	High
Prospects for evolution of locally adapted population	Low	High
Genetic age-structure in local population	Even-aged	Variable
Genetic life span	Long	Variable
Spatial context for including genetic population dynamics	Large scale	Small scale

shorter life spans because they are continually being replaced (Eriksson, 1993). Of course, many species are likely to be located along a continuum between ISR and RSR. Table 5.7 summarizes the life history traits associated with ISR and RSR patterns.

Case history: Plantain pussytoes – a species with agamospermy, clonal reproduction and sexual reproduction

Plantain pussytoes (*Antennaria parlinii*) is a herbaceous perennial in eastern North America that reproduces via both sexual and agamospermic seeds and clonally through stolons. Asexual populations tend to be more prevalent in disturbed early successional sites (fields and pastures), whereas sexually reproducing populations tend to be in less disturbed sites (open woods and old fields).

Michaels and Bazzaz (1986) compared demographic characteristics and resource allocation of sexual and asexual populations of plantain pussytoes (Table 5.8). Asexual individuals had higher fecundity rates because they produced more, but smaller seeds; however, seedling survivorship was lower. Clonal growth was high in asexual populations with more ramets being produced; however, stolon length and survivorship were decreased. Sexual populations produced long-lived wandering stolons that allowed the genet to persist in spatially and temporally unpredictable envi-

Table 5.8. Comparison of seed production, seedling establishment and clonal growth in populations of sexual (female plants only) and asexual (agamospermic) pussytoes (*Antennaria parlinii*) (from data in Michaels and Bazzaz, 1986, 1989).

Stage	Factor	Sexual populations	Agamospermic populations
Seed production	Seed number/inflorescence	252 seeds	389 seeds
	Seed mass[a]	approx. 77 μg	approx. 68 μg
	Inflorescences/plant	Fewer	More
Seedling survivorship	Midsummer	47%	22%
	End of growing season	7%	4%
Ramet demography	Ramet production (no. ramets/genet)	2.3	2.8
	Ramet survivorship	85%	68%
	Stolon length (cm/genet)	8 cm	5 cm
Biomass	Total biomass	More	Less
	Allocation to reproduction	Less	More
	Response to increase in resources	Little change in biomass allocation	More towards reproduction

[a]Estimated from graphed data.

ronments. Sexual populations had higher competitive ability, whereas asexual populations had higher fecundity.

Asexual populations of plantain pussytoes were able to grow in a wider range of nutrient and light levels than sexual populations (Michaels and Bazzaz, 1989). This often is beneficial to colonizing species, especially in environments with unpredictable selection pressures. Furthermore, asexual populations responded to increased resources by allocating proportionally more to increased fecundity and reproductive biomass. Overall, sexual populations of pussytoes allocated more resources to clonal growth and persistence, and were more competitive than asexual populations. Asexual populations had higher fecundity and were able to respond to changes in environmental resources.

Summary

As illustrated by the case study of plantain pussytoes, the relative costs and benefits of sexual and asexual reproduction and, hence the relative success of either reproductive method, are influenced by the selection pressures in the environment. The main benefits of asexual reproduction (apomixis) are similar to those of seed production. Like seed produced via selfing or other forms of inbreeding, the cost of asexual reproduction is that there is little chance for genetic recombination. Without genetic recombination, there is reduced ability to produce new genotypes that can be 'fit' to new or otherwise changing environments. Further, alleles that impede survival may accumulate and ultimately end the ability of an individual to reproduce, asexually or sexually. The trade-off between agamospermy and clonal reproduction is that agamosperms produce seeds to allow long dispersal colonization, while cloning increases colonizing ability and persistence because it eliminates seedlings – the stage of growth when risk of mortality is highest.

Questions

1. Does the plant you selected, reproduce asexually? If so, describe how. Is this an important reason why your species is considered a weed?.
2. Why does clonal reproduction become important in northern environments or in other harsh environments?
3. Explore the available literature and list some examples of weeds that have a phalanx growth form, or a guerrilla growth form.
4. How would asexual reproduction in weeds influence their management?
5. Considering the management regime for your lawn, design a 'perfect weed' for colonization and persistence.

General References

Abrahamson, W.G. (1980) Demographic and vegetative reproduction. In: O.T. Solbrig (ed.) *Demography and Evolution in Plant Populations*, Blackwell Scientific, Oxford, pp. 89–106

de Kroon, H. and van Groenendael, J. (1997) *The Ecology and Evolution of Clonal Plants*. Backhuys Publishers, The Hague.

Oborny, B and Bartha, S. (1995) Clonality in plant communities – an overview. *Abstracta Botanica* 19, 115–127.

Literature Cited

Abrahamson, W.G. (1980) Demography and vegetative reproduction. In: Solbrig, O.T. (ed.) *Demography and Evolution in Plant Populations.* Blackwell Scientific, Oxford, pp. 89–106.

Alpert, P. and Stuefer, J.E. (1997) Division of labour in clonal plants. In: de Kroon, H. and van Groenendael, J. (eds) *The Ecology and Evolution of Clonal Plants.* Backhuys Publishers, Leiden, The Netherlands, pp. 137–154.

Asker, S.E., and Jerling, L. (1992) *Apomixis in Plants.* CRC Press, Boca Raton, Florida, 298 pp.

Barrett, J.P. and Silander, J.A. Jr (1992) Seedling recruitment limitation in white clover (*Trifolium repens*; Leguminosae). *American Journal of Botany* 79, 643–649.

Barrett, S.C.H. (1992) Genetics of weed invasions. In: Jain, S.K. and Botsford, L.W. (eds) *Applied Population Biology.* Kluwer, Dordrecht, pp. 91–119.

Brock, J.H., Child, L.E., de Waal, L.C. and Wade, M. (1995) The invasive nature of *Fallopia japonica* is enhanced by vegetative regeneration from stem tissue. In: Pyšek, P., Prach, K., Rejmánek, M. and Wade, M. (eds) *Plant Invasions – General Aspects and Special Problems.* SPB Academic Publishing, Amsterdam, pp. 131–139.

Callaghan, T.V., Carlsson, B.A., Jonsdittir, I.S., Svensson, B.M. and Jonasson, S. (1992) Clonal plants and environmental change: introduction to the proceedings and summary. *Oikos* 63, 341–347.

Cheplick, G.P. (1995) Life history trade-offs in *Amphibromus scabrivalvis* (Poaceae): allocation to clonal growth, storage, and cleistogamous reproduction. *American Journal of Botany* 82, 621–629.

Cook, R.E. (1985) Growth and development in clonal plant populations. In: Jackson, J.B.C., Buss, L.W. and Cook, R.E. (eds) *Population Biology and Evolution of Clonal Organisms.* Yale University Press, New Haven, Connecticut, pp. 259–296.

Cox, P.A. (1985) Islands and dioecism: insights from the reproductive ecology of *Pandanus tectorius* in Polynesia. In: White, J. (ed.) *Studies in Plant Demography.* Academic Press, London, pp. 359–372.

de Kroon, H. and Hutchings, M.J. (1995) Morphological plasticity in clonal plants: the foraging concept reconsidered. *Journal of Ecology* 83, 143–152.

Eriksson, O. (1989) Seedling dynamics and life histories in clonal plants. *Oikos* 55, 231–238.

Eriksson, O. (1993) Dynamics of genets in clonal plants. *Trends in Ecology and Evolution* 8, 313–317.

Ford, H. (1985) Life history strategies in two coexisting agamospecies of dandelion. *Biological Journal of the Linnean Society* 25, 169–186.

Grant, M.C. (1993) The trembling aspen. *Discover* 14, 82–89.

Hartnett, D.C. and Bazzaz, F.A. (1985) Physiological integration among intraclonal ramets of *Solidago canadensis. Ecology* 64, 779–788.

Hermanutz, L.A., Innes, D.J. and Weis, I.M. (1989) Clonal structure of arctic dwarf birch (*Betula glandulosa*) at its northern limits. *American Journal of Botany* 76, 755–761.

Hughes, J.W., Fahey, T.J. and Bormann, F.H. (1988) Population persistence and reproductive ecology of a forest herb: *Aster acuminatus. American Journal of Botany* 75, 1057–1064.

Hutchings, M.J. and Mogie, M. (1990) The spatial structure of clonal plants: control and consequences. In: van Groenendael, J and de Kroon, H. (eds) *Clonal Growth in Plants.* SPB Academic Publishing, Amsterdam, pp. 57–76.

Jeník, J. (1994) Clonal growth in woody plants: a review. *Folia Geobotanica et Phytotaxis* 29, 291–306.

Jonsdottir, I.S. and Watson, M.A. (1997) Extensive physiological integration: an adaptive trait in resource-poor environments. In: de Kroon, H. and van Groenendael, J. (eds) *The Ecology and Evolution of Clonal Plants.* Backhuys Publishers, Leiden, The Netherlands, pp. 109–136.

Klimeš, L., Klimesova, J., Hendrik, R. and van Groenendael, J. (1997) Clonal plant architecture: A comparative analysis of form and function. In: de Kroon, H. and van Groenendael, J. (eds) *The Ecology and Evolution of Clonal Plants.* Backhuys Publishers, Leiden, The Netherlands, pp. 1–29.

Leakey, R.R. (1981) Adaptive biology of vegetatively regenerating weeds. *Advances in Applied Biology* 6, 57–90.

Lovett Doust, L. (1981) Population dynamics and local specialization in a clonal perennial (*Ranunculus repens*). I. The dynamics of ramets in contrasting habitats. *Journal of Ecology* 69, 743–755.

Michaels, H.J. and Bazzaz, F.A. (1986) Resources allocation and demography of sexual and apomictic *Antennaria parlinii. Ecology* 67, 27–36.

Michaels, H.J. and Bazzaz, F.A. (1989) Individual and population responses of sexual and apomictic plants to environmental gradients. *American Naturalist* 134, 190–207.

Mogie, M. (1992) *The Evolution of Asexual Reproduction in Plants.* Chapman and Hall, London.
Ohkubo, T., Tanimoto, T. and Peters, R. (1996) Response of Japanese beech (*Fagus japonica* Maxim.) sprouts to canopy gaps. *Vegetatio* 124, 1–8.
Peterson, C.J. and Jones, R.H. (1997) Clonality in woody plants: a review and comparison with clonal herbs. In: de Kroon, H. and van Groenendael, J. (eds) *The Ecology and Evolution of Clonal Plants.* Backhuys Publishers, Leiden, The Netherlands, pp. 263–289.
Pyšek, P. (1997) Clonality and plant invasions: can a trait make a difference? In: de Kroon, H. and van Groenendael, J. (eds) *The Ecology and Evolution of Clonal Plants.* Backhuys Publishers, Leiden, The Netherlands, pp. 405–427.
Richards, A.J. (1997) *Plant Breeding Systems,* 2nd edn. Chapman and Hall, London.
Ronsheim, M.L. and Bever, J.D. (2000) Genetic variation and evolutionary trade-offs for sexual and asexual reproductive modes in *Allium vineale* (Liliaceae). *American Journal of Botany* 87, 1769–1777.
van Dijk, P. and van Damme, J. (2000) Apomixis technology and the paradox of sex. *Trends in Plant Science* 5, 81–84.
van Groenendael, J.M., Klimeš, L., Klimesova, J. and Hendriks, R.J.J. (1996) Comparative ecology of clonal plants. *Philosophical Transactions of the Royal Society of London, Series B* 351, 1331–1339.
Waller, D.W. (1988) Plant morphology and reproduction. In: Lovett Doust, J. and Lovett Doust, L. (eds) *Plant Reproductive Ecology: Patterns and Strategies.* Oxford University Press, New York.
Wijesinghe, D.K. and Handel, S.N. (1994) Advantages of clonal growth in heterogeneous habitats: an experiment with *Potentilla simplex. Journal of Ecology* 82, 495–502.
Williams, G.C. (1975) *Sex and Evolution.* Princeton University Press, Princeton, New Jersey.

6
From Seed to Seedling

Concepts

- The number of seeds that a plant produces will depend on many interacting factors including genetic constraints, germination date, plant size and environmental conditions. There is a trade-off between seed size and seed number.
- Seed dispersal moves seeds away from the parent plant and may lead to higher seedling survivorship. Animals, water, wind, humans or machinery may disperse seeds.
- Seeds are dispersed over time by remaining dormant in the seed bank. Dormancy can be innate, induced or enforced.
- Seeds germinate in response to environmental conditions such as light, temperature and moisture.
- Seedling establishment is dependent on environmental conditions. Conditions that are optimal for germination may not be optimal for seedling establishment.

Introduction

Reproduction is but one part of the larger life cycle of plants and there is no start or finish (Fig. 6.1). We started our discussion of plant life cycles with the process of seed production, although we could very well have started with seed dispersal or germination. In this chapter, we will focus on these three topics as well as the subsequent stages of seedling emergence. We discuss how biotic and abiotic processes affect each of these stages, and how events during one stage will influence the progress of subsequent stages.

Seeds are the primary mobile stage of the life cycle. Like the other mobile phase (pollen dispersal), the fate of seeds is dependent on the wind, water or animals that disperse them. Furthermore, because they often contain high levels of nutrients, seeds are a good food source for many animals. In some cases, consumption results in their dispersal to favourable habitats, while in other cases, seeds are destroyed by consumption or end up in hostile environments. Essentially, the seed must find its 'safe site', a set of tolerable or favourable environmental conditions, in order to survive to produce a seedling. Even if seeds survive, the resultant seedlings have

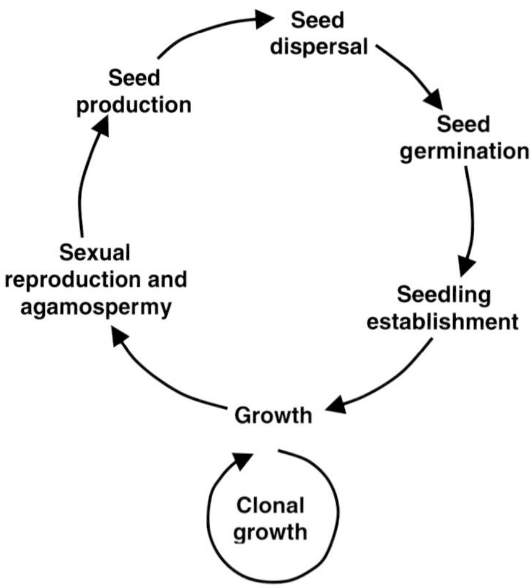

Fig. 6.1. Life cycle of plants has no real beginning or end. One chooses an arbitrary 'start'.

high rates of mortality because they are small and lack long-term nutritional reserves. It seems that the odds are against the survival of seeds and seedlings, and yet, many individuals have strategies that increase their chance of survival.

Fruit and Seed Production

What are seeds and fruits?

A seed develops from the fertilized ovule and contains an embryonic plant surrounded by a protective seed coat. It also contains nutritional reserves in the form of either endosperm or cotyledons. In angiosperms, seeds may be dispersed within a fruit formed from the flower ovary or receptacle. A fruit may contain one or many seeds. Fruits are divided into two basic types: fleshy and dry. Fleshy fruits include peaches, tomatoes and figs; acorns, rice and beans are dry fruits.

Seed set

Since a plant has a limited amount of resources for reproduction, there is a trade-off between the number of seeds an individual can produce and the size of seeds. The number of successful seedlings produced by an individual plant depends on the total number of seeds it produces, and on their size.

Seed number

The number of seeds produced by an individual plant will depend on the number of ovules produced, their rate of fertilization, and on how many fertilized ovules survived to become mature seeds. What determines the actual number of seeds produced? First, there are the genetic constraints over the number (and size) of seeds a species can produce. Orchids produce thousands of dust-sized seeds but cannot produce coconut-size seeds; coconut trees cannot produce as many seeds as orchids do. Within these constraints, seed number is influenced by the availability of resources and by the environmental conditions during pollination and seed development. For example, the number of seeds produced by redroot pigweed (*Amaranthus retroflexus*) decreases as light level decreases (McLachlan *et al.*, 1995).

The benefit of producing many seeds is that they may have more opportunities for colonization because of the sheer numbers of seeds produced and reduced losses from seed herbivores that cannot usually find and destroy all the seeds.

Seed size

Seed size has many repercussions for dispersal and seedling establishment. The main benefit of having large seeds is that resultant seedlings are usually more competitive because they have more nutrient reserves and can survive harsher conditions for longer periods of time. Seedlings from large seeds are better able to withstand drought, defoliation, shade, litter and competition from established vegetation or concurrently emerging seedlings from relatively small seeds (Westoby *et al.*, 1996; Leishman, 2001). Larger seeds, however, require more energy to produce and are more likely to be consumed by seed herbivores in search of easy to find and nutritious meals (Reader,

1993; Thompson *et al.*, 1993; Rees, 1996). There are advantages to having small seeds. Individuals with small seeds produce many more of them than individuals with large seeds and can do so because small seeds require less energy to produce (Leishman, 2001). Though small seedlings (from small seeds) are more likely to die before reaching maturity, they make up for the loss in sheer numbers.

While it was once thought that seed size was a genetically stable trait within a species (Harper, 1977), it is now recognized that environmental variation often causes seed size to vary greatly within species, populations and individuals (Michaels *et al.*, 1988). For example, the seed size of the annual weedy cucumber (*Sicyos deppi*) is dependent on the environment in which the fruit develops (Orozco-Segovia *et al.*, 2000). This weed is a vine that climbs up the stems and trunks of the other vegetation in fields and disturbed forests of Mexico. As a result, some fruits develop in full sunlight while others develop in the shade. Seeds that develop in full sunlight are larger and heavier than shaded seed, but seed viability is the same. Alternatively, the size of redroot pigweed seeds do not vary with light level (McLachlan *et al.*, 1995)

Because seed size is genetically controlled and environmentally influenced, selection pressures could lead to a change in seed size. For example, seed size of the weedy gold-of-pleasure (*Camelina sativa*) has diverged over time, depending on the type of flax crop (*Linum* spp.) the weed grows in. Weed seeds became similar to crop seeds because both pass through the winnowing machine at the same time and thus weed seeds similar in size and weight to flax are selected for. In fibre flax, gold-of-pleasure seeds are flat; in flax grown for oil production, the weed seeds are smaller and plumper.

Examples of the trade-off between seed number and seed size

Producing a very few, nutrient-rich, seeds can be disastrous if all of the seeds die because they end up in unsuitable habitats or are destroyed (e.g. eaten). Likewise, producing many nutrient-poor seeds can be equally disastrous if the seedlings are not able to survive the biotic and abiotic stresses of their environment. So how will a plant resolve this trade-off? Eriksson (2000) proposed a model to explain the dispersal and colonizing ability of a species based on seed

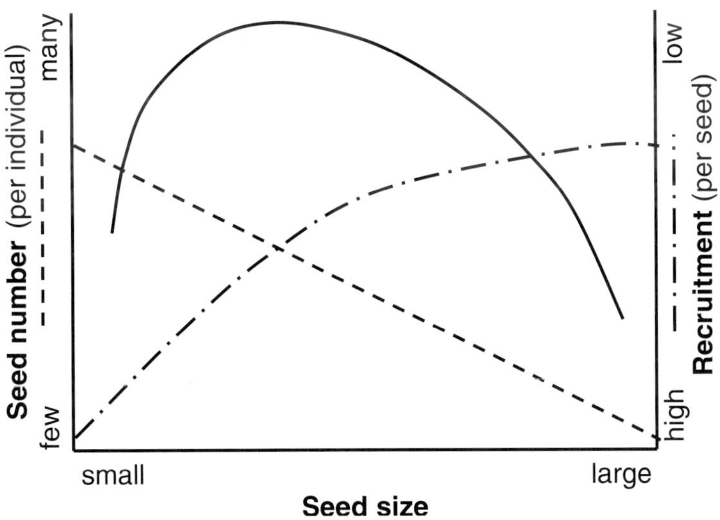

Fig. 6.2. A model of the relationship between seed size, seed number, recruitment and dispersal and colonizing ability (redrawn from Eriksson, 2000).

size (Fig. 6.2). Here, seed size decreases with increased seed production. The number of seedlings that survive to become adults ('recruited') increases as seed size increases but this reaches a limit, i.e. a maximum recruitment threshold. As a result, the maximum combined dispersal and colonizing ability is at an intermediate level of seed size. Eriksson (2000) recognized that this relationship is dependent on habitat and community type. The peak of the dispersal and colonizing ability curve moves leftward in disturbed sites because the benefits of increasing seed number outweigh those of increasing seed size. Additionally, these maxima will vary with short- and long-term environmental variation.

Fruit and seed polymorphisms

In some cases, one plant may produce two or more types of seeds that differ in morphology. Species with two types of seed ('morphs') are dimorphic and those with more are polymorphic. Seed morphs may have different sets of germination requirements or different dispersal mechanisms associated with them. Having two or more morphs is a form of 'bet-hedging'. By producing seeds with different germination and/or dormancy requirements, the plant is likely to have at least *some* seeds germinate. This strategy is advantageous in environmentally variable habitats. The trade-off with bet-hedging is that while it increases the chance of at least some seed germinating in most conditions, fewer seeds will germinate in optimal conditions.

Seeds of common lambsquarters (*Chenopodium album*) are dimorphic for two characters: seed wall and seed texture. There are thin-walled brown seeds that germinate immediately, and thick-walled black seeds that are dormant. In addition, both brown and black seeds have smooth coat and textured coat morphs. The proportion of each of the four seed types varies among populations (Harper *et al.*, 1970). Seed polymorphisms are common in the daisy (*Asteraceae*), goosefoot (*Chenopodiaceae*), grass (*Poaceae*), and mustard (*Brassicaceae*) families; these families also contain many weed species (Harper, 1977).

Wingpetal (*Heterosperma pinnatum*) is a summer annual that has different types of 'achenes' (a dry, single-seeded fruit) within each flowering head; the polymorphisms ensure that some of these short-lived (1 year) seeds will germinate each year. Venable *et al.* (1995) divided these into three morphology types (central, intermediate and peripheral) based on length/width ratio, and the presence of a beak and/or wing (Fig. 6.3). Central achenes are awned (winged) and tend to disperse further. They lose dormancy earlier than other morphs and germinate in the spring. Peripheral achenes do not disperse as far but tend to have higher germination under harsh conditions. Intermediate achenes are longer and skinnier than peripheral achenes but are not awned. The relative proportion of these morphs differs among population; this is a result of selection under different environments. For example, populations in disturbed habitats (Mirador population) have a higher proportion of central achenes as these are more likely to dispersal away from a habitat that may be eliminated. In populations receiving heavy early rains (Tula population), there are more peripheral achenes that germinate late and can withstand this environment.

Getting Away from Your Parents: I. Seed Dispersal (Dispersal in Space)

Seeds are dispersed away from parents for several reasons. First, it avoids seedlings competing with their parents or siblings. Competition among siblings or between parent and offspring is more intense because they tend to have similar resource requirements and strategies to acquire these resources. Being dispersed away from the maternal plant also decreases the likelihood of mating with a sibling, which could lead to inbreeding depression.

Second, dispersal increases colonization opportunities. If all seeds fall directly around their parent plant, then the species has little chance of colonizing new habitats

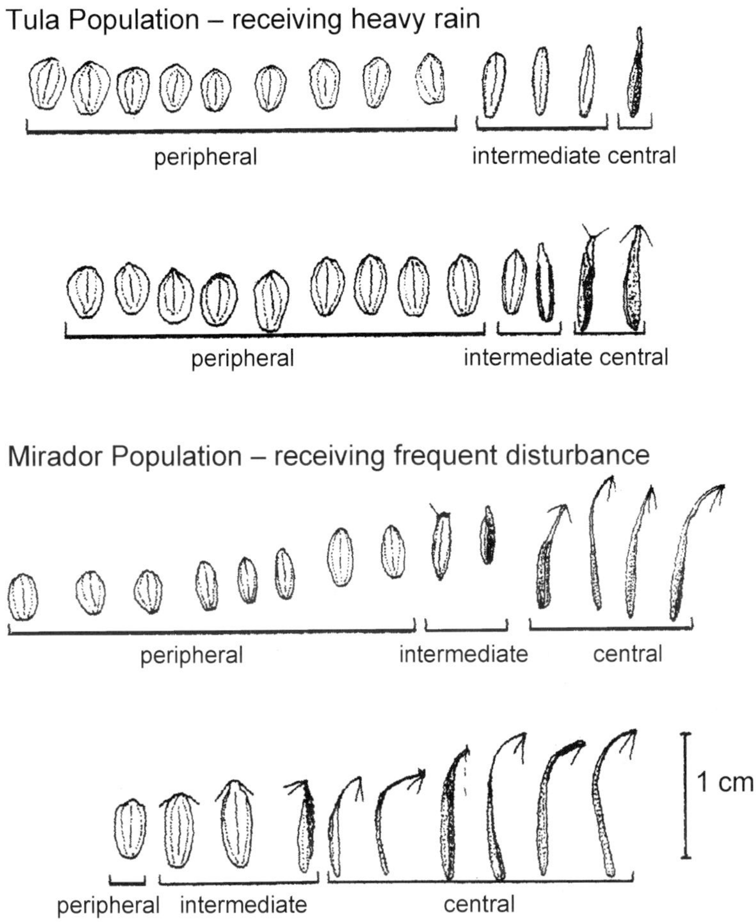

Fig. 6.3. Examples of three morphs (peripheral, intermediate and central) of achenes from two populations of wingpetal (*Heterosperma pinnatum*). The Tula population received heavy rain while the Mirador population was frequently disturbed (Venable *et al.*, 1995; with permission of the Botanical Society of America and the authors).

or expanding its range. Also, seedling establishment is sometimes higher in habitats away from the parent plant, e.g. where a parent plant creates too much shade for the seedling.

Finally, dispersal also reduces the chance of all seeds from an individual being eaten by seed herbivores or otherwise destroyed all at once by parasites or pathogens. This could occur if all seeds were dispersed to the same small area in high densities of closely related individuals. This would increase the intensity of attack by parasites or pathogens attracted to specific genotypes of seeds.

The cost of dispersal is that energy is allocated to dispersal structures (e.g. wings or fleshy fruits) and away from seed production itself; the benefits of seed dispersal must outweigh this cost (Howe and Smallwood, 1982; Willson, 1992; Eriksson and Kiviniemi, 1999). If a plant produces dispersal structures, then less resources are available to produce seeds. Consequently, there is a trade-off between dispersal and seed production, and plants have developed

Table 6.1. Dispersal methods (based on van der Pijl, 1982, and Willson *et al.*, 1990).

Dispersal method	Morphological features	Weed examples
Wind	Wings or plumes that slow rate of fall	Dandelion, *Taraxacum officinalis* Bull thistle, *Cirsium vulgare*
Vertebrate	Fleshy fruit, aril or seeds/fruit with high nutritional content	European buckthorn *Rhamnus cathartica* Common barberry *Berberis vulgaris*
Ant	Elaiosome (food body) to attract ants	Bull thistle, *Cirsium vulgare* Mile-a-minute weed, *Polygonum perfoliatum*
Ballistic	Seed is ejected from fruit by specialized structure	Bur-cucumber, *Sicyos angulatus* Touch-me-not, *Impatiens glandulifera*
External	External barbs and hooks or hairs that adhere to fir and feathers	Bluebur, *Lappula echinata* Common burdock, *Arctium minus*
None	No evident structures	Lambsquarters, *Chenopodium albam*

different dispersal mechanisms or 'syndromes' to maximize dispersal and minimize energy losses.

Dispersal syndromes

The main types of dispersal syndromes are: wind, vertebrate, ant, ballistic and external. Others have no special dispersal device (van der Pijl, 1982; Willson *et al.*, 1990) (Table 6.1). Wind-dispersed seeds are small and light, and tend to have wings or plumes that slow their rate of fall and this increases their dispersal distance. Seeds can be dispersed by animals, either on the animal's exterior or else internally. Seeds dispersed on the outside of animals have features such as barbs, hooks or are sticky causing them to attach to fur, hair or feathers. Seeds that are dispersed internally tend to have fleshy fruit that attracts animals but also have hard seed coats to protect the seed while it passes through the animal's gut. The behaviour of an animal determines the fate and the distribution of a seed; for example, hoarding animals may produce seed caches – clumps of seeds (and seedlings) (Howe and Westley, 1997). A special type of animal dispersal is by ants. These seeds offer external fat tissues (elaisomes) to entice ants. Ballistically dispersed seeds are usually housed in fruit that has a trigger mechanism to propel the seed away from the parent plant. Seeds with no special devices are usually round and have no external morphological features.

This classification system is based on morphological traits of the seeds. Thus, while small light seeds are easily dispersed by wind, they are not considered 'wind-dispersed' unless they have specific structures (Willson *et al.*, 1990). Seed size is associated with dispersal type. Large seeds (>100 mg) tend to be vertebrate dispersed whereas small seeds (<0.1 mg) tend to be unassisted: seeds with other types of dispersal syndromes are intermediate in size (Hughes *et al.*, 1994). Seeds and fruits with specific dispersal syndromes tend to have predictable characteristics (colour, odour, form) and may provide rewards to dispersal agents. Some species may have two or more modes of dispersal.

Humans have become an excellent agent of seed dispersal. While weed seeds do stick to our hair and skin, we are an important seed dispersal agent more because of our technology and mobility across the planet. There are numerous examples of vehicles and native, exotic and domesticated animals dispersing weed seeds, although much of it is rather passive or at least accidental in terms of the motives of the dispersal agent (McCanny and Cavers, 1988; Schmidt, 1989; Mack, 1991; Lonsdale and Lane, 1994). Examples include:

- seeds from burdock (*Arctium minus*) and many other species sticking to clothes and fur;
- seeds from stinging nettle (*Urtica dioica*) and many other species stick to mud affixed to motor vehicles;
- seeds from many agricultural weeds (e.g. proso millet) being combined along with the crops;
- weeds that have been a part of commercial seed mixes for both agriculture and ecological restoration purposes (e.g. dame's rocket, *Hesperis matronalis*).

The last bullet point above is also an example of deliberate dispersal, i.e. species were introduced and used before it was realized that they were prone to being weeds. We have intentionally introduced seeds for landscaping, farming or erosion control.

Weed dispersal may be said to be more 'deliberate' in the sense that native seed dispersers may preferentially consume fruits from exotic weeds. This occurs when weeds have larger, more palatable seeds or fruit than native species; for example, weedy English hawthorn (*Crataegus monogyna*) has seeds that are more attractive to birds than native *Crataegus* species (Sallabanks, 1993). An odder example is that of the seeds of weedy pine species (*Pinus*) introduced to Australia are now dispersed by cockatoos instead of by wind as is normal in native pine habitats (Richardson and Higgins, 1998).

Primary and secondary dispersal

Seeds will not be evenly dispersed throughout a habitat because they are subject to myriad abiotic and biotic factors. These may be categorized as primary dispersal (seed is dispersed from the parent plant to the ground) and secondary dispersal (seed is

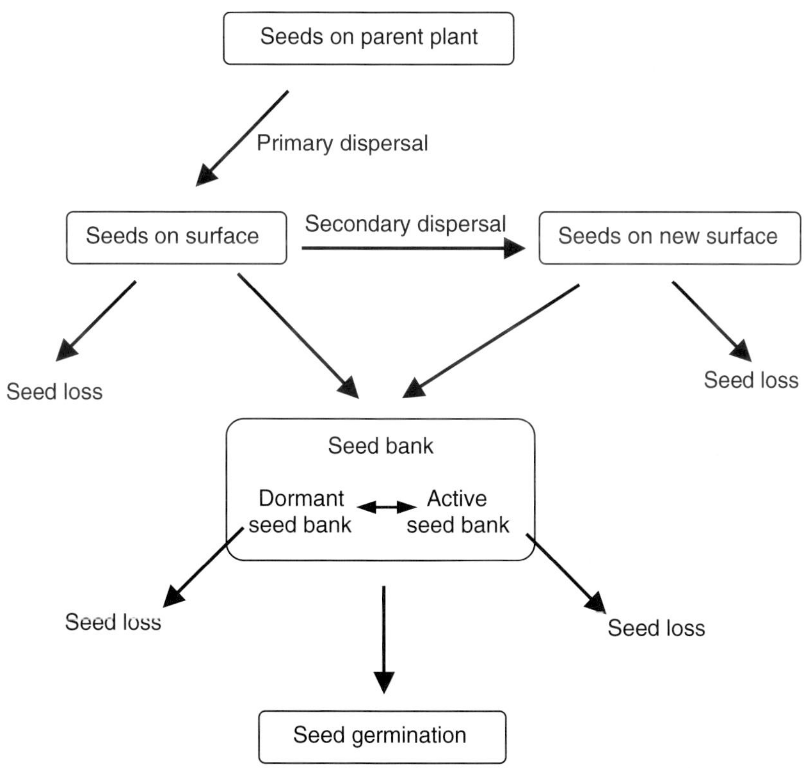

Fig. 6.4. Movement and fates of seeds (based on Chambers and MacMahon, 1994).

subsequently moved to other sites) (Fig. 6.4).

The extent of primary dispersal is determined by the morphological characteristic of the seed. For example, small winged or plumed seeds (e.g. fleabanes, *Erigeron* spp.) travel further in wind than unadorned and relatively heavy seeds (e.g. pigweeds). Secondary dispersal is an important factor because it moves seeds horizontally away from or closer to the parent, or vertically deeper or shallower in the soil. The extent of secondary dispersal is dependent on seed characteristics and how they interact with abiotic factors (gravity, wind, rain and frost heaving) and the physical structure of the community (topography, vegetation and soil) (Chambers and MacMahon, 1994). Even slight changes in the physical structure can change secondary dispersal patterns. A slight change in soil particle size can determine whether a seed moves in response to wind. Secondary dispersal is further influenced by biotic factors. Many types of seed movement are the result of animals (digging, burrowing, scatter-hoarding). For example, the presence of earthworms, increases the depth at which annual bluegrass (*Poa annua*), common chickweed (*Stellaria media*) and shepherd's purse (*Capsella bursa-pastoris*) seeds are buried (van der Reest and Rogaar, 1988; Fig. 6.5). Animals show specific preferences for weed seed types, but their level of consumption is dependent on seed density (e.g. Cromar *et al.*, 1999). Animals also have microsite pref-

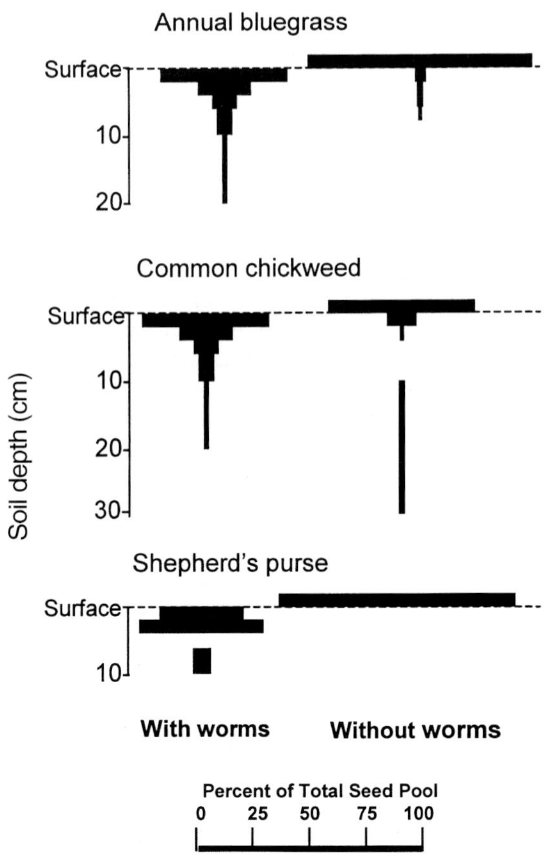

Fig. 6.5. The effect of earthworms on the vertical distribution of annual bluegrass (*Poa annua*), common chickweed (*Stellaria media*), and shepherd's purse (*Capsella bursa-pastoris*) seed (redrawn from van der Reest and Rogaar, 1988).

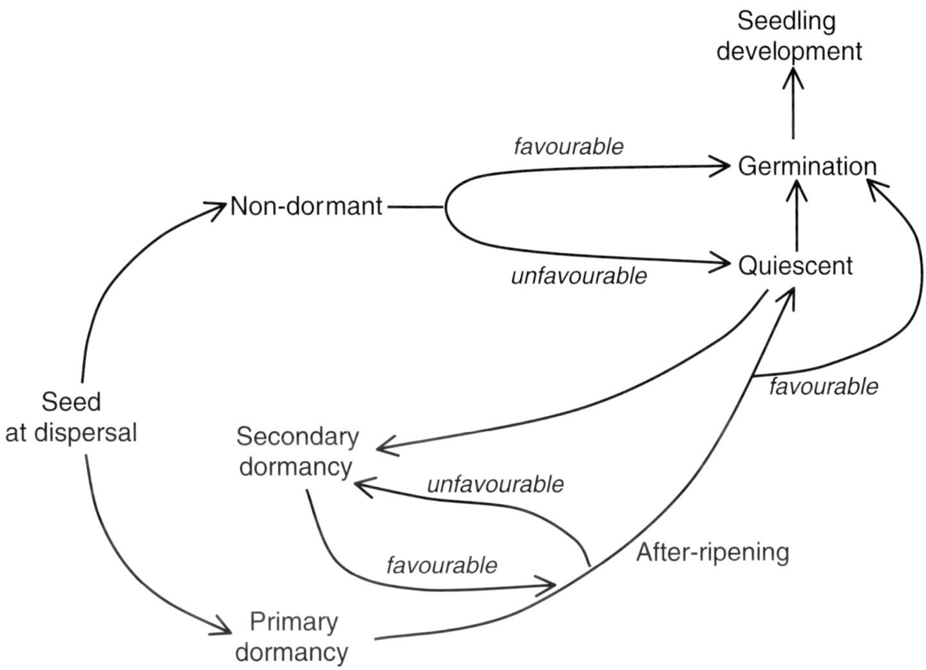

Fig. 6.6. Biotic and abiotic factors affecting seed from the time it is dispersed from maternal parent to its germination (Foley, 2001).

erences, hence this affects the availability of seeds from different weed species, e.g. some animals may prefer depths where seeds of certain weed species are not found.

Getting Away from Your Parents: II. Seed Banks and Seed Dormancy (Dispersal in Time)

When a seed is dispersed away from its parent, it may be either dormant (in a resting state and be unable to germinate) or non-dormant (able to germinate) (Fig. 6.6). Both dormant and non-dormant seeds eventually become incorporated into the soil as part of the 'seed bank' – think of it as a repository of seeds that will be withdrawn over time. Seeds that are able to remain in the seed bank for long periods of time do so because they are dormant. In the following sections we discuss seed banks and seed dormancy. Seed germination is discussed later in this chapter.

Seed banks

The seed bank is referred to as 'dispersal in time' because it provides the same essential benefit as dispersal through space – it increases the chance that at least some seeds will survive to germinate under suitable environmental conditions. Unfortunately for plants, the seed bank is not a benign place and seeds cannot survive indefinitely. Seed survival decreases because of failed germination, physiological death, disease, herbivory, pathogens, adverse soil conditions (pH and moisture) and deep burial (Simpson *et al.*, 1989). Burial in the seed bank offers a brief respite at best for most individuals. Seeds in the seed bank are continually redistributed through secondary dispersal. Human activity is another mechanisms of seed redistribution. For example, tillage can alter the distribution (Fig. 6.7a) and density (Fig. 6.7b) of seeds in the soil, often species specifically (Clements *et al.*, 1996).

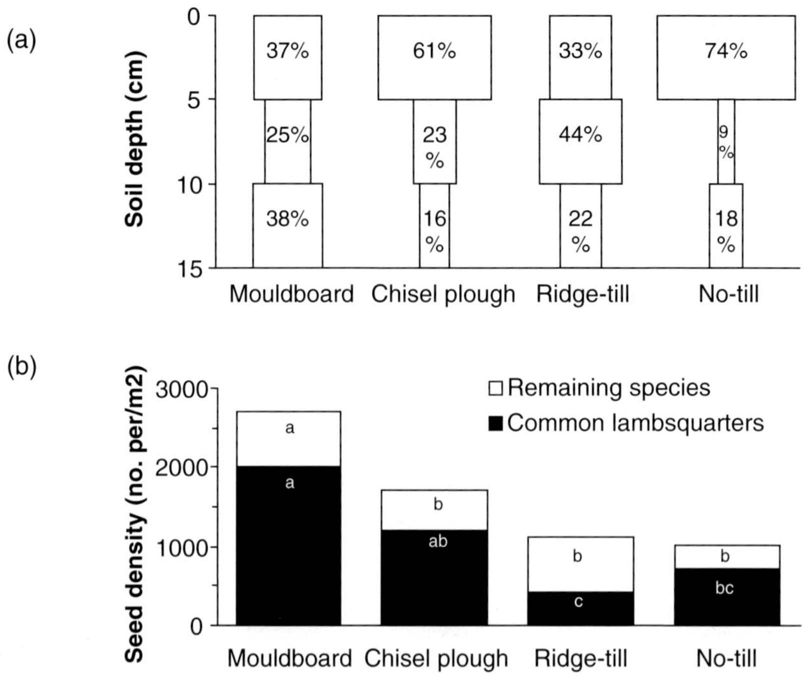

Fig. 6.7. Effect of four types of tillage on (a) seed distribution, and (b) the density of common lambsquarters (*Chenopodium album*) in the seed bank (Clements *et al.*, 1996; with permission of the Weed Science Society of America).

A species' seed bank can be classified based on seasonal variation in germinable seeds (Thompson and Grime, 1979). The two main types are transient and persistent seed banks. These two types are further subdivided. Transient seed banks contain seeds that do not last for more than 1 year; they may contain either autumn-germinating seeds (Type I – e.g. perennial rye grass, *Lolium perenne*) or spring-germinating seeds (Type II – e.g. Himalayan balsam, *Impatiens glandulifera*) (Fig. 6.8).

Persistent seed banks contain seeds that remain viable for more than 1 year. They either contain many seeds that germinate in autumn but maintain a small seed bank throughout the year (Type III – e.g. Rhode Island bentgrass, *Agrostis tenuis*), or they may have a large persistent seed bank year round (Type IV – e.g. common chickweed, *Stellaria media*). Most of the world's worst weeds have large persistent seed banks (Holm *et al.*, 1977). Species that are increasing in abundance are more likely to have a persistent seed bank than those that are decreasing in abundance (s.f. Hodgson and Grime, 1990).

Monocarpic species tend to have more persistent seed banks than polycarpic species. This allows a seed carry-over effect for monocarpic species and prevents them from becoming locally extinct. If all individuals of a monocarpic species die before they can reproduce, there will still be viable seeds that can germinate. In polycarpic perennial species, seed carry-over is not as important because seeds are produced repeatedly during the life span. Species with persistent seed banks tend to have small, rounded seeds (Thompson and Grime, 1979; Thompson *et al.*, 1993). This is likely to occur because small seeds are more likely to become buried either by falling down cracks in the soil or through burial by soil organisms (Thompson *et al.*, 1993). Once buried, seeds are more likely to persist because they have lower predation rates and are less like-

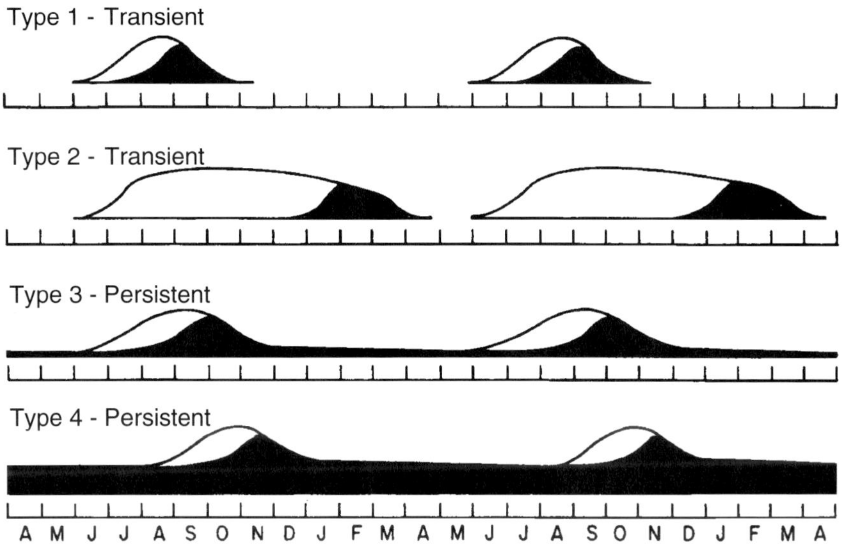

Fig. 6.8. Types of transient (Type 1 and 2) and persistent (Type 3 and 4) seed banks. Shaded areas show seed capable of immediate germination, while unshaded areas show viable seed that are not capable of immediate germination. Examples of each seed bank type are: Type 1: annual and perennial grasses of dry and disturbed habitats. Type 2: annual and perennial herbs colonizing gaps of early spring. Type 3: species mainly germinating in the autumn but maintaining a small persistent seed bank. Type 4: annual and perennial herbs and shrubs with large persistent seed banks. (Thompson and Grime, 1979; with permission of Blackwell Science.)

ly to face germination-inducing environmental conditions, such as light, compared to seeds on the soil surface.

Seed dormancy

The main reason that seeds remain viable in the seed bank is because they are usually dormant, i.e. they cannot germinate until a specific set of environmental and physiological conditions are met. During dormancy the seed exhibits little growth or development and respiration is reduced (Rees, 1997; Benech-Arnold *et al.*, 2000). This allows the seed to persist but expend little resources on maintenance. Dormancy prevents germination while the seed is still on the parent plant and ensures temporal dispersal into environments favourable to seedling survival (Murdoch and Ellis, 1992). Long-lived perennials tend to have short-lived seed

Table 6.2. Definitions of terms associated with seed dormancy.

Term	Definition
After-ripening	A process whereby seeds are gradually able to germinate over a broader range of conditions
Dormant	Seeds unable to germinate even though they have imbibed and are under favourable environmental conditions
Non-dormant	Being able to germinate under favourable environmental conditions
Primary dormancy	Seeds that are unable to germinate when they mature and are either dispersed or still attached to the maternal parent plant
Quiescent	Being unable to germinate due to unfavourable environmental conditions
Secondary dormancy	Dormancy that is imposed on the seed after being dispersed

banks, whereas short-lived species are more dependent on dormancy. Table 6.2 is a list of definitions associated with dormancy.

Primary and secondary dormancy

Seeds that are unable to germinate when they first mature have primary dormancy. When dormancy is imposed after seeds have dispersed, this is called secondary dormancy. Dormancy is usually imposed when environmental conditions are unfavourable for prolonged periods of time. Dormancy is adaptive because it prevents seeds from germination during seasons when environmental conditions are unsuitable for the growth of that species.

Seeds may cycle in and out of dormancy, changing from dormant to conditionally dormant (where they germinate under a smaller range of conditions) to non-dormant; this cycle repeats and can result in annual dormancy cycles observed in many weeds such as barnyardgrass (Honek *et al.*, 1999) (Fig. 6.9). Dormancy of summer annuals such as common ragweed (*Ambrosia artemisiifolia*) and lady's-thumb (*Polygonum persicaria*) is released in the spring by low winter temperatures and re-induced by early summer high temperatures. Winter annuals such as ivy-leaved speedwell (*Veronica hederifolia*) and henbit (*Lamium amplexicaule*) require higher summer temperatures to release dormancy for fall germination. Dormancy cycles ensure that seeds remain viable over seasons (by not expending resources) but are able to germinate when conditions are appropriate for them.

Secondary dormancy is maintained by several mechanisms. Physiological mechanisms in the seed embryo may prevent it from germinating. Physical mechanisms can also enforce dormancy (Foley, 2001). A hard seed coat is one type of secondary dormancy as it prevents water and/or gasses from entering the seed. Seeds with a hard seed coat usually require some sort of physical or chemical abrasion to break dormancy. Alternatively, chemical inhibitors within the seed coat can maintain dormancy. These chemicals must be removed, for example by leaching, before the seed can germinate. Velvetleaf (*Abutilon theophrasti*) and field bindweed (*Convolvulus arvensis*) are examples of weeds with seed coat-imposed dormancy, while wild oat (*Avena fatua*) and annual sunflower (*Helianthus annua*) have embryo dormancy.

Seeds exhibiting dormancy usually have to experience periods of favourable environmental conditions during a period called 'after-ripening' (Fig. 6.6). After-ripen-

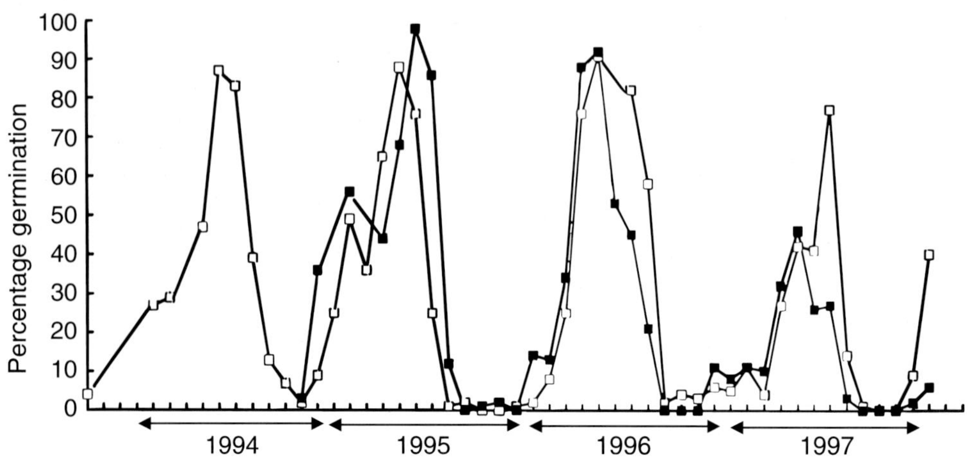

Fig. 6.9. Patterns of cyclic dormancy in barnyardgrass (*Echinochoa crus-galli*). Seeds buried in 1993 (open squares) and 1994 (black squares) were periodically retrieved and germinated in light at 25°C (Honek *et al.*, 1999; with permission of Blackwell Science).

ing is a process whereby seeds are gradually able to germinate over a broader range of conditions (Baskin and Baskin, 1989). The environmental conditions required for after-ripening to occur are specific to individuals (and often broadly to populations and species). For example, common and giant ragweed (*Ambrosia artemisiifolia* and *A. trifida*) require cool, moist conditions (Bazzaz, 1970; Ballard *et al.*, 1996) whereas common cocklebur (*Xanthium strumarium*) requires warm dry conditions (Esashi *et al.*, 1993).

Breaking dormancy

The conditions required to break dormancy often vary among species, but also vary within species or populations. This is especially true if their geographical range or habitats have a high degree of environmental variation (Allen and Meyer, 1998). In species that normally experience wide environmental variation, e.g. beard-tongues (*Penstemon* spp.) and blue flax (*Linum perenne*), only some seeds respond to dormancy-breaking conditions and they maintain a long-term seed bank as a hedge against sudden environmental change within a growing season. Within a given species, populations that experience colder winters have more dormant seeds and require longer periods of cold to break dormancy than populations with milder winters. Intermediate populations often have variable dormancy, with differences occurring either within or among individual plants (Allen and Meyer, 1998).

Even seeds produced from one individual may have different dormancy-breaking requirements. This often occurs in species with polymorphic seeds where seeds with different morphs have different types of dormancy. For example, seeds from the peripheral flowers on tansy ragwort (*Senecio jacobaea*) are larger, heavier and require longer periods of time to break dormancy; whereas seeds from central flowers are small, lighter, disperse further and are less dormant (McEvoy, 1984). Similarly, common lambsquarters produces mainly dormant black seeds but a few (< 1% of total) are non-dormant brown seeds (Roman *et al.*, 2000).

How the cues for breaking dormancy and germination differ

When dormancy is 'released' or 'broken', this does not necessarily lead to seed germination as conditions required to break dormancy are not necessarily the same as those required for germination (Benech-Arnold *et al.*, 2000). A seed that loses dormancy may either germinate, become dormant again, or it may die. Release from dormancy and seed germination generally are considered as two sequential processes.

Seed Germination

Seeds that are not dormant may not germinate if they have not encountered favourable environmental conditions: these are termed 'quiescent' (Foley, 2001). They are 'seeds in waiting'. Quiescent seeds are able to germinate immediately once they encounter favourable environmental conditions, but may revert to secondary dormancy (Fig. 6.6).

The critical factor for seeds is to be able to germinate at an appropriate time, a daunting task since environmental conditions vary on small spatial scales and also are rarely constant from day to day and year to year. For example, in temperate regions, seeds that germinate late often experience intense competition from other individuals and have a shorter growing season to complete their life cycle. Conversely, seeds that germinate early may experience high mortality from unfavourable environmental conditions (e.g. frost). However, the risk may be worth it as the ones that survive are better competitors for light and other resources and have higher fitness (Ross and Harper, 1972; Marks and Prince, 1981; Gross, 1984).

Most weeds are able to germinate throughout the growing season and this is an important reason why they are successful, though each individual will experience differential success for the reasons we just mentioned. For example, redroot pigweed and lambsquarters can germinate from April to October. In addition to the early season weather risks, any individuals in agricultural fields likely will be subject to weed management. Some may escape management and survive to set seed but the chances of success are generally low. However, this may be worth the risk because later germinating weeds (e.g. in mid- to late summer) can be far worse at reproducing successfully. Later germinating weeds may not be subject to management because it is not economically viable for a farmer to do so. These weeds may acquire enough light, water and nutrients to survive under the crop canopy but they may have insufficient time or resources to set seed. While the parent may survive, their offspring may not (or may not even exist) and their fitness is reduced. In fact, it is likely that most of the weed seeds returned to the seedbank will be from those individuals that germinated early, survived weather and management conditions, and produced seeds (Swanton *et al.*, 1999).

The timing of seed germination is triggered by environmental cues. The most common cues are light quality and quantity, temperature, moisture and gases (O_2 and CO_2). These generally vary on large scales (latitude) but they also vary locally. On a local scale, seeds of different species are sensitive to different aspects of their physical microenvironment. For example, Sheldon (1974) found that annual sow-thistle and dandelion seeds that had the attachment end (end that was formerly attached to the seed head) in closest contact with a moist substrate had the highest percentage germination (Fig. 6.10).

Requirements differ among species but, generally, species with small seeds tend to require light for germination more than large seeded species (Milberg *et al.*, 2000). Alternating temperatures are required for the seed germination of many weeds, including common lambsquarters, large crabgrass (*Digitaria sanguinalis*), field bindweed, orchardgrass (*Dactylisglomerata glomerata*) and some species of dock (*Rumex* spp.). The effect of temperature fluctuations depends on the:

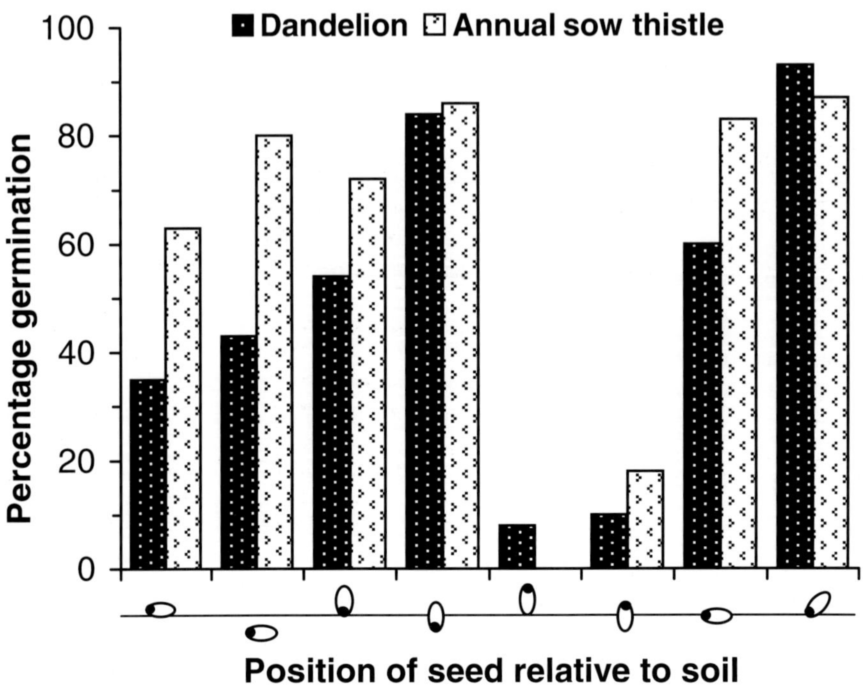

Fig. 6.10. Effect of seed position (in relation to the soil surface) on the germination of dandelion and annual sow thistle. The dark end of the seed is the attachment end (adapted from Sheldon, 1974).

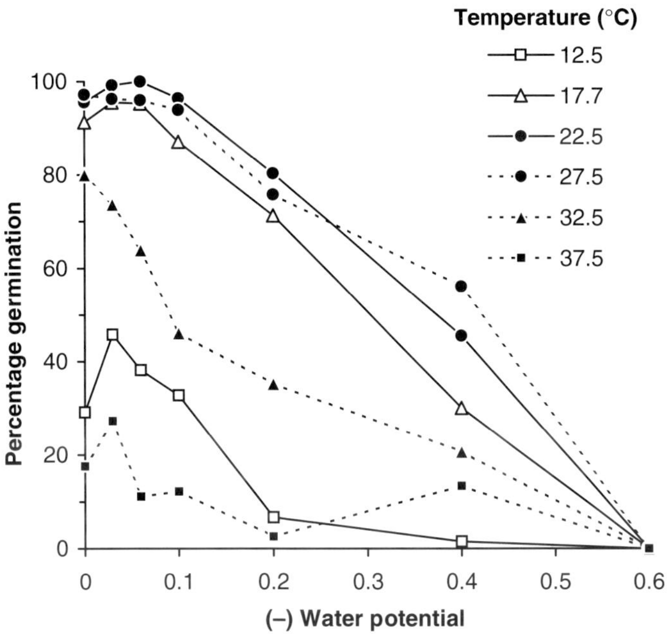

Fig. 6.11. The effect of temperature and moisture (water potential) on the germination of common lambsquarters. The highest percentage germination was achieved at moderate temperatures (17.7–27.5°C) and low water potential (high water availability) (based on data from Roman, 1998).

- amplitude (difference between maximum and minimum temperatures);
- mean temperature;
- thermoperiod (time above mean temperature each day) (Probert, 1992).

In some species, germination requirements are well understood and we can be predict what proportion of them will germinate based on moisture and temperature (Fig. 6.11) (Roman, 1998). We also understand that for some species, germination cues can be influenced by the maternal environmental conditions. For example, the temperatures during seed after-ripening affect germination requirements of redroot pigweed and common lambsquarters (Baskin and Baskin, 1987). When seeds develop at higher temperatures, more seeds germinate at higher temperatures.

Seedling Emergence

The term 'seedling' is simply another way of saying 'young plant' and implies no specific age or stage range. Trees are often referred to as seedlings into their second or third year simply because they are small compared to their adult form, whereas a fast-growing annual may be called a seedling only for a matter of days. The distinction between seed germination, seedling emergence and establishment is not always clear and the terminology can be confusing. Germination normally means that the seed is physiologically active and the embryo is undergoing mitosis to produce a shoot and/or root that emerges from the seed coat. Emergence usually refers to the appearance of a shoot above the soil or a root from the seed. Establishment is generally considered to occur once a seedling no longer depends on seed reserves (endosperm and cotyledons), i.e. it is photosynthetically independent. We discuss establishment in Chapter 7.

Factors affecting seedling emergence

The seedling stage often has the highest mortality rate of a plant's life cycle (Harper, 1977) because seedlings are vulnerable to environmental stress. As with germination, the timing of seedling emergence is important because it determines whether an individual will be able to compete with its neighbours, subject to herbivory or disease, and timing of other life history events (Forcella *et al.*, 2000). The timing of seedling emergence is determined by the interaction of seed size, dormancy, germination, and the rate of stem and root elongation with abiotic factors (e.g. soil temperature, temperature fluctuations, soil moisture, depth of burial, light) (Allen and Meyer, 1998; Forcella *et al.*, 2000; Roman *et al.*, 2000). For example, Benvenuti *et al.* (2001) examined the effect of burial depth on 20 weed species. Time to emergence increased with depth of burial, and few seeds germinated below 10 cm. In addition, they found that the depth at which 50% of a species' seeds germinated was related to seed weight. For example, species with light seeds (<1 g) had at least 50% emergence only when buried less than 5.5 cm, whereas species with large seeds (>8.5 g) had more than 50% emergence at up to 7 cm burial. Therefore, larger seeds were able to emerge from a greater depth.

Summary: Seeds and Seedling Must Find Their Safe Sites

In summary, a seed must find its safe site, i.e. a site that provides all the conditions necessary for the seed to germinate and emerge from the soil. These conditions include the environmental conditions necessary to break dormancy and allow germination to proceed, as well as protecting the seed and seedling from hazards such as herbivores, competitors and disease. The problem is that most seeds do not end up in a safe site. A seed is not guaranteed to find a safe site because it cannot control its own dispersal; dispersal agents like animals, wind and water do not guarantee safe passage and delivery to a good place to germinate and grow.

Even if the seed finds a safe site and then germinates, the seedling may not survive because safe germination sites may not promote seedling emergence. A seedling may require different environmental conditions, or the environmental conditions may change by the time a seedling emerges.

Environmental conditions are the main barrier to finding a safe site. The environmental conditions needed for a safe site vary among individuals, populations and species. In response to different selection pressures wrought by environmental variation, plants have adapted. Adaptations include increasing the number of seeds produced, increasing the resources allocated to seeds (parents 'pack a bigger lunch' in bigger seeds to sustain them), having protective seed coats, and using dormancy to wait in the seed bank until a safe site is available. Once a safe site is available, a seed may germinate and produce a seedling. In the next chapter, we will discuss how plants respond to the challenges of life as they grow up.

Questions

1. How are seeds and/or fruits dispersed in your selected species, and what are its safe site conditions? What type of seed bank does it have?
2. When trying to reduce the seed bank of a weed, how would the type of dormancy (primary vs. secondary) affect control strategies?
3. Crops have been selected for seeds with no dormancy. Given that most weeds have some type of dormancy, how will this influence management of agricultural and forestry weeds.
4. Explain why in the Eriksson (2000) model (Fig. 6.2), the hump of the dispersal and colonizing ability curve moves to the left in disturbed sites.
5. Explain the benefits of dispersing in time and/or space.

General References

Benech-Arnold, R.L., Sánchez, R., Forcella, F., Kruk, B.C. and Ghersa, C.M. (2000) Environmental control of dormancy in weed seed banks. *Field Crops Research* 67, 105–122.

Forcella, F., Benech Arnold, R.L., Sánchez, R. and Ghersa, C.M. (2000) Modeling seedling emergence. *Field Crops Research* 67, 123–139.

Grubb, J.P. (1977) The maintenance of species-richness in plant communities: the importance of the regeneration niche. *Biological Review* 52, 107–145.

Howe, H.F. and Miriti, M.N. (2000) No question: seed dispersal matters. *Trends in Ecology and Evolution* 15, 434–435.

Literature Cited

Allen, P.S. and Meyer, S.E. (1998) Ecological aspects of seed dormancy loss. *Seed Science Research* 8, 183–191.

Ballard, T.O., Foley, M.E. and Bauman, T.T. (1996) Germination, viability, and protein changes during cold stratification of giant ragweed (*Ambrosia trifida* L.) seed. *Journal of Plant Physiology* 149, 229–233.

Baskin, J.M. and Baskin, C.C. (1987) Temperature requirements for after-ripening in buried seeds of four summer annual weeds. *Weed Research* 27, 385.

Baskin, J.M. and Baskin, C.C. (1989) Physiology of dormancy and germination in relation to seed bank ecology. In: Leck, M.A., Parker, V.T. and Simpson, L.R. (eds) *Soil Seed Banks*. Academic Press, San Diego, California, pp. 53–66.

Bazzaz, F.A. (1970) Secondary dormancy in the seeds of the common ragweed *Ambrosia artemisiifolia*. *Bulletin of the Torrey Botanical Club* 97, 302–305.

Benech-Arnold, R.L., Sánchez, R.A., Forcella, F., Kruk, B.C. and Ghersa, C.M. (2000) Environmental control of dormancy in weed seed banks. *Field Crops Research* 67, 105–122.

Benvenuti, S., Macchia, M. and Miele, S. (2001) Quantitative analysis of emergence of seedlings from buried weeds with increasing soil depth. *Weed Science* 49, 528–535.

Chambers, J.C. and MacMahon, J.A. (1994) A day in the life of a seed: movements and fates of seeds and their implications for natural and managed systems. *Annual Review of Ecology and Systematics* 25, 263–292.

Clements, D.R., Benoit, D.L., Murphy, S.D. and Swanton, C.J. (1996) Tillage effects on weed seed return and seedbank composition. *Weed Science* 44, 314–322.

Cromar, H.E., Murphy, S.D. and Swanton, C.J. (1999) Influence of tillage and crop residue on post-dispersal predation of weed seeds. *Weed Science* 47, 184–194.

Eriksson, O. (2000) Seed dispersal and colonization ability of plants – assessment and implications for conservation. *Folia Geobotanica* 35, 115–123.

Eriksson, O. and Kiviniemi, K. (1999) Evolution of plant dispersal. In: Vuorisalo, T.O. and Mutikainen, P.K. (eds) *Life History Evolution in Plants*. Kluwer, Dordrecht, pp. 215–238.

Esashi, Y., Ogasawara, M., Gorecki, R. and Leopold, A.C. (1993) Possible mechanisms of after-ripening in *Xanthium* seeds. *Physiological Plant* 87, 359–364.

Foley, M.E. (2001) Seed dormancy: an update on terminology, physiological genetics, and quantitative trait loci regulating germinability. *Weed Science* 49, 305–317.

Forcella, F., Benech Arnold, R.L., Sánchez, R. and Ghersa, C.M. (2000) Modeling seedling emergence. *Field Crops Research* 67, 123–139.

Gross, K.L. (1984) Effects of seed size and growth form on seedling establishment of six monocarpic perennials. *Journal of Ecology* 72, 369–388.

Harper, J.L. (1977) *Population Biology of Plants*. Academic Press, London.

Harper, J.L., Lovell, P.H. and Moore, K.G. (1970) The shapes and sizes of seeds. *Annual Review of Ecology and Systematics* 1, 327–356.

Hodgson, J.G. and Grime, J.P. (1990) The role of dispersal mechanisms and regenerative strategies and seed banks in the vegetation dynamics of the British landscape. In: Bunce, R.G.H. and Howard, D.C. (eds) *Species Dispersal in Agricultural Habitats*. Belhaven, London, pp. 65–81.

Holm, L.G., Plucknett, D.L., Pancho, J.V. and Herberger, J.P. (1977) *The World's Worst Weeds: Distribution and Biology*. University of Hawaii, Honolulu.

Honek, A., Martinkova, Z. and Jarosik, V. (1999) Annual cycles of germinability and differences between primary and secondary dormancy in buried seeds of *Echinochloa crus-galli. Weed Research* 39, 69–79.

Howe, H.F. and Smallwood, J. (1982) Ecology of seed dispersal. *Annual Review of Ecology and Systematics* 13, 201–228.

Howe, H.F. and Westley, L.C. (1997) Ecology of pollination and seed dispersal. In: Crawley, M.J. (ed.) *Plant Ecology*. Blackwell Scientific, Oxford, pp. 262–283.

Hughes, L., Dunlop, M., French, K., Leishman, M.R., Rice, B., Rodgerson, L. and Westoby, M. (1994) Predicting dispersal spectra: a minimum set of hypotheses based on plant attributes. *Journal of Ecology* 82, 933–950.

Leishman, M.R. (2001) Does the seed size/number trade-off model determine plant community structure? An assessment of the model mechanisms and their generality. *Oikos* 93, 294–302.

Lonsdale, W.M. and Lane, A.M. (1994) Tourist vehicles as vectors of weed seeds in Kahada National Park, northern Australia. *Biological Conservation* 69, 277–293.

Mack, R.N. (1991) The commercial seed trade: early dispersers of weeds in the United States. *Economic Botany* 45, 257–273.

Marks, M. and Prince, S. (1981) Influence of germination date on survival and fecundity in wild lettuce *Lactuca serriola. Oikos* 36, 326–330.

McCanny, S.J. and Cavers, P.B. (1988) Spread of proso millet (*Panicetum milaceum* L.) in Ontario, Canada. II. Dispersal by combines. *Weed Research* 28, 67–72.

McEvoy, P.B. (1984) Dormancy and dispersal in dimorphic achenes of tansy ragwort, *Senecio jacobaea* L. (Compositae). *Oecologia* 61, 160–168.

McLachlan, S.M., Murphy, S.D., Tollenaar, M., Weise, S.F. and Swanton, C.J. (1995) Light limitation of reproduction and variation in the allometric relationship between reproductive and vegetative biomass in *Amaranthus retroflexus* (redroot pigweed). *Journal of Applied Ecology* 32, 157–165.

Michaels, H.J., Benner, B., Hartgerink, A.P., Lee, T.D., Rice, S., Willson, M.F. and Bertin, R.I. (1988) Seed size variation: magnitude, distribution and ecological correlates. *Evolutionary Ecology* 2, 157–166.

Milberg, P.J., Andersson, L. and Thompson, K. (2000) Large-seeded species are less dependent on light for germination than small-seeded ones. *Seed Science Research* 10, 99–104.

Murdoch, A.J. and Ellis, R.H. (1992) Longevity, viability and dormancy. In: Fenner, M. (ed.) *Seeds: the Ecology of Regeneration in Plant Communities.* CAB International, Wallingford, UK, pp. 193–229.

Orozco-Segovia, A., Brechú-Franco, A.E., Zambrano-Polano, L., Osuna-Fernández, R., Languna-Hernández, G., Sánchez-Coronado, M.E. (2000) Effects of maternal light environment on germination and morphological characteristics of *Sicyos deppei* seeds. *Weed Research* 40, 495–506.

Probert, R.J. (1992) The role of temperature in germination ecophysiology. In: Fenner, M. (ed.) *Seeds: the Ecology of Regeneration in Plant Communities.* CAB International, Wallingford, UK, pp. 285–325.

Rathcke, B. and Lacey, E.P. (1985) Phenological patterns of terrestrial plants. *Annual Review of Ecology and Systematics* 16, 179–214.

Reader, R.J. (1993) Control of seedling emergence by ground cover and seed predation in relation to seed size for some old-field species. *Journal of Ecology* 81, 169–175.

Rees, M. (1996) Evolutionary ecology of seed dormancy and seed size. *Philosophical Translations of the Royal Society of London, Series B* 351 1299–1308.

Rees, M. (1997) Seed dormancy. In: Crawley, M.J. (ed.) *Plant Ecology*, 2nd edn. Blackwell Science, Oxford, pp. 214–238.

Richardson, D.M. and Higgins, S.I. (1998). Pines as invaders in the southern hemisphere. In: Richardson, D.M. (ed.) *Ecology and Biogeography of Pinus.* Cambridge University Press, Cambridge, pp. 450–473.

Roman, E.S. (1998) Modelling seedling emergence of common lambsquarters in corn. PhD thesis, University of Guelph, Ontario.

Roman, E.S., Murphy, S.D. and Swanton, C.J. (2000) Simulation of *Chenopodium album* seedling emergence. *Weed Science* 48, 217–224.

Ross, M.A. and Harper, J.A. (1972) Occupation of biological space during seedling establishment. *Journal of Ecology* 60, 77–88.

Sallabanks, R. (1993) Fruiting plant attractiveness to avian seed dispersers: native vs. invasive *Crataegus* in Western Oregon. *Madrono* 40, 108–116.

Schmidt, W (1989) Plant dispersal by motor cars. *Vegetatio* 80, 147–152.

Sheldon, J.C. (1974) The behaviour of seeds in soil. III. The influence of seed morphology and the behaviour of seedlings on the establishment of plants from surface-lying seeds. *Journal of Ecology* 62, 47–66.

Simpson, L.R., Leck, M.A. and Parker, V.T. (1989) Seed banks: general concepts and methodological issues. In: Leck, M.A., Parker, V.T. and Simpson, L.R. (eds) *Soil Seed Banks.* Academic Press, San Diego, California, pp. 3–8.

Swanton, C.J., Chandler, K. and Shrestha, A. (1999) Weed seed return as influenced by the critical weed-free period in corn (*Zea mays* L.). *Canadian Journal of Plant Science* 79, 165–167.

Thompson, K. and Grime, J.P. (1979) Seasonal variation in the seed banks of herbaceous species in ten contrasting habitats. *Journal of Ecology* 67, 893–921.

Thompson, K., Band, S.R. and Hodgson, J.G. (1993) Seed size and shape predict persistence in soil. *Functional Ecology* 7, 236–241.

van der Pijl, L. (1982) *Principles of Dispersal in Higher Plants*, 3rd edn. Springer-Verlag, Berlin.

van der Reest, P.J. and Rogaar, H. (1988) The effect of earthworm activity on the vertical distribution of plant seeds in newly reclaimed polder soils in the Netherlands. *Pedobiologia* 31, 211–218.

Venable, D.L., Dyreson, E. and Morales, E. (1995) Population dynamic consequences and evolution of seed traits of *Heterosperma pinnatum* (Asteraceae). *American Journal of Botany* 82, 410–420.

Westoby, M., Leishman, M. and Lord, J. (1996) Comparative ecology of seed size and dispersal. *Philosophical Translations of the Royal Society of London, Series B* 351, 1309–1318.

Willson, M.F. (1992) The ecology of seed dispersal. In: Fenner, M. (ed.) *Seeds: the Ecology of Regeneration in Plant Communities.* CAB International, Wallingford, UK, pp. 61–86.

Willson, M.F., Rice, B.L. and Westoby, M. (1990) Seed dispersal spectra: a comparison of temperate plant communities. *Journal of Vegetation Science* 1, 547–562.

7
Growing Up, Getting Old and Dying

Concepts

- Phenology is the study of regularly occurring life cycle events. A plant's phenology is determined by the interaction of its genetic make-up and its biotic and abiotic environment.
- Plants allocate their limited resources to the functions of growth, maintenance and reproduction. Species will do this in different ways, and allocation patterns change with the environment.
- Weeds are often phenotypically plastic and this allows them to take advantage of environmental heterogeneity.
- Their genetic make-up and the abiotic environment (e.g. temperature and light) determine a plant's response via phenology, resource allocation and phenotypic plasticity. This is known as the genotype × environment interaction.
- Senescence is the natural deterioration of plants (or plant parts) leading to death at the end of the plant's life cycle.

Introduction

In the previous chapter, we discussed the importance of the seed and early seedling stages of a plant's life cycle. In this chapter we continue the theme of plant phenology, from the stage of seedling establishment (i.e. when a seedling becomes independent of its seed reserves) through to death. We focus on how abiotic factors influence these life cycle events. We also look at how plants allocate resources to their growth, maintenance and reproduction and how allocation changes over the life of a plant. Finally we discuss plant senescence, the natural process of plant death.

Phenology

A seedling becomes independent of its maternal parent when it has used up all the endosperm nutrient resources (when cotyledons drop off). Following this a plant will grow, producing stems, leaves and roots. Eventually, a plant will flower and produce seeds, and then either die (monocarpic) or have repeated periods of flowering (polycarpic) throughout its life. The study of these regular life cycle events is called phenology. To a certain extent, a plant's phenology is determined by its genetics (a maple tree will never be an annual); however, plants also respond to their environment and therefore the environment influences phenology.

Effect of abiotic factors on phenology

The abiotic environment will have direct and indirect effects on phenology. A phenological event may be directly triggered by a specific environmental cue; for example, a seed may not germinate until a specific temperature is reached. A plant's direct responses may be immediate or delayed (inductive effect). In addition, environmental cues may be qualitative (absolute), where they are necessary for the effect to occur at all, or they may be quantitative (facultative), where they increase the rate of occurrence but are not necessary for it to occur. In some cases, a series of environmental cues may be required before a phenological event is triggered.

Indirectly, the environment can influence phenology because it can influence the rate of important physiological processes (e.g. photosynthesis, cell division). Generally, a plant's response increases as the level of factor increases until an optimal level is reached and then begins to decrease (Fig. 7.1). Beyond an upper limit, elevating the level of the factor will inhibit growth and yield. For example, most plants will not survive in either very cold or very hot temperatures and will have optimum growth at intermediate temperatures. Every species (population or individual) will have a different set of minimum, maximum and optimum responses to a factor. Because plants respond to their environment, the rate at which they pass through their life cycle can vary. Thus, a plant growing in optimal conditions will grow faster than one under stressful conditions, and then be larger when it flowers.

Factors that influence phenology include light, temperature, water, nutrients, gases (oxygen and carbon dioxide), and soil characteristics (pH and texture). In the next sections we look at how temperature and

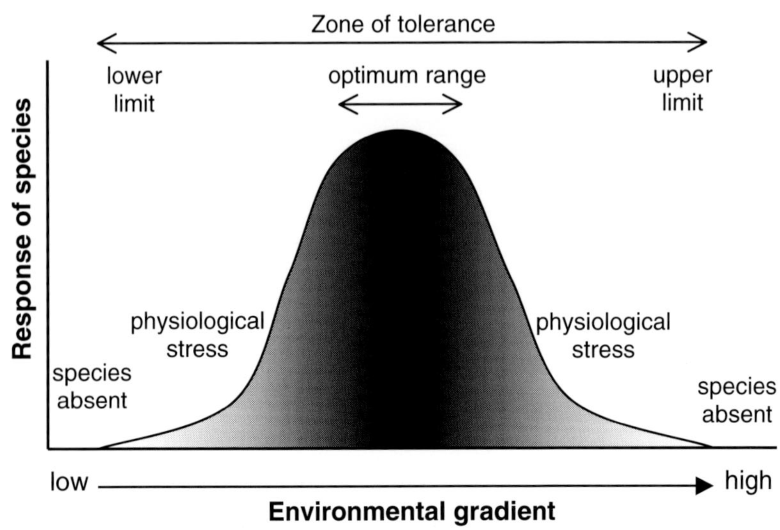

Fig. 7.1. Effect of environment factors (e.g. temperature and light) on the rate of important physiological processes (e.g. photosynthesis, cell division). A plant's response increases as the level of factor increases until an optimal level is reached and then begins to decrease. Beyond an upper and lower limit of the factor, the plant will not survive.

light influence phenology, and then go on to discuss how abiotic factors interact. Note that plants are influenced by other factors, but the effects of these are beyond the scope of this book.

Temperature

Temperature has a strong influence on plant growth rate. We see this when garden plants appear to stop growing on cold spring days, and then 'grow before our eyes' on the next warm day. Below the threshold level, plants will die from freezing damage (or chilling in warmer climates). As temperature increases the metabolic rate increases and growth increases; however, a few degrees beyond the optimum, proteins begin to be denatured (destroyed) and the plant dies. The critical thresholds will change depending on the plant's life stage and environment. For example, a seed will have different limits from those of a seedling or mature plant. Furthermore, individuals growing in colder climates may be adapted or acclimated to lower temperatures than those in warmer climates.

Temperature influences other processes such as dormancy, germination, and bud and flower initiation. Canada thistle, for example, produces new shoots from over-wintering roots only when temperatures reach 5°C (Sheley and Petroff, 1999). Some species require a cold period to promote flowering; if kept in warm temperatures, they will continue to grow vegetatively but will not flower (Salisbury and Ross, 1985). Dalmatian toadflax (*Linaria dalmatica*), for example, requires a winter dormancy period and then exposure to temperatures between 10 and 20°C to produce floral stems (Sheley and Petroff, 1999). Temperature is often an important factor controlling the overall distribution of a species, because it influences many important processes.

The critical temperatures will differ among various processes in the plant. For example, seedling emergence of wild mustard (*Sinapis arvensis*) is fastest under temperatures of 29/19°C and 35/25°C, while plants grow taller at 23/13°C (Huang *et al.*, 2001) (Fig. 7.2). For shoot elongation the optimum temperature (and high and low

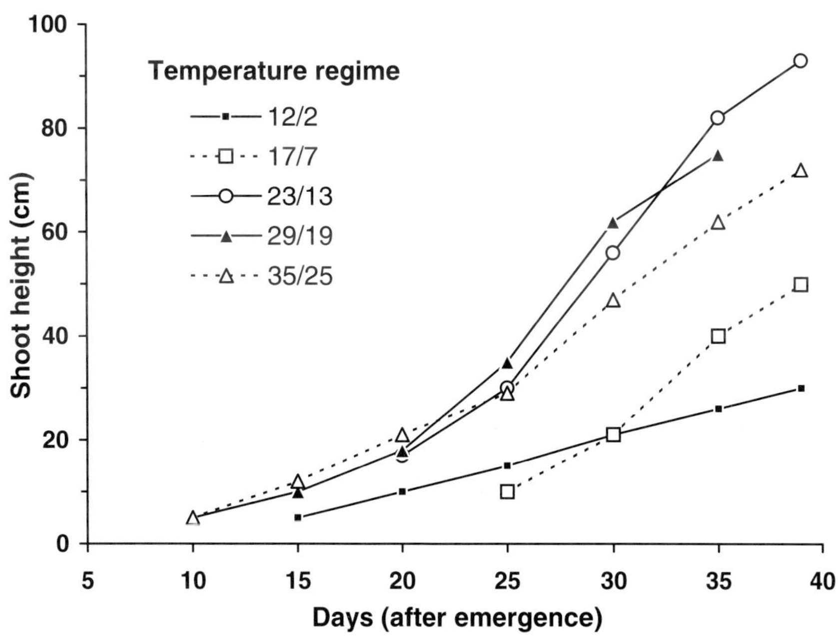

Fig. 7.2. Change in shoot height of wild mustard (*Sinapis arvensis*) over time at five temperature regimes (redrawn from Huang *et al.*, 2001).

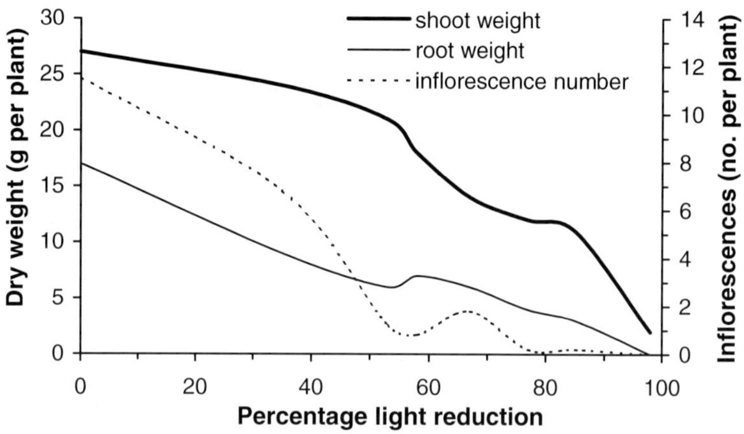

Fig. 7.3. Effect of light intensity on: (a) shoot and root dry weight, and (b) inflorescence number of Canada thistle (adapted and redrawn from Zimdahl *et al.*, 1991).

thresholds) is 24.5°C (5.5–40.0°C), whereas for leaf appearance it is 27°C (1.5–48°C) (Huang *et al.*, 2001).

Many weeds are particularly well adapted to low temperatures and this provides a potential advantage over non-weedy plants. For example, seed germination, growth, flowering and seed set all occur at 12/2°C in wild mustard (Huang *et al.*, 2001). Weeds are often less sensitive to short periods of cold temperature in the early spring. For example, chilling reduced the level of leaf area expansion of velvetleaf (*Abutilon theophrasti*) and spurred anoda (*Anoda cristata*) in cotton (*Gossypium hirsutum*); however, the weeds recovered faster and subsequently gained a competitive advantage over the delayed cotton (Patterson and Flint, 1979). While specific temperatures are required for some processes, temperature fluctuations are required for others such as seed germination (see Chapter 6).

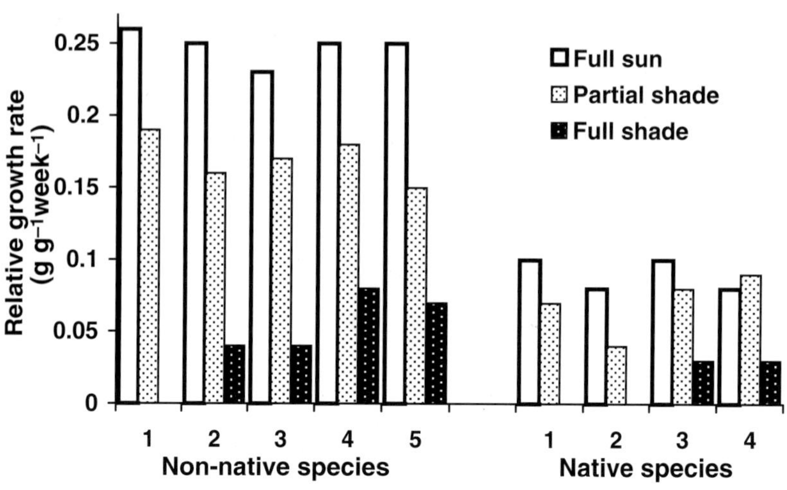

Fig. 7.4. Effect of shading on the relative growth rate (g g^{-1} week^{-1}) on four native and five non-native species of Hawaiian rainforest (redrawn from Pattison *et al.*, 1998).

Light

Light (irradiance) has a direct effect on many plant processes, such as growth, flowering, stem elongation, seed dormancy, formation of storage organs, leaf fall and on plant morphology such as leaf, number and shape (Salisbury and Ross, 1985; Lambers *et al.*, 1998). Plants respond to changes in light quality (spectral composition), quantity (intensity), photoperiod (periodicity of light and dark cycles) and direction (phototropism). It is sometimes difficult to determine what type of light effect is occurring because they interact; for example, as light quantity decreases the spectral composition changes.

Reduced light quantity will have species specific effects. In Canada thistle (*Cirsium arvense*), for example, shading reduces shoot and root production, and the number of inflorescences produced (Zimdahl *et al.*, 1991) (Fig. 7.3). Conversely, itchgrass (*Rottboellia exaltata*) plants growing in shade grow taller than those in full sunlight; this is thought to be why itchgrass is competitive with maize (Patterson, 1985). Pattison *et al.* (1998) compared the relative growth rates of native and non-native species grown in full sun, partial sun (30% of full sun) and full shade (5% of full sun) of Hawaiian rainforests. As light decreased, all species had lower relative growth rates, but native species were less affected (Fig. 7.4). This, incidentally, shows why non-native species are more successful in open disturbed habitats.

As light passes through the leaf canopy the quality of light is also altered. The spectrum of light wavelengths changes because plants reflect green and far-red wavelengths while absorbing blue and red. This causes the ratio of red to far-red (R:FR) light to decrease in shaded conditions (Fig. 7.5). Many species respond to this change in the R:FR ratio. For example, higher levels of FR light triggers internode extension in white mustard (*Sinapis alba*), Chinese datura (*Datura ferox*), and lambsquarters (*Chenopodium album*) causing plants to grow taller into better higher light situations (Alm *et al.*, 1991). A high-

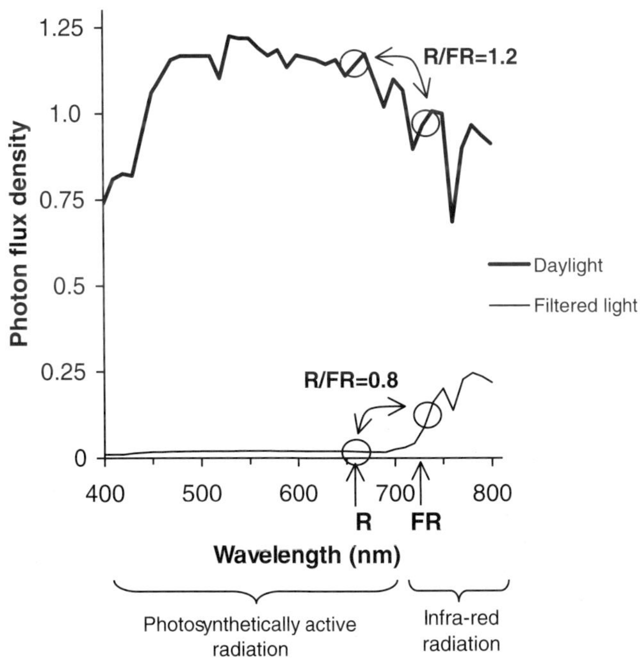

Fig. 7.5. Spectral photon distribution in the 400–800 nm wavelength region on an overcast day in: (a) full daylight, and (b) filtered through a canopy of ash (*Fagus*) (redrawn and adapted from Pons, 2000).

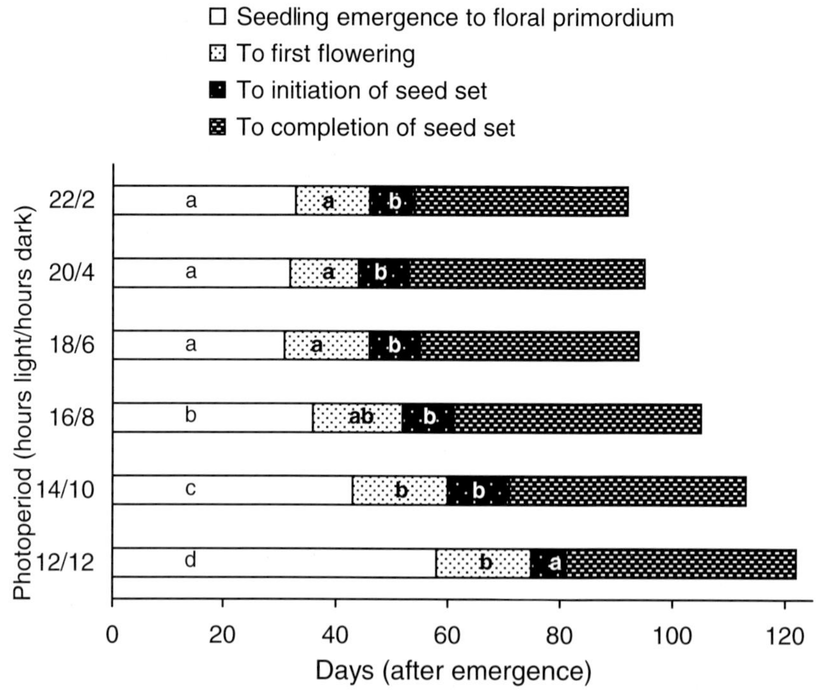

Fig. 7.6. The effect of photoperiod on the duration from emergence, to floral primordia, to first flowering, to seed set initiation and to completion of seed set in wild mustard (from data in Huang *et al.*, 2001). Different letters within each developmental stage indicate significant differences. The lengths of the final developmental stage were not different.

er R:FR ratio promotes flowering in some species; for example, in shaded conditions, oxeye daisy allocates more resources to reproduction, resulting in higher seed production than individuals in sun (Olson and Wallander, 1999).

Photoperiod is what synchronizes many of the seasonal events observed in nature. Alternating light and dark cycles give an accurate cue as to the time of year; temperature can vary unpredictably from day to day, and light quantity and quality is altered by the surrounding vegetation. Thus, photoperiod is often a trigger for important phenological events such as reproduction. In wild mustard, for example, photoperiod influences many reproductive events (Huang *et al.*, 2001). As photoperiod shortens, the time from seedling emergence to floral primordia increases, that from flower primordia to flowering increases, but the time from flowering to initiation of seed set decreases (Fig. 7.6). When photoperiod increases above 18 h, the timing of phenological events is not affected. Photoperiod also effects the morphology of some species. In wild mustard, plants grown in 12–20 h of day length are taller than those grown at 22 h (Huang *et al.*, 2001). Furthermore, plants at 12 and 14 h of day length have more leaves, while plants in 22 and 24 h day length have larger inflorescences than plants grown in other day lengths.

The flowering response to photoperiod can take many forms. Species responses may be triggered by short days, long days, long days and then short days, or by other variations in the sequence. Table 7.1 summarizes flowering responses to day length and gives some weed examples. A response to photoperiod typically requires several cycles to occur. Some species, however, only require

Table 7.1. Types of photoperiod responses and examples of representative species (adapted from Salisbury and Ross, 1985).

Photoperiod response	Examples
Short-day	Red goosefoot, *Chenopodium rubrum*
	Goosefoot, *Chenopodium polyspermum*
	Common cocklebur, *Xanthium strumarium*
	Redroot pigweed, *Amaranthus retroflexus*
Long-day	Scarlet pimpernel, *Anagallis arvensis*
	White mustard, *Sinapis alba*
	Henbane, *Hyoscyamus niger*
Short–long-day	White clover, *Trifolium repens*
	Kentucky bluegrass, *Poa pratensis*
Intermediate-day	Purple nutsedge, *Cyperus rotundus*
Ambiophotoperiod	Hooked bristlegrass, *Setaria verticillata*
Day-neutral	Onion, *Allium cepa*
	Wild carrot, *Daucus carota*
	Barnyardgrass, *Echinochloa crus-galli*
	Indian goosegrass, *Eleusine indica*
	Portulaca, *Portulaca oleracea*
	Itchgrass, *Rottboellia exaltata*

one cycle to trigger a response. For example, a single short-day cycle will induce flowering in redroot pigweed. While we refer to the importance of day length, it is actually the length of dark period that usually triggers a response. For example, a 1-h interruption of fluorescent light during the dark period will inhibit flowering of redroot pigweed (Gutterman, 1985). Plants that respond to photoperiod generally go through three stages of sensitivity: a pre-inductive stage where photoperiod has no effect, an inductive phase where photoperiod triggers reproductive response, and a post-inductive phase where reproduction will continue irrespective of photoperiod (Patterson, 1995).

Because photoperiod changes with latitude, there are often 'biotypes' of species. A biotype is a group of individuals that have similar genetic structure but respond to their environment in different ways than other biotypes of the same species. For example, there are biotypes of common lambsquarters and cocklebur (*Xanthium strumarium*) which respond differently to photoperiod. Northern biotypes usually require shorter nights to initiate flowering. Patterson (1993) examined the differences between Mississippi and Minnesota populations of velvetleaf. The Minnesota population produced more vegetative growth than the Mississippi in short days (12 h light), but the reverse was true for long days (16 h light). In the longest photoperiods, Minnesota plants allocated resources to reproduction, thereby limiting further vegetative growth. Northern populations of side-oats grama (*Bouteloua curtipendula*) are long-day ecotypes while southern populations are short-day ecotypes (Olmsted, 1944).

Interaction of abiotic factors

We have looked at how plants respond to changes in temperature and light separately. Data of this type are usually derived from controlled experiments where one variable is changed and all others remained constant. Such results do not necessarily reflect real situations for several reasons. First, environmental variables tend to fluctuate in tandem (when the sun comes out, light and temperature increase); therefore manipulation of one variable is not realistic. Secondly, plants respond in a complex fashion to the array of environmental factors they face and this may not be predictable by looking at one factor at a time. The response to one environmental factor will affect how an individual responds to another factor.

Reproduction is often determined by an interaction between photoperiod and temperature. For example, at low temperatures, poinsettia (*Euphorbia pulcherrima*) and morning glory (*Ipomoea purpurea*) are long-day plants but they flower in short days at high temperatures. At intermediate temperatures they are day neutral. Interactions of temperature and light occur for other processes. Kikuyugrass *(Pennisetum clandestinum*) exemplifies the interactions of environmental factors and how they can affect weeds (Wilen and Holt, 1996). Kikuyugrass is an introduced turfgrass from the tropics that became a weed in temperate climates. The reason is that it is able to maximize photosynthesis during warm temperatures (25–40°C), as expected from a tropical grass, but it also has high photosynthetic rates during the still-long day lengths of the Mediterranean zone in spring and autumn despite cooler weather. The physiology of kikyugrass is such that lower temperatures do not act to inhibit photosynthesis. We might expect that the result of an interaction of lower temperatures with abundant light would inhibit a tropical grass but this is not the case. Most successful weeds are able to accommodate and maximize growth over a wide range of light and temperature interactions (see also Plowman and Richards, 1997; Roche *et al.*, 1997; Kibbler and Bahinsch, 1999).

Water stress can effect the plant's ability to respond to other environmental triggers. For example, water stress can limit or prevent flowering in single-cycle photoperiod species such as common cocklebur and rye grass (*Lolium temulentum*) (Chiariello and Gulmon, 1991). Conversely, water stress may promote flowering in some species (e.g. siratro, *Macroptilium atropurpureum*, a tropical pasture legume).

Before a plant can respond to an environmental cue, it must reach a phenological state where it is able to sense the cue. Cocklebur, for example, must reach a certain size before it responds to photoperiod cues. In this species, an individual leaf must be at least 1 cm long before it will respond to light, and the most sensitive leaves are the fastest growing ones that are half-expanded (Salisbury and Ross, 1985). The cotyledons of lambsquarters (*Chenopodium* spp.) will respond to photoperiod and thus even tiny seedlings may flower given the appropriate light conditions. The long-day plant scarlet pimpernel (*Anagallis arvenis*) is most sensitive as a seedling; new leaves on older plants will respond, but they are less sensitive (Salisbury and Ross, 1985).

The same environmental cue will trigger different phenological events depending on the species. Mahal and Bormann (1978) identified four phenological patterns of understorey species in New Hampshire northern hardwood forests. Spring ephemerals germinate and complete their life cycles early in the spring, taking advantage of higher light conditions before tree canopy closure. Summer green species emerge soon after the death of spring ephemerals and maintain their leaves until the autumn. They are shade-tolerant species. Late summer species develop slowly over the summer but persist into the autumn taking advantage of higher light conditions after the tree canopy opens. Finally, evergreen species are perennials that maintain their leaves for up to 3 years. They produce new leaves in the spring under higher light conditions and these are then maintained throughout the summer and into the following years. These four phenological patterns are subject to the same environmental triggers, but respond to them in different ways. Weeds also exhibit similar phenological patterns. In North America, dandelion (*Taraxacum officinales*) and coltsfoot (*Tussilago farfara*) flower in the spring, ox-eye daisy (*Chrysanthemum leucanthemum*) and vetch (*Vicia* sp.) in the summer, and common ragweed (*Ambrosia artemisiifolia*) and goldenrods (*Solidago* spp.) in the autumn.

Resource Allocation

Every plant has a limited amount of resources (e.g. carbon, nitrogen) that it allots to the functions of growth, maintenance and

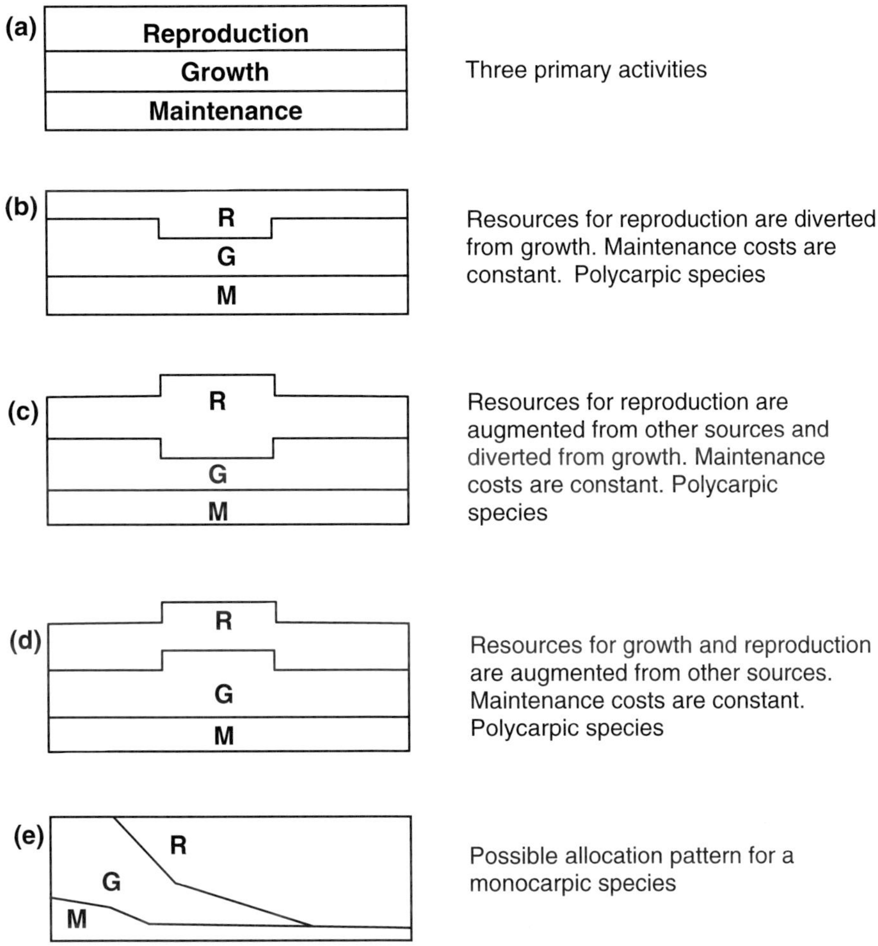

Fig. 7.7. Theoretical resource budgets of plants over one growing season. Resources are allocated to maintenance (M), growth (G) and reproduction (R) (redrawn from Willson, 1983).

reproduction (Fig. 7.7a). The 'principle of allocation' states that plants have a limited supply of resources and that this is allocated to various structures in a way that maximizes lifetime fitness (Bazzaz, 1996; Barbour *et al.*, 1999). Obviously, plants do not make conscious decisions on where to allocate resources. This is determined by the interaction of their genotype and their environment. How a plant allocates resources is important because if too much is spent on one function, then other functions may suffer. For example, if a perennial species allocates too many resources to reproduction and not enough to storage, then it may not be able to survive a harsh winter.

The amount of resources allocated to various functions will vary with the life history strategy and will change over the course of a plant's life cycle. Early on, plants accumulate biomass/nutrients in roots, shoots and leaves. In annual species, reproduction events require expenditures of resources towards the production of reproductive structures (gametes, protective tissue, attraction structures) and towards the care of maturing embryos (Willson, 1983); therefore, as the season progresses, more resources will be devoted to reproduction

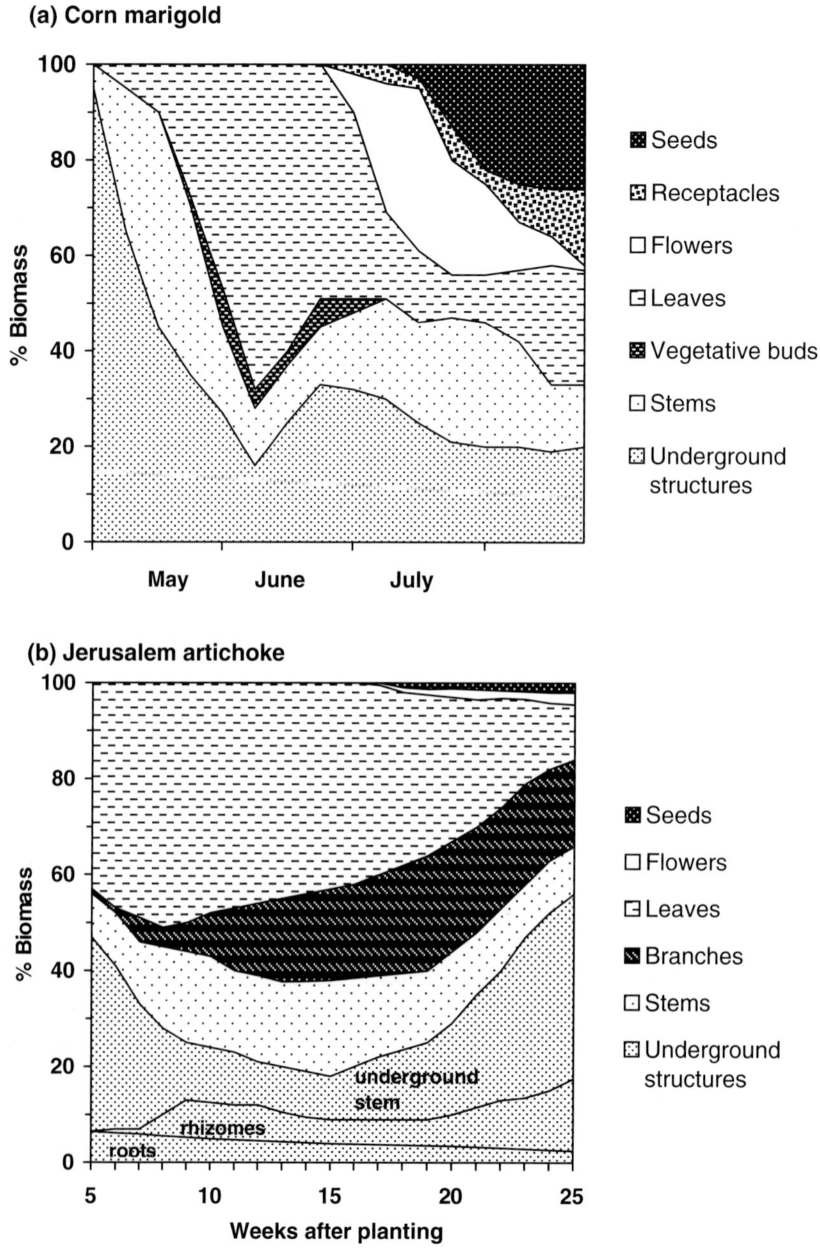

Fig. 7.8. Allocation of dry weight to vegetative and reproductive structures of: (a) an annual species, corn marigold and (b) a perennial species, Jerusalem artichoke (adapted and redrawn from Harper, 1977, and Swanton and Cavers, 1989).

and less to vegetative structures (Fig. 7.8a). If resources are limiting, then the individual may not be able to reproduce, or may reproduce, but at the cost of future fitness or survival. Allocation patterns differ among species. Common groundsel (*Senecio vulgaris*), for example, allocates proportionally more resources to stems and less to flowers than corn marigold (*Chrysanthemum segetum*), and yet both are annuals (Harper, 1977). Within species, allocation patterns will vary with the environment. In common groundsel, for example, more resources are allocated to roots in stressful environments.

In perennials, allocation patterns differ primarily because fewer resources are allocated to reproduction. For example, the Jerusalem artichoke (*Helianthus tuberosus*) is a herbaceous perennial well adapted to invading open areas, particularly cultivated fields (Swanton and Cavers, 1989). Here, a relatively large proportion of biomass is allocated to structural organs such as stems, leaves and branches (Fig. 7.8b). Over the season the allocation to storage organs such as roots, rhizomes and tubers increases and is much larger than biomass allocated to flowers and seeds. This pattern of allocation ensures long-term survival through clonal structures, as well as seed production.

Each reproductive event comes at a cost that must either be compensated for through the accumulation of new resources, or through a trade-off within the plant. In polycarpic species, for example, the plant diverts only some of its resources towards each reproductive event (Fig. 7.7b–d). In some cases the cost occurs at the expense of growth (Fig. 7.7b), while at other times the cost may be covered by the uptake of additional resources which will partially or totally compensate (Fig. 7.7c and d). This increase in resource supply occurs when reproductive structures take up more resources (e.g. flower and fruit may be photosynthetic) or when they enhance the uptake of resources through vegetative structures. During reproduction, for example, leaf photosynthesis of quackgrass may increase or decrease, depending on the plant's genotype and nutrient status (Reekie and Bazzaz, 1987). Where the change in leaf photosynthesis is positive, the cost of reproduction is offset. In monocarpic species, the plant diverts most of its stored resources towards reproduction at the end of its life cycle (Fig. 7.7e).

Phenotypic Plasticity

Phenotypic plasticity is the ability of an individual to respond in the short term to environmental conditions. It is not genetically based. Species that are phenotypically plastic are often able to survive and reproduce under many environmental conditions. Plasticity can be used to take advantage of a sudden, temporary improvement in environmental conditions or to avoid stressful environments (Meerts, 1995; Sans and Masalles, 1997; Wulff *et al.*, 1999; Weinig, 2000).

In general, weeds are thought to have a high degree of phenotypic plasticity, although there are relative few studies that clearly demonstrate this (D.R. Clements *et al.*, unpublished observations). Phenotypically plastic weeds are able to mature and reproduce under a broad range of environmental conditions. For example, showy crotalaria (*Crotalaria spectablis*) can reproduce in heavy shade even though it is substantially smaller in the shade (Patterson, 1982). This species can also produce seed under a range of temperatures in spite of decreased size and biomass. Barnyardgrass (*Echinochloa crus-galli*) showed extreme plasticity among six cohorts planted from March to September. While vegetative biomass was over 3000 g in early cohorts and less than 25 g in a later cohort, late cohorts still flowered – although the number of flowers/plant was reduced from 10,000 to less than 100. Although the allocation of resources to reproduction may be genetically controlled, it is influenced strongly by the environment (Bazzaz *et al.*, 2000).

Other weeds exhibiting phenotypic variation are dandelion, jimsonweed (*Datura stramonium*) and velvetleaf (Clements *et al.*, 2002). Two similar weeds, wild oat (*Avena fatua*) and slender wild oat (*Avena barbata*), rely on different mechanisms to respond to environmental heterogeneity (Marshall and Jain, 1968). Wild oat is more genetically variable, and this allows it to persist in many types of environments. Slender wild oat is more genetically uniform but is able to persist in a variety of habitats because it is phenotypically plastic. Wild oat is genetically variable and thus populations contain many genotypes, some of which will be able to survive and reproduce in the ambient environment. Thus genotypic variation and phenotypic plasticity are two mechanisms to deal with environmental stress and heterogeneity.

Senescence: Programmed Death

It is easy to understand why plants die when they are eaten, trampled or run out of water. It is not as easy to understand why plants die naturally, unthreatened by external forces at the end of their life cycle. 'Senescence' is the programmed process of deterioration that leads to the natural death of a plant (or plant part). It is an internally controlled process that determines the life span of a whole organism or its parts. Knowledge of when and why a weed senesces has implications for weed management. Here, we discuss whole plant senescence. For reviews of the mechanisms of leaf senescence, see Nooden *et al.* (1997) and Chandlee (2001).

We do not always know the cause of senescence (Crawley, 1997). It may be due to the accumulation of deleterious mutations, or it may be that the soil becomes unsuitable when trace nutrients are depleted or when allelopathic chemicals accumulate. Alternatively, senescence may be caused by the negative affects of earlier life stages. For example, a high allocation of resources to early reproduction may inevitably lead to senescence because not enough has been allocated to the maintenance of structures. The mechanisms of senescence may differ between monocarpic and polycarpic species.

Senescence in monocarpic species

Wilson (1997) reviewed several hypothesized mechanisms for senescence of monocarpic species. The 'death hormone' hypothesis suggests that the rapid death (plant suicide) of monocarpic plants is pre-programmed and likely hormonally controlled. Most biological evidence supports this hypothesis, although the actual mechanisms are not clear. The alternative hypothesis to the 'death hormone' is that senescence in monocarps is caused by nutrient starvation following flowering. Here it is suggested that the act of reproduction diverts stored photosynthate away from vegetative tissue, causing it to die. For example, senescence is delayed in the annual weed and crop (used in traditional medicine to treat malaria) beggar's tick (*Bidens pilosa*) when flowers are removed (Zobolo and van Staden, 1999). Wilson (1997) points out that, while this may occur in some species, the biological evidence does not generally support this alternative hypothesis. Among other reasons, he cites:

- leaves do not generally exhibit nutrient deficiency before death nor are reduced minerals evident in leaf tissue;
- the addition of nutrients does not delay senescence;
- senescence is rapid rather than prolonged which is typical of nutrient deficiency;
- other triggers such as long days have been identified as necessary for senescence to occur.

Senescence in polycarpic species

The death of polycarpic perennials is somewhat harder to explain, partly because it is difficult to separate the effects of age and size. As a plant (say a tree) gets larger, it begins to decline physiologically because it has more respiratory tissue to support, it must transport water and nutrients further, and it is more susceptible to herbivores and pathogens (Watkinson, 1992). Thus the effects of size are similar to those of age; however, they do not fully explain senescence of old plants. Senescence of woody plants could be caused by the collapse of structural tissue, but this does not explain why some woody plants that regenerate from dormant meristems (e.g. when a tree resprouts from its base after being cut) often senesce after temporary regrowth (Watkinson, 1992). Alternatively, toxins accumulating in cells may lead to senescence. The two evolutionary mechanisms explaining senescence in long-lived plants are (Orive, 1995; Pedersen, 1995):

- deleterious mutations accumulate over time and eventually become lethal to the individual;
- genes that increase fitness early in the life cycle may decrease fitness later on (antagonistic pleiotropy).

The senescence of clonal organisms (i.e. the entire clone) is even harder to explain. Some clonal plants *appear* to have escaped senescence altogether, by evidence of their extremely old age. In other cases, an entire clone can senesce at the same time (Watkinson, 1992).

Conclusions

We have now completed our look at the dynamics of individual plant populations. Until now, we have looked at population dynamics as being separate from the dynamics of other species. Common sense tells us that this does not reflect real life. Even a 'monoculture' maize field has hundreds (thousands?) of other species including weeds, insects, soil fungi and bacteria, mammals, and birds, many of whom will influence the population dynamics of maize. To begin our examination of how populations interact, we look first at competition in Chapter 8, and then herbivory and other types of interactions in Chapter 9. From there, we have a chapter on how populations and their interactions are studied (Chapter 10).

Questions

1. Describe the phenology and resource allocation patterns of your selected weed species. What environmental factors trigger important phenological events?
2. Explain why it would be adaptive to have a longer time from emergence to seed set at photoperiods of 12/12 (Fig. 7.6; Huang *et al.*, 2001).
3. Why will knowledge about a plant's phenology and senescence aid in its management?
4. Explain the difference between 'death' and 'senescence'.

General References

Huang, J.Z., Shrestha, A., Tollenaar, M., Deen, W., Rajcan, I., Rahimian, H. and Swanton, C.J. (2001) Effect of temperature and photoperiod on the phenological development of wild mustard (*Sinapis arvensis* L.). *Field Crops Research* 70, 75–86.

Sultan, S.E. (2001) Phenotypic plasticity for components of *Polygonum* species of contrasting ecological breadth. *Ecology* 82, 328–343.

Wilson, J.B. (1997) An evolutionary perspective on the 'death hormone' hypothesis in plants. *Physiologia Plantarum* 99, 511–516.

Literature Cited

Alm, D.M., McGiffen, M.E., Jr and Hesketh, J.D. (1991) Weed phenology. In: Hodges, T. (ed.) *Predicting Crop Phenology*. CRC Press, Boca Raton, Florida, pp. 191–218.

Barbour, M.G., Burk, J.H., Pitts, W.D., Gilliam, F.S., and Schwartz, M.W. (1999) *Terrestrial Plant Ecology*, 3rd edn. Addison Wesley Longman, Menlo Park, California.

Bazzaz, F.A. (1996) *Plants in Changing Environments: Linking Physiological, Population and Community Ecology*. Cambridge University Press, New York.

Bazzaz, F.A., Ackerley, D.D. and Reekie, E.G. (2000) Reproductive allocation in plants. In: Fenner, M. (ed.) *Seeds: the Ecology of Regeneration in Plant Communities*, 2nd edn. CAB International, Wallingford, UK, pp. 1–30.

Chandlee, J.M. (2001) Current molecular understanding of the genetically programmed process of leaf senescence. *Physiologia Plantarum* 113, 1–8.

Chiariello, N.R. and Gulmon, S.L. (1991) Stress effects on plant reproduction. In: Mooney, H.A., Winner, W.E. and Pell, E.J. (eds) *Response of Plants to Multiple Stresses*. Academic Press, San Diego, pp. 161–188.

Crawley, M.J. (1997) Life history and environment. In: Crawley, M.J. (ed.) *Plant Ecology*, 2nd edn. Blackwell, Oxford, pp. 73–131.

Gutterman, Y. (1985) Flowering, seed development, and the influences during seed maturation on seed germination of annual weeds. In: Duke, S.O. (ed.) *Weed Physiology*, Vol. I. *Reproduction and Ecophysiology*. CRC Press, Boca Raton, Florida, pp. 1–26.

Harper, J.L. (1977) *Population Biology of Plants*. Academic Press, Toronto.

Huang, J.Z., Shrestha, A., Tollenaar, M., Deen, W., Rajcan, I., Rahimian, H. and Swanton, C.J. (2001) Effect of temperature and photoperiod on the phenological development of wild mustard (*Sinapis arvensis* L.). *Field Crops Research* 70, 75–86.

Kibbler, H. and Bahnisch, L.M. (1999) Distribution of *Hymenachne acutigluma* (Steudel) Guilliland in ponded pasture is limited by photosynthetic response to temperature. *Australian Journal of Experimental Agriculture* 39, 437–443.

Lambers, H., Chapin, F.S. III, and Pons, T.L. (1998) *Plant Physiological Ecology*. Springer-Verlag, New York.

Mahal, B. and Bormann, F.H. (1978) A quantitative description of the vegetative phenology of herbs in a northern hardwood forest. *Botanical Gazette* 139, 467–481.

Marshall, D.R. and Jain, S.K. (1968) Phenotypic plasticity of *Avena fatura* and *A. barbata*. *American Naturalist* 102, 457–467.

Meerts, P. (1995) Phenotypic plasticity in the annual weed *Polygonum aviculare*. *Botanica Acta* 108, 414–424.

Nooden, L.D., Guiamet, J.J. and John, I. (1997) Senescence mechanisms. *Physiologia Plantarum* 101, 746–753.

Olmsted, C.E. (1944) Growth and development in range grasses. IV. Photoperiodic responses in twelve geographic strains of side-oats grama. *Botanical Gazette* 106, 46–74.

Olson, B.E. and Wallander, R.T. (1999) (eds) Oxeye daisy. In: Shely, R.L. and Petroff, J.K. (eds) *Biology and Management of Noxious Rangeland Weeds*. Oregon State University Press, Corvallis, Oregon, pp. 282–289.

Orive, M.E. (1995) Senescence in organisms with clonal reproduction and complex life-histories. *American Naturalist* 145, 90–108.

Patterson, D.T. (1982) Effects of shading and temperature on showy crotalaria (*Crotalaria spectabilis*). *Weed Science* 30, 692–697.

Patterson, D.T. (1985) Comparative ecophysiology of weeds and crops. In: Duke, S.O. (ed.) *Weed Physiology* Vol. I, *Reproduction and Ecophysiology*. CRC Press, Boca Raton, Florida, pp. 101–129.

Patterson, D.T. (1993) Effects of temperature and photoperiod on growth and development of sicklepod (*Cassia obtusifolia*). *Weed Science* 41, 574–582.

Patterson, D.T. (1995) Effects of photoperiod on reproductive development in velvetleaf (*Abutilon theophrasti*). *Weed Science* 43, 627–633.

Patterson, D.T. and Flint, E.P. (1979) Effects of chilling on cotton (*Gossypium hirsutum*), velvetleaf (*Abutilon theophrasti*), and spurred anoda (*Anoda cristata*). *Weed Science* 27, 473.

Pattison, R.R., Goldstein, G. and Ares, A. (1998) Growth, biomass allocation and photosynthesis of invasive and native Hawaiian rainforest species. *Oecologia* 117, 449–459.

Pedersen, B. (1995) An evolutionary theory of clonal senescence. *Theoretical Population Biology* 47, 292–320.

Plowman, A.B. and Richards, A.J. (1997) The effect of light and temperature on competition between atrazine susceptible and resistant *Brassica rapa*. *Annals of Botany* 80, 583–590.

Pons, T.L. (2000) Seed responses to light. In: Fenner, M. (ed.) *Seeds: the Ecology and Regeneration of Plant Communities*, 2nd edn. CAB International, Wallingford, UK, pp. 237–260.

Reekie, E.G. and Bazzaz, F.A. (1987) Reproductive effort in plants. I. Carbon allocation to reproduction. *American Naturalist* 129, 876–896.

Roche, C.T., Thill, D.C. and Shafi, B. (1997) Prediction of flowering in common crupina (*Crupina vulgaris*). *Weed Science* 45, 519–528.

Salisbury, F.B. and Ross, C.W. (1985) *Plant Physiology*, 3rd edn. Wadsworth Publishing, Belmont, California.

Sans, F.X. and Masalles, R.M. (1997) Demography of the arable weed *Diplotaxis erucoides* in central Catalonia, Spain. *Canadian Journal of Botany* 75, 86–95.

Sheley, R.L. and Petroff, J.K. (1999) *Biology and Management of Noxious Rangeland Weeds*. Oregon State University Press, Corvallis, Oregon.

Swanton, C.J. and Cavers, P.B. (1989) Biomass and nutrient allocation patterns in Jerusalem artichoke (*Helianthus tuberosus*). *Canadian Journal of Botany* 64, 2280–2287.

Watkinson, A. (1992) Plant senescence. *Trends in Ecology and Evolution* 7, 417–420.

Weinig, C. (2000) Limits to adaptive plasticity: temperature and photoperiod influence shade-avoidance responses. *American Journal of Botany* 87, 1660–1668.

Wilen, C.A. and Holt, J.S. (1996) Physiological mechanisms for the rapid growth of *Pennisetum clandestinum* in Mediterranean climates. *Weed Research* 36, 213–225.

Willson, M.F. (1983) *Plant Reproductive Biology*. John Wiley & Sons, New York.

Wilson, J.B. (1997) An evolutionary perspective on the 'death hormone' hypothesis in plants. *Physiologia Plantarum* 99, 511–516.

Wulff, R.D., Causin, H.F., Benitez, O. and Bacalini, P.A. (1999) Intraspecific variability and maternal effects in the response to nutrient addition in *Chenopodium album*. *Canadian Journal of Botany* 77, 1150–1158.

Zimdahl, R.L., Lin, J. and Dall'Arellina, A.A. (1991) Effect of light, watering frequency, and chlorsulfuron on Canada thistle (*Cirsium arvense*). *Weed Science* 39, 590–594.

Zobolo, A.M. and van Staden, J. (1999) The effects of deflowering and defruiting on growth and senescence of *Bidens pilosa* L. *South African Journal of Botany* 65, 86–88.

Part II

Interactions Between Populations

8
Interactions Between Populations I: Competition and Allelopathy

Concepts

- Interactions can have positive, negative or neutral effects on the individuals involved.
- Competition is a negative interaction where individuals make simultaneous demands that exceed limited resources and, while both suffer, one individual suffers less.
- Competition may be less important in habitats where resources are so poor and scattered that populations never get dense or large enough to cause simultaneous demands that exceed limiting resources.
- Competition can be difficult to test because weeds can compete for more than one resource at a time (above and below ground). Competition is influenced by other interactions like herbivory and parasitism (see Chapter 9), and environmental and genetic variation.
- Individuals also may deny access of resources to others (interference competition); it is difficult to tell whether this occurs in weeds. Allelopathy, possibly a form of interference competition, involves the production of toxins that can make tissues or habitats unsuitable for use.
- The outcome of competition is often related to the size of individuals and density of populations.

Interactions in Populations and Communities

Like humans, individual weeds live in contact and interact with a myriad of individuals of their own species and other species. We have already alluded to this in previous chapters where we mentioned pollination and occasionally mentioned the influence of other organisms via competition and herbivory (especially in Chapter 7), though we have yet to explore what this all means. Perhaps the most important implication is that the fate of an individual cannot really be disentangled from its interactions with others. Interactions are complex as they benefit, harm, or have no effect and will vary amongst the many, varied individuals found in populations and communities. While based on the environmental variation and

Table 8.1. Summary of interactions that might occur between two species.

Interaction	Species A	Species B	Explanation
Neutralism	0	0	Neither species affected
Competition (amensalism)	0/–	–	Both species inhibited, or one species affected, the other not
Allelopathy	0	–	Species A releases a chemical then inhibits species B
Herbivory	+	–	Species A (animal) consumes part of species B (plant)
Mutualism	+	+	Both species benefit
Commensalism	+	0	Species A benefits while species B is not affected
Parasitism	+	–	Species A (parasite) exploits species B (host) by living on or in it

individual characteristics like germination, growth rate and death rate, the sheer number of possible interactions and outcomes complicates the dynamics and structures of populations and communities beyond what we have discussed so far.

Because of the complexity of interactions, they may be categorized and presented as if only two individuals were interacting at one time (Table 8.1). Again, in reality, many individuals (too numerous to actually count) are involved but it is easier for us to visualize and discuss interactions between a pair of individuals. In this chapter, we focus on interactions where at least one individual is negatively affected. These are called competition (both individuals are negatively affected, but one individual less so) and amensalism (only one individual is negatively affected and the other neither benefits nor is harmed). Because determining whether an interaction inhibits one or both of the individuals is difficult, we group amensalism with competition (Lawton and Hassel, 1981; Connell, 1983).

Defining Competition

Competition requires that individuals make demands on a common pool of resources (e.g. water, light, nutrients and space) that are *limited*. Because this situation occurs frequently, competition is a key interaction that determines population and community structures and dynamics. It would be better for organisms to avoid competition because it harms all involved. Being a better competitor just means you suffer less and have a greater relative fitness; if you could avoid allocating resources to characteristics good only for competition and, for example, allocate resources instead to produce more seeds, then your absolute fitness would increase. Because most plants require the same types of resources, avoiding competition may not be feasible during a plant's life cycle. Thus, individuals are often selected for characteristics that allow them to outcompete others in order to increase their relative fitness (by surviving and reproducing). Outcompeting others requires being better at acquiring more of the common resource pool. It requires being able to tolerate a lack of resources when others acquire more. Hence, the competitive ability of a plant can be measured in two ways: its ability to suppress other individuals (competitive effect), and its ability to avoid being suppressed (competitive response) (Aarssen, 1989; Goldberg, 1990; Goldberg and Landa, 1991).

Competition involving plants has been dissected by many authors (Zimdahl, 1980; Keddy, 1989; Grace and Tilman, 1990; Bengtsson *et al.*, 1994; Casper and Jackson, 1997). We focus on the main issues related to competition since these are especially relevant to weeds:

- What is the importance of competition in different environments?

- What general mechanisms of competition for different resources exist?
- What determines the outcome of competition?

The Importance of Competition

Is competition more likely to occur in resource-rich or resource-poor environments?

It is generally agreed that in resource-rich environments, plants are less likely to compete for nutrients because they will not be limiting, but that plants still compete for light because more individuals will survive and develop leaves that reduce available light (Goldberg, 1990). The discussion on whether plants compete in resource-poor environments has been more polarized. Grime (1979) maintains that competition is not important in resource-poor environments because resources cannot be depleted further. He argues that plants adapted for stress tolerance will dominate in these environments, while competitors will be favoured in resource-rich environments. Tilman (1988, 1990) argues that competition occurs in resource-poor environments because, while some plants do adapt generally to tolerate these stressful environments, many individuals also deplete resources more than others by having a high efficiency of nutrient uptake. Grace (1990) believes that both theories are compatible and reflect different aspects of competition, i.e. Grime focuses on the long-term competitive effect on a community, while Tilman focuses on the shorter-term competitive response of individuals. Furthermore, if the ability to tolerate being denied resources is considered a competitive trait (Aarssen, 1989, 1992), then stress tolerance is actually a competitive trait. Indeed, most plants do compete in response to resource-poor envi-

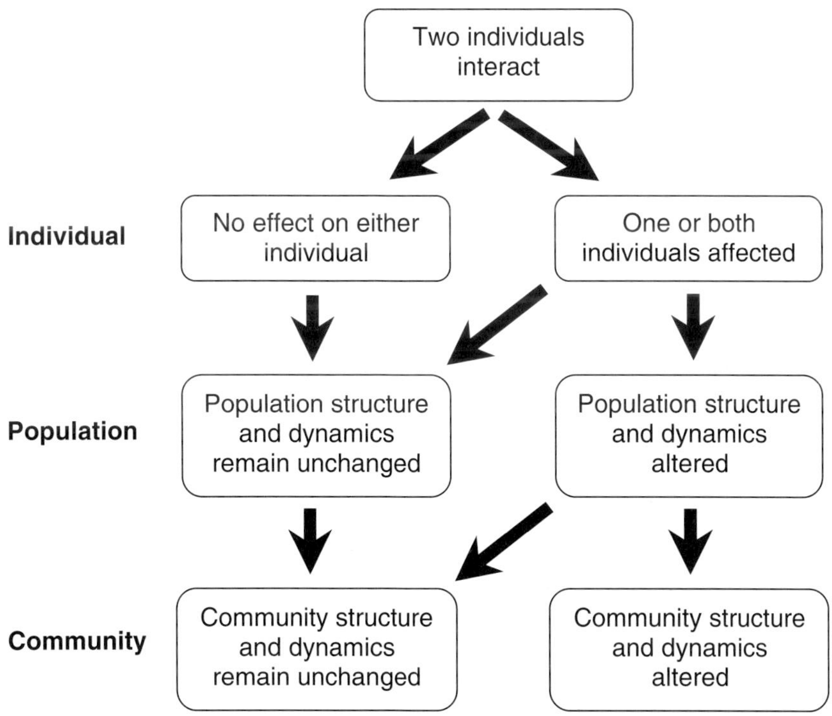

Fig. 8.1. Series of events that must occur before competition between two individuals will influence populations or communities.

ronments – recall that one of the conditions for competition was the existence of limited resources.

Like so many concepts in ecology, the issue is generally one of semantics – in this case it is about what we define as 'resource-rich' and 'resource-poor'. There is no good way to quantify the division between these two. Hence, most studies find evidence for both theories, though a harsh environment (in terms of its local climate, for example) will create resource-poor conditions. Weeds and other plants in these environments may be limited by abiotic conditions rather than biotic interactions like competition. This is because they may not be able to grow in large numbers and densities to create an excess demand on available resources (see Grime, 1979; Goldberg, 1985; Grace, 1990; Tilman, 1990; Aarssen, 1992, for detailed discussions).

The effect of competition on populations and communities

Establishing the relative importance of competition in real world habitats requires evidence that competition affects population size or biomass, or that it affects community composition (Goldberg and Barton, 1992; Goldberg *et al.*, 1995); to do this, a series of events must occur (Fig. 8.1). Even if we can establish that there is an effect on populations and communities, it may be ephemeral and the population or community may return to its original state. Complicating matters further is that even if observed patterns suggest competition is causing them, there can be other explanations, e.g. current interactions such as predation or environmental changes. Patterns also may be the result of unknown past (historical) events or interactions. Attempting to determine the exact cause (i.e. was it caused by competition) is like chasing the ghost of these past events (Connell, 1980). What ultimately makes competition at the population and larger scales difficult to examine is that the pattern we see results from millions of competitive interactions between individuals. Hence, ecologists often focus on individual-scale experiments and extrapolate to populations and beyond. By examining the variation in the competitive interactions between individuals, it is possible to determine what is likely to happen at larger scales, though we caution again that populations, communities and ecosystems are the result of more than just competition and individual interactions.

What evidence do we have that competition occurs?

When you observe a habitat, how can you tell if individual weeds (for example) are competing or not? Competition involving weeds is subtle and difficult to observe but perhaps easier to test than for whole populations or communities. Still, in a meadow, you do not see weeds throwing punches at each other. Thus, ecologists often use controlled experiments to test whether inferences made in the field about competition are real. The easiest way to detect competition experimentally is to grow plants together (in competition) and apart (without competition), and compare their growth or survival. There are many variations on this basic experimental design (see Chapter 10).

While these types of paired experiments are useful, they are limited in their broad-scale application because they ignore all other species normally found in a habitat. The design of experiments can be more complex. For example, if we wished to look at the competitive effect of weeds on an unmanaged forest ecosystem, we would have to decide what to measure (e.g. survivorship or growth), the number and types of species to measure it on (e.g. one weed species or many species, weeds and otherwise), and what time span to cover (e.g. 1 year or 10).

One example of how competition is assessed is from Wilson and Tilman (1995). They planted three species of native grasses into prairie soil and measured their growth in three competition treatments: competition for light and nutrients by allowing roots and shoots to interact, competition for nutrients by excluding the shoots from interactions, and no competition by allowing neither roots or shoots to interact. They found that excluding the shoots of neighbours had no effect on plant growth, i.e. they prevented any possible competition for light but competition for nutrients still was possible. Excluding both shoots and roots caused a drastic increase in growth of all three species, i.e. when the possibility of competition for nutrients also was prevented, the plants benefited. Tilman (1988, 1990) also added a variety of soil nutrients and found that plants responded only when nitrogen was no longer limiting; therefore, he concluded that competition was for nitrogen. This study illustrates that:

- some grasses are better than others at 'exploiting' nitrogen while in competition with one another;
- there were several types of resources (including nitrogen) for which grasses might have competed.

The pattern of adult plants reflects the competition for nitrogen between individuals.

Weed competition studies generally use similar approaches and reach similar conclusions. Bergelson (1996) reviewed her experiments on competition between two weeds, annual bluegrass (*Poa annua*) and common groundsel (*Senecio vulgaris*) (see also Bergelson, 1990, for example). By comparing the growth of each species alone with situations where groundsel was surrounded by annual bluegrass individuals, the outcome of competition could be studied. The fundamental results reinforce the general patterns that competition involving weeds often exhibit. The main pattern is that even small differences in seed germination and seedling emergence matter – earlier ones often are better competitors. In theory, getting a head start on competitors means that a weed is competing with others of a similar age (i.e. all start out as seedlings). If this is true, then 'neighbourhood competition' is important (Pacala and Silander, 1990). This means that competitive ability is useful in densely vegetated environments but it also means that some individuals avoid competition because of chance. Avoidance is not a strategy – some individuals happen to colonize a 'neighbourhood' area that is relatively free of competitors of similar ages and have higher fitness regardless of their competitive ability. 'Gap colonization' may be equally or more important in influencing whether competition between individuals, i.e. competition is not always influenced by interactions between co-generational individuals but by the gaps left open or still occupied by the previous generation. Competition can be affected by the influence of the previous or older generation. In

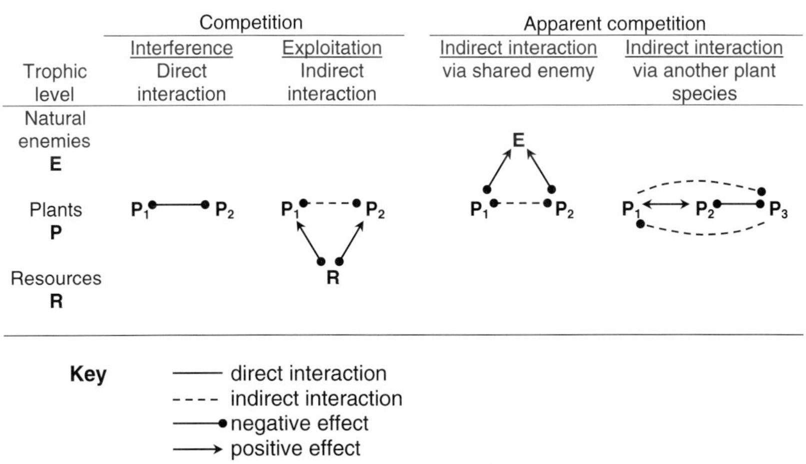

Fig. 8.2. Types of traditional competition and apparent competition showing direct and indirect interactions (redrawn from Connell, 1990).

Bergelson's (1996) review, she explains how the main influence on competition's outcome was that the litter from dead bluegrass prevented groundsel seedlings from emerging. Though this is *not* what Connell (1980) meant by the 'ghost of competition past' (see previous section), the analogy here is similar sounding because competition is between live groundsel and the 'ghost' (the litter) of dead bluegrass.

Exploitation and Interference Competition

There are two general mechanisms of competition. In 'exploitation competition', weeds (as with all plants) compete with one another by being better at exploiting the resources, i.e. obtaining more of them than others (Fig. 8.2). In 'interference competition', there can be a direct removal of a resource (taking it away from another who already captured it) or the occupation of a resource that may not be needed immediately, but denying it to others raises relative fitness (Murphy, 1999). In weeds, interference competition does not involve direct removal of a resource because this situation applies more to animals that physically confront one another over resources like nests or caches of food. Thus, interference competition in plants is generally the denial of resources to other plants.

Exploitation competition

While we will discuss general trends in exploitation competition for different types of resources, we caution that the relationship between the amount and type of resources available and competition is not simple because:

- weeds can compete for more than one resource at a time and may do so above and below ground;
- the outcome of competition is modified by abiotic factors (e.g. temperature, pH) and other biotic interactions (e.g. predation, parasitism) that influence the rate and efficiency at which the resources are consumed;
- weeds are genetically variable and thus all individuals of a species will not respond to competition in the same way.

Light

Both light quality and quantity are important aspects of competition (e.g. Novoplansky, 1991). Weeds can compensate in a number of ways to avoid poor light conditions, all of these related to the basic processes of germination, emergence and adult phenology (Chapters 6 and 7). Many weeds germinate early and grow taller at a faster rate to acquire as much light as possible, e.g. self-heal (*Prunella vulgaris*) (Miller *et al.*, 1994). Since the presence of dense leaf canopies reduces the quantity and quality of light available to weeds, competition for light is greatest when plant density is highest. The plasticity of weeds explains why:

- they tend to be taller when grown in high densities (Nagashima *et al.*, 1995);
- the position and orientation of leaves changes to intercept more light (Alphalo *et al.*, 1999);
- stems may elongate so that leaves are positioned above competing vegetation (Alphalo *et al.*, 1999).

Nutrients

Plants compete mostly for nitrogen, phosphorus and potassium (but there are many others). Phosphorus is usually the most limited nutrient in aquatic ecosystems. Nitrogen is usually the most limited nutrient in terrestrial habitats. Potassium is often overlooked but some terrestrial weeds (e.g. dandelions) might be managed better if potassium-poor fertilizers were used because some weeds are limited by this nutrient (Tilman *et al.*, 1999). In general, weeds compete directly for nutrients; however, weeds that germinate and emerge prior to other plants are able to deplete the available nutrient resource pool first. Because nitrogen is also one of the most mobile and absorbable nutrients, adding fertilizers may benefit weeds that are poor nitrogen com-

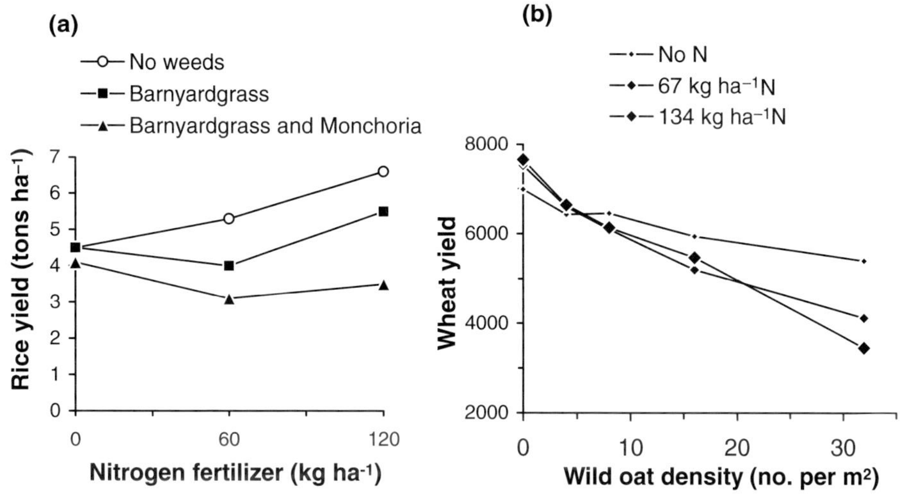

Fig. 8.3. (a) Effect of nitrogen addition and two weeds (barnyardgrass and the annual broadleaf monchoris) on rice yield (redrawn from data from Moody, 1981 in Zimdahl, 1999) and (b) the effect of nitrogen and increasing wild oats density on wheat yield (redrawn from data in Carlson and Hill, 1986).

petitors at a cost to crops (Fig. 8.3). However, crops such as maize have been shown to outcompete weeds for the extra nutrients and as a result suffer less from weed competition (Tollenaar *et al.*, 1994). Agricultural weeds may be poor competitors at low levels of nitrogen, although there are exceptions (Iqbal and Wright, 1997; Hashem *et al.*, 2000).

Water

Competition for water occurs below ground between roots. The ability to absorb water is related to rooting volume. However, not only are the dimensions (breadth and depth) of rooting zones important: so is the degree of water extraction. The ability of a plant to use water depends on its water use efficiency (WUE). This is the ability to minimize water use for a given amount of carbon assimilation. When water is abundant, plants with low WUE (and flood tolerance) are more competitive, whereas in arid conditions, plants with high WUE (and drought tolerance) have the advantage (Di Tomaso, 1998; Gealy, 1998; Hunt and Beadle, 1998; Iqbal and Wright, 1998; Walch *et al.*, 1999; Schillinger and Young, 2000). The examples of weedy saltcedars (*Tamarix* spp.) and their impact on the southwestern USA are particularly illustrative as these weeds have invaded 600,000 ha, in part, because of their superior WUE (Di Tomaso, 1998). Similarly, trees in forest plantations of Tasmania, Australia, suffer because some weeds have a high WUE and can outcompete the tree for water (Hunt and Beadle, 1998). Plants in more temperate climates are less likely to compete for water because rain and snowmelt are usually abundant. However, we should not ignore the importance of WUE, because this may be an important adaptation needed to survive climate changes and a tendency for temperate areas to receive less precipitation (Norris, 1996).

Physical space

It is easier to understand how weeds compete for a resource such as nutrients, water or light because we can imagine this being consumed by the plant. It is harder to imagine space as something that is competed for, and yet the lack of physical space creates consequences for individuals. Root restriction experiments are one way to test the effect of limiting space. Although not all species react equally, restricting a plant's rooting space generally decreases shoot biomass, height and/or growth rate, even when ample water and nutrients are supplied (Richards and Rowe, 1977; Gurevitch *et al.*, 1990; McConnaughay and Bazzaz, 1991; Matthes-Sears and Larson, 1999). Schenk *et al.* (1999) argue that plants are 'territorial' because they defend space for their exclusive use. A territorial individual may not be better at exploiting limited space; it may be better at preventing others from using it. This might apply to forestry weeds competing with willows (*Salix*) (Sage, 1999) and pasture weeds like Paterson's curse (*Echium plantagineum*) (Grigulis *et al.*, 2001).

Exploiting different resources: competition above and below ground

Above- and below-ground competition may be considered separately because plants use different structures (e.g. roots vs. leaves) to compete for different resources (e.g. nutrients vs. light) above and below ground. Below-ground root competition is more complex because individuals compete for space, water and many nutrients, and these all differ in distribution, mobility, molecular size and other aspects. In above-ground competition, individuals normally compete only for light or space (Casper and Jackson, 1997). Below-ground competition usually reduces plant performance more than above-ground competition, except in weed–crop competition where shoot competition tends to be more intense (Wilson, 1988). Root competition is more prevalent in arid and nutrient-poor systems because plants tend to be spaced further away from neighbours and therefore do not shade each other; however, root zones may overlap.

From a physiological perspective, roots and shoots are so integrally related that it is practically impossible to separate them. In weeds, this creates problems because they must trade-off allocating resources between tissues involved in above- and below-ground competition. For example, if the weedy rice cockspur (*Echinochloa oryzoides*) increases leaf area to compete for light, it reduces allocation of resources to roots and is vulnerable to competition for nutrients (Gibson and Fischer, 2001). Many weed species experience such trade-offs (e.g. McLachlan *et al.*, 1995).

The relationship between above- and below-ground competition is not usually additive: that is, the total competitive effect is not simply the above-ground effect plus the below-ground effect, although this is often presumed (Wilson, 1988). Root and shoot competition may have opposing effects, or be subject to complex interactions, and this may not be evident when measured together. One species may benefit from below-ground interaction and this may counter any negative effect of above-ground competition (Wilson and Tilman, 1995).

Interference competition

While the term 'interference competition' is still used, all plant competition might be considered exploitation if denial of resources is caused by overexploiting a common resource pool (Schoener, 1983). For example, Rebele (2000) grew weedy feathertop (*Calamagrostis epigeios*) and Canada goldenrod (*Solidago canadensis*) in mixed stands for several years. For the first 3 years, the much taller and large leaved goldenrod dominated feathertop. Subsequently, Canada goldenrod seedlings or regenerating shoots began to be buried by the accumulation of litter from feathertop. Eventually feathertop dominated because it interfered with the space that Canada goldenrod needed. However, the outcome could be interpreted as exploitation competition if the key process was that the litter prevented Canada goldenrod from competing for light.

Allelopathy

'Allelopathy' is an interaction that *might* be considered to be interference competition (Aarssen, 1989, 1992; Murphy, 1999). Allelopathy is the direct effect of one individual on others through the release of chemical compounds from roots, shoots, leaves or flowers (Rice, 1995). Allelopathy can require resources in the form of the chemicals used or can create autotoxic effects that harm the allelopathic plant; since neither the allelopathic plant nor its targets gain absolute benefits, allelopathy can fit this part of the definition of competition. However, because no resource is being exploited and because the release of beneficial compounds can occur, allelopathy is sometimes described as a unique interaction that is not related to competition (Rose *et al.*, 1984; Aarssen, 1989; Inderjit and del Moral, 1997; Murphy, 1999; Olofsdotter *et al.*, 1999). It is even unclear how often allelopathy occurs since many chemicals released by plants only become toxic after being transformed by other species in soil or water. Such indirect effects may be considered to be allelopathy but whether the plant that exuded the original chemical benefits at all is unclear (Connell, 1990; Williamson, 1990).

While claims that allelopathy exists in many weeds (like velvetleaf, quackgrass, redroot pigweed) should be treated sceptically, it does occur (Rice, 1984, 1995). Generally, allelopathy seems to be easier to demonstrate and more ecologically important in species-poor habitats where one species may 'dominate the biochemistry of the soil' (Wardle *et al.*, 1998). Additionally, if some species adapt and resist allelopathic chemicals, then these 'resistant' species may dominate, e.g. pink flower (*Lantana camara*) (Gentle and Duggin, 1997; Mallik and Pellissier, 2000). When nodding thistle (*Carduus nutans*) dies, its decomposing leaves release allelopathic chemicals that suppress or kill only white clover (*Trifolium repens*). However, white clover increases soil nitrogen, and loss of white clover reduces nitrogen available to all other species. Nodding thistle may be less affected by loss of nitrogen and benefits from the reduced competition by all other species, hence nodding thistle populations increase greatly (Wardle *et al.*, 1998). In pollen allelopathy, the chemicals are released from pollen that lands on the stigma of the 'wrong' species (see Chapter 4). This reduces the success of sexual reproduction in recipient species, and can allow the allelopathic species to dominate. Parthenium weed (*Parthenium hysterophorus*), hawkweeds (*Hieracium* species) and timothy (*Phleum pratense*) are the only weedy species demonstrated to have allelopathic pollen (Murphy and Aarssen, 1995a,b; Murphy, 1999).

Apparent competition

Even with well-designed experiments, an observer may conclude that competition is occurring between individuals, even though some other reason may explain the outcome. For example, common sage (*Salvia officinalis*) attracts small herbivores because it provides shelter for them. The herbivores, in turn, consume other vegetation around the sage plant. The gross result is that other species do poorly while sage is healthy; if you did not look for the herbivores, you might erroneously conclude that sage was outcompeting other species. This is called 'apparent competition' (Holt, 1977; Connell, 1990) (Fig. 8.2). In another example, the survival, growth and reproduction of the parsley fern (*Botrychium australe*), native to New Zealand decreased after the introduction of the non-native colonial bentgrass (*Agrostis capillaris*). This occurred because the colonial bentgrass provided habitat for non-native slugs (*Deroceras reticulatum*), which then predated on the parsley fern (Sessions and Kelly, 2002)

What Determines the Outcome of Competition?

One of the reasons weeds are so successful is because they adapt rapidly to new environmental conditions, including the 'competitive neighbourhood' of other weeds, crops and plants in general. Weeds do not 'know' how competitive others are – if others are much better competitors, the weed simply dies without reproducing. If a weed is at a competitive disadvantage but still produces offspring, there should be selection for the offspring to develop better competitive abilities (as long as the genes are available). What complicates the situation is that weeds are subject to selection from other types of interactions (herbivory, for example – see Chapter 9) and this makes it difficult to

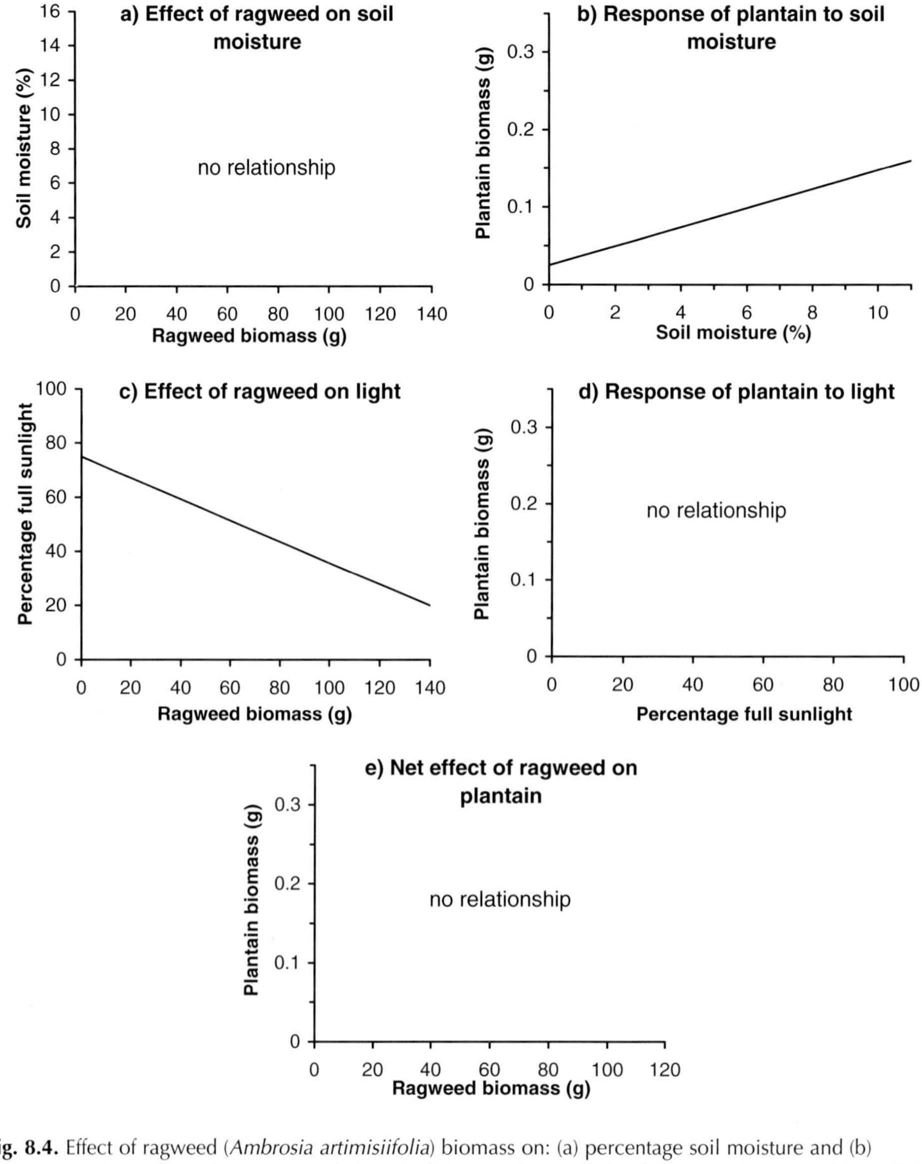

Fig. 8.4. Effect of ragweed (*Ambrosia artimisiifolia*) biomass on: (a) percentage soil moisture and (b) percentage full sunlight, the response of plantain (*Plantago lanceolata*) to (c) soil moisture and (d) sunlight, and (e) the net effect of ragweed on plantain (redrawn and adapted from Goldberg, 1990).

determine what abilities are a response to the need to compete.

Competitive traits

Despite the complex nature of interactions, the outcome of competition depends on the same processes and structures that influence all aspects of a weed's existence, i.e. selection on the traits that each individual weed has. Weeds have traits that will suppress neighbours or avoid being suppressed by them, but it is rare to find weeds that do both well. For example, Goldberg (1990) found that the presence of common ragweed

Table 8.2. List of characteristics associated with competitive plants (not in rank order) (adapted from Zimdahl, 1999).

Characteristics
Shoot characteristics
Rapid expansion of tall, foliar canopy
Horizontal leaves under overcast conditions and obliquely slanted leaves (plagiotropic) under sunny conditions
Large leaves
A C_4 photosynthetic pathway and low leaf transmissivity of light
Leaves forming a mosaic leaf arrangement for best light interception
A climbing habit
A high allocation of dry matter to build a tall stem
Rapid extension in response to shading
Root characteristics
Early and fast root penetration of a large soil area
High root density/soil volume
High root–shoot ratio
High root length per root weight
High proportion of actively growing roots
Long and abundant root hairs
High uptake potential for nutrients and water

(*Ambrosia artemisiifolia*) decreased the percentage sunlight available but did not affect soil moisture. Because its competitor, narrow-leaved plantain (*Plantago lanceolata*), responded to decreases in moisture, but not to decreases in light, ragweed was not competitively superior to plantain (Fig. 8.4).

Generally, weeds may adapt to a competitive environment and develop traits that allow them to specialize in being relatively superior at competing for one (or a few) resources. Zimdahl (1999) listed traits associated with highly competitive agricultural weeds (Table 8.2). Possessing any one or a few of these traits does not guarantee competitive success as traits vary in their effectiveness in different populations and communities. For example, the Australian native Sydney golden wattle (*Acacia longifolia*) has a high photosynthetic rate, but is still outcompeted by weedy tick berry (*Chrysanthemoides monilifera*). This is because tick berry has a more efficient leaf arrangement to intercept light and therefore is able to outcompete the native species (Weiss and Nobel, 1984).

Weed ecologists may study 'size' as being the most important trait in competition. 'Size', however, is really not a trait itself, it is more of a general description. We say this because size could mean a weed has adapted (or is phenotypically plastic enough) to grow taller, branch out more or produce more roots to capture resources (Goldberg and Werner, 1983; Schoener, 1983; Goldberg, 1987). Hence, size results from interacting traits like rate of cell division, leaf expansion, seed germination and seedling emergence time and speed. One of the implications of the Bergelson (1996) review we discussed earlier is that larger individuals are often those that germinated or emerged first and captured more resources. In agriculture and forestry, early emerging weeds are the ones that cause crop losses because they compete with the young and vulnerable crops for nutrients and light, depending on the planting conditions used (Forcella, 1993; Van Acker *et al.*, 1993; Knezevic *et al.*, 1994; Chikoye *et al.*, 1995; Weinig 2000). The same principle applies to competition in non-crop ecosystems (Gerry and Wilson, 1995; Tremmel and Bazzaz, 1995). For example, garlic mustard (*Alliaria petiolata*) is probably competitive because it germinates in the autumn, over-winters (and perhaps photosynthesizes) and quickly grows tall as soon as the temperature, moisture and light allow, usually before native spring plants. In this manner, garlic mustard captures early season light, nutrients, moisture and space at the expense of individuals of native species.

The actual impact of size on the outcome of competition can be difficult to quantify. You might expect that a weed that is (initially) twice the size of a competitor would imply that the weed is then twice as competitive. Indeed, this may happen if competition is 'size-symmetric'. In other cases, competition is 'size-asymmetric', meaning that an individual that is (initially) twice the size of a competitor may be (for example) four times as competitive. In practice, you might measure the outcome of competition by examining the relative weights of the weed and its competitor: (i) in competition and (ii) not in competition.

Size, however, is not always a determinant of competitive success (Wilson, 1988; Gerry and Wilson, 1995). To test for size advantage, Grace *et al.* (1992) grew six grasses alone and in pairs. During the first 2 years the initial plant size was correlated to competitive success measured as relative yield (a comparison of yield when grown alone and when grown in competition). In the third year, however, the initial size did not confer an advantage. Individuals with higher relative growth rate were at an advantage rather than those that were initially bigger. Weigelt *et al.* (2002) suggest that size is more important during the seedling stage, whereas species-specific traits such as biomass allocation patterns are more important during the adult stages of a plant's life cycle. Size is less likely to be advantageous in situations of low nutrients and high light, where size does not improve an individual's chance of obtaining resources.

Below-ground competition (for nutrients or water) is more likely to be size-symmetric, while above-ground competition (for light) is more likely to be size-asymmetric (Casper and Jackson, 1997; Schwinning and Weiner, 1998). This is because a weed that successfully outcompetes others for early-season light often has accelerated growth, leading to faster suppression of competitors, capture of increasingly available light as daylength increases, further suppression of competitors and so on. This type of 'feedback' is what leads to size-asymmetric competition. Below ground, weeds that have more roots (or more efficient roots) will capture more resources, but the process is much slower as the water and nutrients are less ubiquitous than light and harder to find. A lack of accelerated capture of resources means that the competitive advantage of a weed with a large root system is restricted to being closely equivalent to its size advantage.

Effect of the environment on competition

Selection pressures change such that a trait may be advantageous in some locations at a given time but may be less advantageous under different environmental circumstances. We have already emphasized that the existence of spatial and temporal variability in the environment is the reality that plants must survive. The more unpredictable the environmental variability, the more risk to existence. In terms of competitive traits, a genotype may survive for years with a suite of traits but if the environment changes drastically, then the genotype may be quickly placed at a competitive disadvantage. This is actually a principle of any weed management: how to outcompete the weeds. The problem again is that weeds tend to adapt more quickly to change and produce a wide variety of genotypes that can be fit to a range of environments.

How adaptable are weeds in changing environments? Generally, weeds can only adapt if they have the genes available. For most weeds, this is rarely a problem as they reproduce sexually and recombine genes constantly (see Chapter 4). When the environment changes, some weed genotypes will die or at least be disadvantaged, but other genotypes will survive to ensure the population and species of weeds will survive. For example, genotypes of lambsquarters were variable enough to live in different concentrations of nutrients and outcompeted the less well adapted carrot crop (*Daucus carota*) (Li and Watkinson, 2000).

Pickett and Bazzazz (1978) examined how a resource gradient (soil moisture) affected competition among six weed

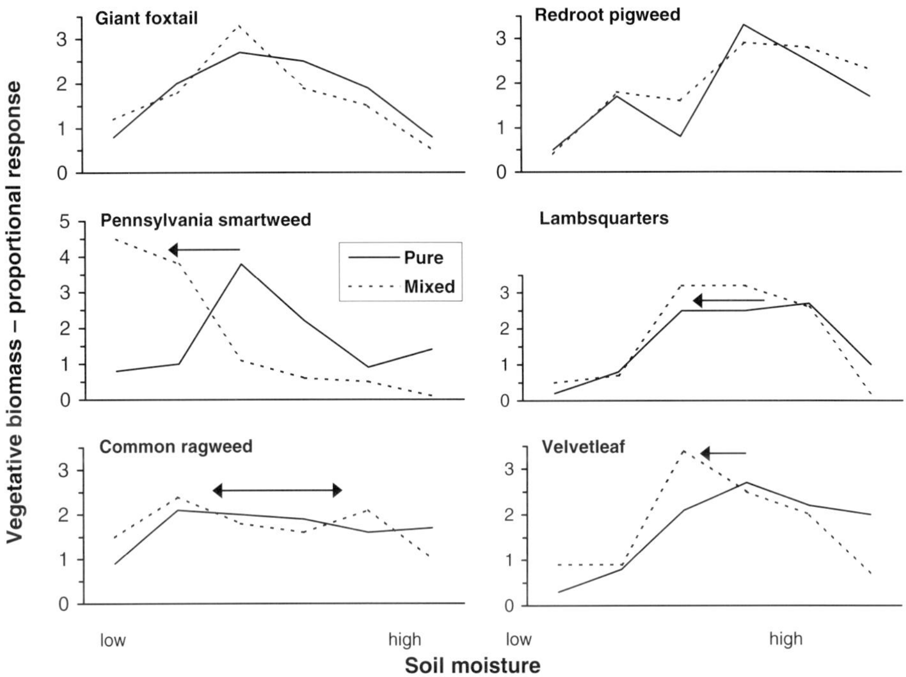

Fig. 8.5. Proportional response of six weeds to a soil moisture gradient when grown alone (pure) and in competition (mixed) (redrawn from data in Pickett and Bazzaz, 1978).

species by growing them alone and in competition with each other (six species together). When grown alone, all species had a broad tolerance to a water gradient (Fig. 8.5). When grown in competition, however, peak biomass tended to shift in four species while the most competitive species, redroot pigweed (*Amaranthus retroflexus*) and giant foxtail (*Setaria faberii*) did not shift. Species primarily responded through phenotypic plasticity rather than higher mortality.

Both of these examples illustrate the effects of spatial heterogeneity on weed competition. Spatial distributions of weeds are the patterns of locations of weed species that we see, for example, in a maizefield or a meadow. Weed distributions are often scattered or 'patchy', in part, because in some areas the weed can outcompete other plants but in other areas the weed is excluded or suppressed. As discussed in Chapter 6, such patchiness also relates to weed seed dispersal. In non-crop habitat, patchiness occurs since variation in topography and substrate will create local micro-environments favouring some weeds and not others. The patchiness may be less obvious in these non-crop habitats since placement of nutrients and water is more precise in crop habitats. In crop fields, the patchiness is exacerbated because weeds will colonize areas where farmers add nitrogen (for example) and where machinery disperses the seeds (Casper and Cahill, 1998; Dieleman and Mortensen, 1999).

One of the implications of clumped population of weeds is that there will be a lot of crowding; in other words, the local density of weeds is often high. This has implications for competition involving weeds. In theory, the more weeds that exist in a given area, the greater the demands placed on local patches of limited resources. Each weed competes for its own benefit. Therefore, when a weed competes it is as likely to harm other weeds (conspecific or otherwise) as it is to harm crops or native plants. Conspecific weeds may be more likely to compete because they should have similar resource demands and probably have similar traits.

Because of the potential importance of plant densities, ecologists and crop scientists often study how density quantitatively affects the outcome of intra- and interspecific competition (e.g. Cousens *et al.*, 1987; Lonsdale, 1990; Kropff and Lotz, 1992; Kropff and Spitters, 1992; Kropff *et al.*, 1992a,b; Cousens and O'Neill, 1993; Frantik, 1994; Knezevic *et al.*, 1994; Chikoye *et al.*, 1995; Ives, 1995; Lindquist and Kropff, 1996; Lutman *et al.*, 1996). It appears that competition is often 'density dependent'. This means exactly what it says – as density changes, so too does the outcome of competition. The change may be direct for a while: every time another weed germinates and starts to grow, there is one more demand on the limited common resource pool and competition increases by an amount directly related to the extra demand.

Most studies, again, tend to examine competition between two species only or, perhaps, between a crop and weed populations comprised of a few weed species. In these studies, density-dependent effects can be illustrated as in Fig. 8.6. Notice that as pigweed density increases beyond a relatively low value of 0.5 pigweed per m^2, competition, as measured by yield loss in maize, slows. This is because the pigweed individuals start to compete *intra*specifically whereas before this they compete mainly with maize (*inter*specifically). While this example does not show it, and it is difficult to demonstrate this experimentally, there also can be a period when the weed density is not yet high enough to cause significant impacts.

Density-dependent competition is typically important but density is neither the only important factor nor is it independent of other factors. The time of emergence of weeds, their morphology (e.g. big leaves or small leaves, tall plants or short plants), and other density-dependent interactions like allelopathy, herbivory (Chapter 9) and parasitism (Chapter 9), can influence competition separately, synergistically or antagonistically (Weidenhamer *et al.*, 1989; Bergelson, 1990; Molofsky, 1999). They are all influenced by genetic and environmental variation, and all of these will not only vary spatially, but temporally. The competitive outcome of a changing environment will be dependent on the timing of the change and on its interaction with the plant's phenology, especially seed germination and seedling emergence.

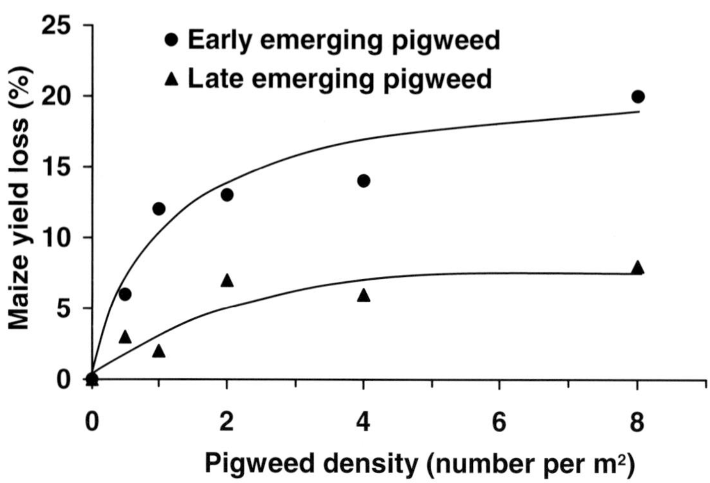

Fig. 8.6. Yield loss in maize as a function of redroot pigweed (*A. retroflexus*) density and date of pigweed emergence (redrawn from Knezevic *et al.*, 1994).

Even the timing of fertilizer application will determine the competitive outcome. For example, two cultivars of sugarbeet (*Beta vulgaris* var. *sacchariferal*) were more competitive with wild mustard (*Sinapsis arvensis*) if there was a late season nitrogen fertilizer application, and more competitive with lambsquarters if nitrogen was applied earlier in the season (Paolini *et al.*, 1999). Similar results were found when nitrogen was applied later in winter wheat (*Triticum aestivum*), thereby reducing the competitive effect of ivy-leaved speedwell (*Veronica hederifolia*) (Agnonin *et al.*, 1996).

Summary

Interactions can have positive, negative or neutral effects on the individuals involved, and these effects may or may not influence population and/or community dynamics. When individuals compete, they may exploit resources such as nutrients, water and space, making them unavailable to competitors (exploitation competition), or they may deny access of resources to others (interference competition). In this chapter we looked at competition and allelopathy: interactions where at least one partner was negatively affected. There is no question that competition can play an important role in shaping populations and communities; however, it is a complex interaction, and to say that 'competition is important' really gives us very little information. Individuals can compete for more than one resource at a time and they can compete above and below ground. The outcome of a competitive interaction will depend on many things, including the relative size and growth rates of individuals as well as the abiotic environment. Allelopathy, possibly a form of interference competition, involves the production of toxins that can make tissues or habitats unsuitable for use. In the next chapter we examine types of interactions where at least one individual benefits.

Questions

1. Would you consider the weed species you selected to be an effective competitor? In what situations will your species be competitive and not be competitive?
2. What is competition, and why is it so difficult to define?
3. What type of competition would you expect to find in an agricultural field? How would this be altered by changing light patterns within the crop canopy or by the addition of fertilizer? Why?
4. Briefly explain Grime's and Tilman's views of competition. In what ways are they similar? In what ways do they differ?
5. Is allelopathy likely to be more important in agricultural or natural systems? Explain why.
6. Compare Baker's (1956, 1974) list of weed traits (Box 1.1) with Zimdahl's (1999) list of competitive weeds (Table 8.2). Explain their similarities and differences.
7. Design an allelopathy experiment.
8. Why is it so difficult to determine how competition and allelopathy occur (i.e. define the mechanism)?

Literature Cited

Aarssen, L.W. (1989) Competitive ability and species coexistence: a "plant's-eye" view. *Oikos* 56, 386–401.

Aarssen, L.W. (1992) Causes and consequences of variation in competition ability in plant communities. *Journal of Vegetation Science* 3, 165–174.

Agnonin, C., Caussanel, J.-P. and Meynard, J.-M. (1996) Competition between winter wheat and *Veronica hederifolia*: influence of weed density and the amount and timing of nitrogen application. *Weed Research* 36, 175–187.

Alphalo, P.J., Ballaré, C.L. and Scopel, A.L. (1999) Plant-plant signalling, the shade-avoidance response and competition. *Journal of Experimental Botany* 50, 1629–1634.

Bengtsson, J., Fagerstrom, T. and Rydin, H. (1994) Competition and coexistence in plant communities. *Trends in Ecology and Evolution* 9, 246–250.

Bergelson, J. (1990) Spatial patterning in plants: opposing effects of herbivory and competition. *Journal of Ecology* 78, 937–948.

Bergelson, J. (1996) Competition between two weeds. *American Scientist* 84, 579–584.

Carlson, H.L. and Hill, J.E. (1986) Wild oat (*Avena fatura*) competition with spring wheat: effects of nitrogen fertilization. *Weed Science* 34, 29–33.

Casper, B. and Jackson, R.B. (1997) Plant competition underground. *Annual Review of Ecology and Systematics* 28, 545–570.

Casper, B.B. and Cahill, J.F. Jr (1998) Population-level responses to nutrient heterogeneity and density by *Abutilon theophrasti* (Malvaceae): an experimental neighborhood approach. *American Journal of Botany* 85, 1680–1687.

Chikoye, D., Weise, S.F. and Swanton, CJ (1995) Influence of common ragweed (*Ambrosia artemisiifolia*) time of emergence and density on white bean (*Phaseolus vulgaris*). *Weed Science* 43, 375–380.

Connell, J.H. (1980) Diversity and the evolution of competitors, or the ghost of competition past. *Oikos* 35,131–138

Connell, J.H. (1983) On the prevalence and relative importance of interspecific competition: evidence from field studies. *American Naturalist* 122, 661–696.

Connell, J.H. (1990) Apparent versus 'real' competition in plants. In: Grace, J.B. and Tilman, D. (eds) *Perspectives on Plant Competition*. Academic Press, San Diego, California, pp. 9–26.

Cousens, R. and O'Neill, M. (1993) Density dependence of replacement series experiments. *Oikos* 66, 347–352.

Cousens, R., Brain, P., O'Donovan, J.T. and O'Sullivan, F.A. (1987) The use of biologically realistic equations to describe the effects of weed density and relative time of emergence on crop yield. *Weed Science* 35, 720–725.

Di Tomaso, J.M. (1998) Impact, biology, and ecology of saltcedar (*Tamarix* spp.) in the southwestern United States. *Weed Technology* 12, 326–336.

Dieleman, J.A. and Mortensen, D.A. (1999) Characterizing the spatial pattern of *Abutilon theophrasti* seedling patches. *Weed Research* 39, 455–467.

Forcella, F. (1993) Seedling emergence model for velvetleaf. *Agronomy Journal* 85, 929–933.

Frantik, T. (1994) Interference of *Chenopodium suecicum* J. Murr. and *Amaranthus retroflexus* L. in maize. *Weed Research* 34; 45–53.

Gealy, D. (1998) Differential response of palmleaf morning glory (*Ipomoea wrightii*) and pitted morning glory (*Ipomoea lacunosa*) to flooding. *Weed Science* 46, 217–224.

Gentle, C.B. and Duggin, J.A. (1997) Allelopathy as a competitive strategy in persistent thickets of *Lantana camara* L. in three Australian forest communities. *Plant Ecology* 132, 85–95.

Gerry, A.K. and Wilson, S.D. (1995) The influence of initial size on the competitive responses of six plant species. *Ecology* 76, 272–279.

Gibson, K.D. and Fischer, A.J. (2001) Relative growth and photosynthetic response of water-seeded rice and *Echinochloa oryzoides* (Ard.) Fritsch to shade. *International Journal of Pest Management* 47, 305–309.

Goldberg, D.E. (1985) Effects of soil pH, competition, and seed predation on the distribution of two tree species. *Ecology* 66, 503–511

Goldberg, D.E. (1987) Neighbourhood competition in an old field community. *Ecology* 68, 1211–1223.

Goldberg, D.E. (1990) Components of resource competition. In: Grace, J.B. and Tilman, D. (eds) *Perspectives on Plant Competition*. Academic Press, San Diego, California, pp. 27–49.

Goldberg, D.E. and Barton, A.M. (1992) Patterns and consequences of interspecific competition in natural communities: a review of field experiments with plants. *American Naturalist* 139, 771–801.

Goldberg, D.E. and Landa, K. (1991) Competitive effect and response: hierarchies and correlated traits in the early stages of competition. *Journal of Ecology* 79, 1013–1030.

Goldberg, D.E. and Werner, P.A. (1983) Equivalence of competitors in plant communities: a null hypothesis and a field experiment approach. *American Journal of Botany* 70, 1098–1104.

Goldberg, D.E., Turkington, R. and Olsvig-Whittaker, L. (1995) Quantifying the community-level consequences of competition. *Folia Geobotanica Phytotaxis* 30, 231–242.

Grace, J.B. (1990) On the relationship between plant traits and competitive ability. In: Grace, J.B. and Tilman, D. (eds) *Perspectives on Plant Competition*. Academic Press, San Diego, California, pp. 51–66.

Grace, J.B. and Tilman, D. (eds) (1990) *Perspectives on Plant Competition*. Academic Press, San Diego, California.

Grace, J.B., Keough, J. and Guntenspergen, G.R. (1992) Size bias in traditional analyses of substitutive competition experiments. *Oecologia* 90, 429–434.

Grigulis, K., Sheppard, A.W., Ash, J.E. and Groves, R.E. (2001) The comparative demography of the pasture weed *Echium plantagineum* between its native and invaded ranges. *Journal of Applied Ecology* 38, 281–290.

Grime, J.P. (1979) *Plant Strategies and Vegetational Processes*. John Wiley & Sons, New York.

Gurevitch, J., Wilson, P., Stone, J.L., Teese, P. and Stoutenburgh, R.J. (1990) Competition among old-field perennials at different levels of soil fertility and available space. *Journal of Ecology* 78, 727–744.

Hashem, A., Radosevich, S.R. and Dick, R. (2000) Competition effects on yield, tissue nitrogen, and germination of winter wheat (*Triticum aestivum*) and Italian ryegrass (*Lolium multiflorum*). *Weed Technology* 14, 718–725.

Holt, R.D. (1977) Predation, apparent competition, and the structure of prey communities. *Theoretical Population Ecology* 12, 197–229.

Hunt, M.A. and Beadle, C.L. (1998) Whole-tree transpiration and water-use partitioning between *Eucalyptus nitens* and *Acacia dealbata* weeds in a short-rotation plantation in northeastern Tasmania. *Tree Physiology* 18, 557–563.

Inderjit and del Moral, R. (1997) Is separating resource competition from allelopathy realistic? *Botanical Review* 63, 221–230.

Iqbal, J. and Wright, D. (1997) Effects of nitrogen supply on competition between wheat and three annual weed species. *Weed Research* 37, 391–400.

Iqbal, J. and Wright, D. (1998) Effects of water deficit and competition on net photosynthesis of spring wheat (*Triticum aestivum* L.) and two annual weeds (*Phalaris minor* Retz. and *Chenopodium album* L.). *Cereal Research Communications* 26, 81–88.

Ives, A.R. (1995) Measuring competition in a spatially heterogeneous environment. *American Naturalist* 146, 911–936.

Keddy, P.A. (1989) *Competition*. Chapman and Hall, London.

Knezevic, S.Z., Weise, S.F. and Swanton, C.J. (1994) Interference of redroot pigweed (*Amaranthus retroflexus*) in maize (*Zea mays*). *Weed Science* 42, 568–573.

Kropff, M.J. and Lotz, L.A.P. (1992) Systems approaches to quantify crop–weed interactions and their application in weed management. *Agricultural Systems* 40, 265–282.

Kropff, M.J. and Spitters, C.J.T. (1992) An eco-physiological model for interspecific competition, applied to the influence of *Chenopodium album* L. on sugar beet I. Model description and parameterization. *Weed Research* 32, 437–450.

Kropff, M.J., Spitters, C.J.T., Schneiders, B.J., Joenje, W. and De Groot, W. (1992a) An eco-physiological model for interspecific competition, applied to the influence of *Chenopodium album* L. on sugar beet II. Model evaluation. *Weed Research* 32, 451–463.

Kropff, M.J., Weaver, S.E. and Smits, M.A. (1992b) Use of ecophysiological models for crop-weed interference: relations amongst weed density, relative time of weed emergence, relative leaf area and yield loss. *Weed Science* 40, 296–301.

Lawton, J.H. and Hassell, M.P. (1981) Asymmetrical competition in insects. *Nature (London)* 289, 793–795.

Li, B. and Watkinson, A.R. (2000) Competition along a nutrient gradient: a case study with *Daucus carota* and *Chenopodium album*. *Ecological Research* 15, 293–306.

Lindquist, J.L. and Kropff, M.J. (1996) Applications of an ecophysiological model for irrigated rice (*Oryza sativa*)–*Echinichloa* competition. *Weed Science* 44, 52–56.

Lonsdale, W.M. (1990) The self-thinning rule: dead or alive? *Ecology* 71, 1373–1388.

Lutman, P. Risiott, J.W.R. and Ostermann, P. (1996) Investigations into alternative methods to predict the competitive effects of weeds on crop yields. *Weed Science* 44, 290–297.

Mallik, A.U. and Pellissier, F. (2000) Effects of *Vaccinium myrtillus* on spruce regeneration: testing the notion of coevolutionary significance of allelopathy. *Journal of Chemical Ecology* 26, 2197–2209.

Matthes-Sears, U. and Larson, D.W. (1999) Limitations to seedling growth and survival by the quantity and quality of rooting space: implications for the establishment of *Thuja occidentalis* on cliff faces. *International Journal of Plant Science* 160, 122–128.

McConnaughay, K.D.M. and Bazzaz, F.A. (1991) Is physical space a soil resource? *Ecology* 72, 94–103.

McLachlan, S.M., Murphy, S.D., Tollenaar, M., Weise, S.F. and Swanton, C.J. (1995) Light limitation of reproduction and variation in the allometric relationship between reproductive and vegetative biomass in *Amaranthus retroflexus* (redroot pigweed). *Journal of Applied Ecology* 32, 157–165.

Miller, T.E., Winn, A.A. and Schemske, D.W. (1994) The effects of density and spatial distribution on selection for emergence time in *Prunella vulgaris* (Lamiaceae). *American Journal of Botany* 81, 1–6.

Molofsky, J. (1999) The effect of nutrients and spacing on neighbor relations *in Cardamine pensylvanica. Oikos* 84, 506–514.

Moody, K. (1981) Weed fertilizer interactions in rice. International Rice Research Institute Thursday Seminar, 9 July 1981, cited in Zimdahl (1999).

Murphy, S.D. (1999) Pollen allelopathy. In: Inderjit, Dakshini, K.M.M. and Foy, C.L. (eds) *Principles and Practices in Plant Ecology: Allelochemical Interactions.* CRC Press, Boca Raton, Florida, pp. 129–148.

Murphy, S.D. and Aarssen, L.W. (1995a) Allelopathic pollen extract from *Phleum pratense* L. (Poaceae) reduces seed set in sympatric species. *International Journal of Plant Science* 156, 435–444.

Murphy, S.D. and Aarssen, L.W. (1995b) Reduced seed set in *Elytrigia repens* caused by allelopathic pollen from *Phleum pratense. Canadian Journal of Botany* 73, 1417–1422.

Nagashima, H., Terashima, I. and Ketoh, S. (1995) Effect of plant density on frequency distributions of plant height in *Chenopodium album* stands: analysis based on continuous monitoring of height-growth of individual plants. *Annals of Botany* 75, 173–180.

Norris, R.F. (1996) Water use efficiency as a method for predicting water use by weeds. *Weed Technology* 10, 153–155

Novoplansky, A. (1991) Developmental responses of portulaca seedlings to conflicting spectral signals. *Oecologia* 88, 138–140.

Olofsdotter, M., Navarez, D., Rebulanan, M. and Streibig, J.C. (1999) Weed suppressing rice cultivars – does allelopathy play a role? *Weed Research* 39, 441–454.

Pacala, S. and Silander, J.A. Jr (1990) Field tests of neighbourhood population dynamic models of two annual weed species. *Ecological Monographs* 60, 113–134.

Paolini, R., Principi, M., Froud-Williams, R.J., Del Pugila, S. and Biancardie, E. (1999) Competition between sugarbeet and *Sinapis arvensis* and *Chenopodium album*, as affected by timing of nitrogen fertilization. *Weed Research* 39, 425–440.

Pickett, S.T.A. and Bazzaz, F.A. (1978) Organization of an assembly of early succession species on a soil moisture gradient. *Ecology* 59, 1248–1255.

Rebele, F. (2000) Competition and coexistence of rhizomatous perennial plants along a nutrient gradient. *Plant Ecology* 147, 77–94.

Rice, E.L. (1984) *Allelopathy.* Academic Press, Toronto.

Rice, E.L. (1995) *Biological Control of Weeds and Plant Diseases: Advances in Applied Allelopathy.* University of Oklahoma Press, Norman, Oklahoma.

Richards, D. and Rowe, R.N. (1977) Effects of root restriction, root pruning, and 6-benzyl-aminopurine on the growth of peach seedlings. *Annals of Botany* 41, 729–740.

Rose, S.J., Burnside, O.C., Specht, J.E. and Swisher, B.A. (1984) Competition and allelopathy between soybeans and weeds. *Agronomy Journal* 76, 523–528.

Sage, R.B. (1999) Weed competition in willow coppice crops: the cause and extent of yield losses. *Weed Research* 39, 399–411.

Schenk, H.J., Callaway, R.M. and Mahall, B.E. (1999) Spatial root segregation: are plants territorial? *Advances in Ecological Research* 28, 146–180.

Schillinger, W.F and Young, F.L. (2000) Soil water use and growth of Russian thistle after wheat harvest. *Agronomy Journal* 92, 167–172.

Schoener, T.W. (1983) Field experiments on interspecific competition. *American Naturalist* 122, 240–285.

Schwinning, S. and Weiner, J. (1998) Mechanisms determining the degree of size asymmetry in competition among plants. *Oecologia* 113, 447–455.

Sessions, L. and Kelly, D. (2002) Predator-mediated apparent competition between an introduced grass, *Agrostis capillaris*, and a native fern, *Botrychium australe* (Ophioglossaceae), in New Zealand. *Oikos* 96, 102–109.

Tilman, D. (1988) *Plant Strategies and the Dynamics and Structure of Plant Communities.* Princeton University Press, Princeton, New Jersey.

Tilman, D. (1990) Mechanisms of plant competition for nutrients: the elements of a predictive theory of competition. In: Grace, J.B. and Tilman, D. (eds) *Perspectives on Plant Competition*. Academic Press, San Diego, California, pp. 117–141.

Tilman, E.A., Tilman, D., Crawley, M.J. and Johnston, A.E. (1999) Biological weed control via nutrient competition: potassium limitation of dandelions. *Ecological Applications* 9, 103–111.

Tollenaar, M., Nissanka, S.P., Aguilera, A., Weise, S.F. and Swanton, C.J. (1994) Effect of weed interference and soil nitrogen on four maize hybrids. *Agronomy Journal* 86, 596–601.

Tremmel, D.C. and Bazzaz, F.A. (1995) Plant architecture and allocation for different neighbourhoods: implications for competitive success. *Ecology* 76, 262–271.

Van Acker, R.C., Swanton, C.J. and Weise, S.F. (1993) The critical period of weed control in soybean [*Glycine max* (L.) Merr]. Weed Science 41, 194–200.

Walch, J.L., Baskin, J.M. and Baskin, C.C. (1999) Relative competitive abilities and growth characteristics of a narrowly endemic and geographically widespread *Solidago* species (Asteraceae). *American Journal of Botany* 86, 820–828.

Wardle, D.A., Nilsson, M.-C., Gallet, C. and Zackrisson, O. (1998) An ecosystem-level perspective of allelopathy. *Biological Review* 73, 305–319.

Weidenhamer, J.D., Hartnett, D.C. and Romeo, J.T. (1989) Density-dependent phytotoxicity: distinguishing resource competition and allelopathic intereference in plants. *Journal of Applied Ecology* 26, 613–624.

Weigelt, A., Steinlein, T. and Beyschlag, W. (2002) Does plant competition intensity rather depend on biomass or on species identity? *Basic and Applied Ecology* 3, 85–94.

Weinig, C. (2000) Differing selection in alternative competitive environments: shade-avoidance strategies and germination timing. *Evolution* 54, 124–136.

Weiss, P.W. and Noble I. R. (1984) Interactions between seedlings of *Chrysanthemoides monilifera* and *Acacia longifolia*. *Australian Journal of Ecology* 9, 107–115

Williamson, G.B. (1990) Allelopathy, Koch's postulates, and the neck riddle. In: Grace, J.B. and Tilman, D. (eds) *Perspectives on Plant Competition*. Academic Press, San Diego, California, pp. 143–162.

Wilson, J.B. (1988) Shoot competition and root competition. *Journal of Applied Ecology* 25, 279–296.

Wilson, S.D. and Tilman, D. (1995) Competitive responses of eight old-field plant species in four environments. *Ecology* 76, 1169–1180.

Zimdahl, R.L. (1980) *Weed–Crop Competition: a Review*. International Plant Protection Centre, Corvallis, Oregon.

Zimdahl, R.L. (1999) *Fundamentals of Weed Science*, 2nd edn. Academic Press, San Diego, California.

9

Interactions Among Populations II: Herbivory, Parasitism and Mutualisms

Concepts

- Herbivory is the consumption of plant tissue by animals.
- Immobile (sessile) plants have developed a variety of ways to avoid, tolerate or defend against herbivory.
- Seed predation is a special type of herbivory that results in significant and often complex effects on plant population dynamics.
- Parasitic plants obtain nutrients, shelter and support from other plants.
- Mutualistic interactions provide benefits to both individuals and are especially important in stressed environments.
- The net effect of one individual on another is the result of all beneficial effects minus all negative effects.

Introduction

There are many types of interactions where one individual gains an absolute benefit over another. For example, in herbivory, parasitism and mutualism (refer to Table 8.1), at least one of the individuals benefits from the interaction. Rather than being distinct processes, there is a gradation of positive to negative effects (Bronstein, 1994a) (Fig. 9.1). In this chapter we look at herbivory, parasitism and mutualisms as separate interactions, and then consider the net effect of all types of interactions on populations.

Herbivory

Herbivory is the consumption of plant tissue by animals. Plants, in general, are a low-quality food because they are low in nitrogen (needed by animals for protein) and high in complex carbohydrates, which are hard for most animals to digest. Herbivores preferentially consume roots, young leaves, flowers, fruits and seeds because these are higher in nutrients and more digestible. In terrestrial plant communities, about 20% of above-ground plant biomass is consumed by herbivores but this varies considerably with community type, and the number and type of herbivores present (Cyr and Pace, 1993).

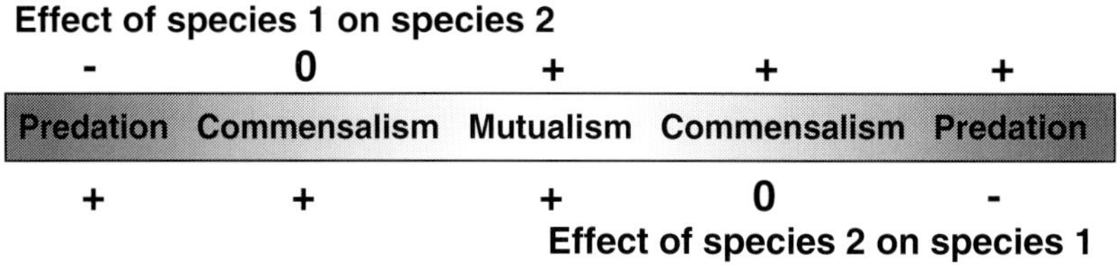

Fig. 9.1. The continuum of species interactions (redrawn from Bronstein, 1994a). If we use the 'cat and mouse' example, the cat (species 1) has a negative effect on the mouse (species 2), and the mouse has a positive effect on the cat.

Herbivores include mammals such as grazing deer and zebra, sap-sucking insects, seedling-eating molluscs, root-feeding larvae, leaf-eating grasshoppers, and seed-eating mice and beetles. Plant herbivores range from specialists (typically invertebrates), that eat only one or a few types of food, to generalists (typically vertebrates) that are unselective feeders (Crawley, 1989). Some herbivores can be both specialist and generalist because they change their feeding behaviour depending on the relative density of the plants.

Herbivory is usually non-lethal unless all of the plant or an essential tissue (e.g. all roots) is eaten, or if the plant is already stressed. If a plant is not able to compensate for the loss of tissue, such as when a tree is girdled by a deer, it will die. The more immediate effects of herbivory are the loss of photosynthetic tissue, stored nutrients and meristems (i.e. sites of active growth) (Noy Meir, 1993). These losses may lessen a plant's ability to photosynthesize, absorb nutrients and water, and grow. There are, however, mechanisms that allow plants to compensate for herbivory. Plants respond by mobilizing stored nutrients, increasing their rate of photosynthesis, stimulating dormant meristems, or growing more roots or shoots (depending on what was eaten) (Crawley, 1992) (Box 9.1). The net effect of long-term herbivory will depend on the age of the foliage, the distribution of damage on the plant, and the stage of development and seasonal timing (Crawley, 1992). For example, the loss of young leaves from herbivores may have more long-term repercussions than the loss of older leaves. No plant species is immune to herbivory completely, but many have evolved ways to defend or compensate for it.

Plant defences and compensation to herbivory

Plants cannot run away from potential herbivores, but they do have ways to avoid being consumed. The two types of defences

Box 9.1. Mechanisms by which plants respond to herbivory (adapted from Crawley, 1992).

- Increased light intensity for surviving leaf area
- Increase in the rate of carbon fixation at a given light intensity
- Improved water and nutrient availability to the surviving leaf tissue
- Delayed senescence (plus rejuvenation) of leaves
- Increased duration of the growing period
- Redistribution of the photosynthate to the production of new leaves and away from roots, flowers, fruits or storage
- Reduced rate of flower abortion
- Production of new shoots from dormant buds or newly produced epicormic buds
- Ungrazable reserve (e.g. storage in roots and woody stems)
- Not regrowing while the herbivore is still around

Table 9.1. Types of herbivore avoidance and tolerance strategies used by plants to defend against herbivory.

Defence	Type	Example	Explanation
Avoidance	Structural	Hairs, spines, trichomes	Make it harder to consume tissue
		Digestibility reducers	Cellulose, lignin, cutins, tannins effect digestion by blocking digestive enzymes
		Protective coating	Shells of fruits make them difficult to eat
	Chemical	Chemical defences	Reduce the palatability of plants or makes them poisonous
	Phenological	Rare or ephemeral	Makes plants hard to find when herbivores are active
		Early growth or reproduction	Plant escapes later emerging herbivores
Tolerance	Compensate	Increase net photosynthetic rate	Plants produce more above-ground biomass after being grazed than non-grazed individuals
		High relative growth rates	
		Increase branching or tillering	After apical dominance is released
		High carbon storage	Pre-existing stores of carbons in roots can be re-allocated for above-ground production
		Carbon re-allocation	Can quickly move carbon from roots to shoots

are direct (i.e. avoidance and tolerance) and indirect (i.e. defend themselves by recruiting bodyguards to protect them from herbivores).

Avoidance

Avoidance defences reduce the chance that a plant will be eaten or reduce the performance of the herbivore. Plants have three strategies to avoid herbivory: structural, phenological and chemical (Table 9.1). Structural and phenological defences are 'constituent' traits present in the plant whether or not a herbivore is present. Constituent defences deter a herbivore from damaging the plant. Chemical defences, however, may not exist until induced by herbivory. These chemicals can inhibit digestion, deter feeding or intoxicate the herbivore, thereby reducing herbivore damage. However, some herbivores adapt and can detoxify chemical defences. For example, saliva from the corn earworm caterpillar (*Helicoverpa zea*) suppresses the induction of the toxin nicotine in tobacco (*Nicotiana tabacum*) (Musser *et al.*, 2002). Herbivores may even use the plant chemicals to their own benefit. The classic example of this is the monarch butterfly (*Danaus plexippus*) and common milkweed (*Asclepias syriaca*) association. Milkweed produces a glycoside that is toxic to the heart and circulatory system in most herbivores. Monarch larvae, however, are able to consume milkweed leaves, and the glycoside, in turn, makes them more unpalatable to their predators.

Tolerance

Tolerance is the ability to minimize damage from herbivory (Belsky *et al.*, 1993; Strauss and Agrawal, 1999). While tolerance may be an alternative to avoidance under certain types of herbivory, both tolerance and avoidance may be used simultaneously. For example, if slugs start eating ginger (*Asarum caudatum*), its palatability decreases (avoidance) and its growth and seed production decreases (tolerance) (Cates, 1975).

The degree to which a plant tolerates herbivory is called 'compensation'. There are several mechanisms used by plants to compensate and increase their tolerance to herbivory or other types of damage (Table 9.1) (Belsky *et al.*, 1993). Regardless of the mechanism, however, the success of any compensation is related to the cost the individual incurs because it has to allocate resources away from growth or reproduction. Ultimately, this re-allocation of resources can reduce the fitness of an individual; for example, jimsonweed (*Datura stramonium*) experienced 15–25% reductions in fitness after defoliation (Foroni and Nunez-Farfan, 2000). As long as the reduction in fitness does not lead to complete extinction of a genotype, tolerance is a good strategy.

Tolerance responses vary with the type of herbivory and the constraints on the plant, i.e. what is morphologically or physiologically possible? For example, cotton (*Glossypium hirsutum*) responds to phloem-sucking aphids by decreasing axillary branching; however, if buds are eaten, branching increases (Sandras, 1996). The weedy leafy spurge (*Euphorbia esula*) seems to be more constrained as it only tolerates defoliation by allocating more resources to its large root system that, after herbivores leave, will produce more shoots (Olson and Wallander, 1999). The internal mechanisms that allow a plant to respond to herbivory are moderated by the timing of herbivory, nutrient, light and water availability, and the presence of plant competitors (Strauss and Agrawal, 1999).

A special type of tolerance is 'over-compensation', as found in weeds such as purple loosestrife (Venecz and Aarssen, 1998). This occurs when herbivory actually benefits plants and increases fitness (Aarssen, 1995). The mechanism is relatively simple: if the apical (shoot) meristem is eaten, the lateral meristems (on the branches) are signalled chemically to grow. As a result, the extra branches and all their leaves may allow the damaged plant to increase photosynthesis and carbohydrate production relative to undamaged plants. If this extra production allows for more sexual or asexual success, then fitness increases (Venecz and Aarssen, 1998; Agrawal, 2000). Over-compensation may only occur under certain conditions such as low competition, and optimum light and moisture (Strauss and Agrawal, 1999).

Indirect plant defences

Indirect plant defences mean that an individual uses another organism to defend itself against herbivory. For example, a plant may be protected when it grows near an unpalatable species. The unpalatable species can mask features, such as scent, that otherwise would attract herbivores (Price *et al.*, 1980). Some plants decrease herbivore damage by recruiting 'bodyguards'. This means an individual provides pollen, nectar, habitat or other rewards that increases the foraging effectiveness of species that harm herbivores (Price *et al.*, 1980; Sabelis *et al.*, 1999; Elliot *et al.*, 2000). In some cases, bodyguards are recruited only as an induced defence, e.g. predatory mites are attracted by chemicals released when herbivorous spider mites damage leaves. It appears that few weeds use bodyguards, though weeds can have the general characteristic of induced defences (e.g. Jennings *et al.*, 2000).

Seed predation: a special case of herbivory

When herbivores eat seeds, it is called 'seed predation'. The term 'predation' is used because the whole individual (as a seed) is consumed just as with the consumption of a whole animal. While seed predation can be severe, the rates are variable because of interactions among plants, herbivores and the environment. It is also difficult to detect the effect of seed predation on population dynamics (Crawley, 1992; Cromar *et al.*, 1999).

Table 9.2. The effect of pre-dispersal flower and seed predation, post-dispersal seed predation and seedling competition with established vegetation (switchgrass – *Paricum virgatum*) on the recruitment of Platte thistle seedlings at Arapaho Prairie, Nebraska (adapted from Louda *et al.*, 1990).

(a) Pre-dispersal flower and seed predation

Stage	With insects, water-spray control (number per plant)	Without insects, insecticide spray (number per plant)
Number seeds initiated	577	716
Number viable seeds	41	105
Seedlings established	0.5	3.0
Adults matured	0.07	0.37

(b) Post-dispersal seed predation

Habitat type	With herbivores, no cage (number per 30×30cm plot)	Without herbivores, with cage (number per 30×30cm plot)
Blowout (open)	0.01	0.02
Grass	2.0	0.01

(c) Seedling survival

Time after transplant	No competition (% survival)	In competition (% survival)
9 weeks	43	5
1 year	19	5
2 years	10	0

Seed predation can occur while seeds are still on the plant ('pre-dispersal') or after they have abscised ('post-dispersal'). Pre-dispersal seed predators are usually invertebrates (mainly insects) with a narrow host range or specialized feeding habits. There is a wider variety of post-dispersal seed predators including vertebrates (birds or rodents), insects (ants and carabid beetles) and molluscs (snails and slugs) (Crawley, 1989; Blaney and Kotanen, 2001).

Louda *et al.* (1990) compared the effect of pre-dispersal flower and seed predation, post-dispersal seed predation, and seedling competition with established vegetation on the recruitment of Platte thistle (*Cirsium canescens*) seedlings (Table 9.2). Pre-dispersal predation by insects reduced seed production and seedling establishment. Post-dispersal seed predation had a more pronounced effect on seedling establishment in exposed open sites than in protected grassy sites. Competition of Platte thistle seedlings with grasses greatly reduced seedling survival, with no seedlings surviving after 2 years when planted in switchgrass (*Panicum virgatum*). The authors concluded that competition confined the distribution of Platte thistle to open areas, while predation limited its abundance.

Some species have defences against seed predators. For example, velvetleaf seeds that are small and have a hard seed coat can often survive intact after being ingested and defaecated. In this case, the seed predator actually serves to disperse seeds. Giant Parramatta grass (*Sporobolus indicus* var. *major*), a major weed of pastures and disturbed areas in Australia, can pass through the gut of cattle (Andrews, 1995). Therefore, to prevent invasion of the weed into uninfested fields, cattle must be isolated for 7 days after grazing in an infested field. Other seeds are unpalatable or are protected by a fruit that is difficult to consume (e.g. cocklebur).

Producing large seed crops at irregular intervals (masting) is a further method of defence against seed predators, because it reduces the chances that all seeds will be destroyed by seed predators. Usually, one to a few years of heavy seed production (mast years) are followed by a period of low seed production. During mast years, the number of seeds consumed by predators is high but the probability of any one seed escaping predation is also high; therefore new seedlings are produced through the sheer quantity of seeds available (Silvertown, 1980). This appears to be a more effective strategy against invertebrate than vertebrate predators because vertebrates are more likely to migrate towards areas of mast seeding. Vertebrates also have other food sources to support them during non-mast years (Crawley, 1989). Weeds that mast include Norway maple (*Acer plantanoides*) and Monterey pine (*Pinus radiata*).

Effect of herbivory on populations

As in the case of testing for competition (Chapter 8), it is relatively simple to determine if herbivory will affect the performance of an individual plant, but less easy to determine if it leads to population or community-level effects. Nevertheless, herbivory can influence the distribution and abundance of populations. The effect will generally depend on the type of herbivore involved, the intensity and frequency of herbivory, the plant species, the type and age of tissue consumed, and the abiotic environment.

We have already presented the example of herbivores controlling the abundance and distribution of a native weed species, i.e. the Platte thistle. In contrast, when weedy exotics are introduced to a habitat, there may be few herbivores that consume them and, consequently, their population distribution and abundance may increase. This is termed 'herbivore release' because the plant species is released from the pressure of herbivory, e.g. garlic mustard (*Alliaria petiolata*). Consequently, one approach to managing weeds is introducing specialist herbivores (often insects) as 'biological control agents'. These herbivores may be exotics themselves but are found in the same area of origin as the weed. While weed species usually are not eradicated, even if the biological control agent consumes only a small proportion of biomass, it may be enough to alter the competitive balance between it and otherwise less-competitive native species.

The biological control of prickly pear cacti (*Opuntia* spp.), introduced from Mexico and the southern USA into Australia, exemplifies both successful biological control and the risks involved. The cactus moth borer (*Cactoblastis cactorum*) was introduced from Argentina as a biological control agent to help control prickly pear cacti in 1926. As a caterpillar, the cactus moth borer consumes the tissue and this introduces bacterial soft rot and other pathogens into the cactus. In 1925, the cactus covered 24 million hectares (Mha) of Australia, but by 1930, the cactus was under control. The cactus still survives in small populations that last only until they are detected by the moths. Recently, however, the cactus moth borer was accidentally introduced to eastern North America from Argentina. Unlike in Australia, prickly pear cacti are native to eastern North America and are not considered weeds in general. Eventually, the moth may spread to Mexico and southwestern North America, where it would find many more species of native cacti and cause serious damage (Cory and Myers, 2000). The cactus moth borer is a native part of the ecosystem in Argentina, a saviour in Australia and a serious pest in North America. This is why using herbivores as biological control agents must be tested carefully – it is not always clear whether the herbivore will do more harm than good and this varies from location to location, e.g. introducing beetles to control purple loosestrife (Blossey *et al.*, 2001a,b).

Parasitism

A parasite is an organism that depends on another organism (its 'host') for nutrition, support or shelter; parasitic plants do this by physically infecting and/or climbing on their hosts (Table 9.3). Plants that are entirely dependent on their host (holoparasites) are usually white because they lack chloro-

Table 9.3. Definitions of the various types of parasitism.

Type of parasitism	Explanation
Holoparasite	Entirely dependent on their host for carbon, water and nutrients
Hemiparasite	Rely on their host for some resources, but are self-sufficient in others
Obligate hemiparasite	Can survive only when associated with the appropriate host
Facultative hemiparasite	Can survive without host but are usually associated with the host
Epiphyte	Rely on host for physical support

phyll and cannot photosynthesize. Plants that rely on their host for only some resources (hemiparasites) form either obligate or facultative relationships. Some parasitic species are dependent on their host for physical support (epiphytes). Orchids, ferns, bromeliads, lichens, mosses and many mistletoes are epiphytic. Epiphytes live upon other plants and may or may not have a negative effect on their host. One example of a parasitic epiphyte is the strangler fig (*Ficus leprieurii*) (Fig. 9.2).

Similarly to herbivory, natives may be more chemically or morphologically suitable than exotic weeds to native parasites. There are exceptions. Some weed species are parasitized by the same species as natives or crops, and weeds may be used as a 'trap' crop to reduce the impact of a parasite on desirable species – the same approach can be used in using weeds to attract herbivores (Scholte, 2000; Rambert *et al.*, 2001). Weeds can be parasitized by fungi that are somewhat akin to a 'sexually transmitted disease' (Kaltz and Schmid, 1995); this essentially describes fungi that attack male sexual structures like anthers, e.g. in white cockle (*Silene latifolia*) (Shykoff and Kaltz, 1998). Many weed species are parasitized by fungi like 'fusarium wilt fungi' – though the weeds do respond with inducible defences as they might to herbivores (Jennings *et al.*, 2000). As with herbivores, biological control of weeds may involve introduction or augmentation of parasites, though using multiple types of biological control can create very complex interactions that are difficult to manage (Rosenheim *et al.*, 1995).

Weeds themselves can be parasitic

Fig. 9.2. Establishment of the parasitic strangler fig (*Ficus leprieurii*) on the palm (*Elaeis quineensis*). The fig germinates in the canopy of the palm and sends aerial shoots downward. Once the fig's roots reach the ground, the shoots increase in size. Eventually the fig overgrows the palm and the palm dies but the fig remains. (Longman and Jenik, 1974; with permission of Pearson Education.)

Table 9.4. Examples of important agricultural parasitic weeds.

Common name	Genera	Host species	Comments
Dodder	*Cuscuta*	Non-specific	Twines around stems the seedling makes contact with, and inserts haustoria into stem
Mistletoe	*Loranthus* *Arceuthobium* *Viscum*	Coniferous trees	Shoots hemiparasite
Broomrape	*Orobanche* *Aeginatia*	Carrots, broad beans, tomatoes, sunflowers, red clover,	Root holoparasite
Witchgrass	*Striga* *Alectra*	Species specific: sorghum, millet, maize, cowpeas, groundnuts, other crops	Root hemiparasite, called witchgrass because it harms crop before the parasite has emerged from the soil

(Table 9.4). Parasitic agricultural weeds are more important in developing countries (Zimdahl, 1999), e.g. witchweeds (*Striga* spp.) infest approximately 44 Mha in Africa and infestations can reduce crop yield by more than 20%. North Americans are more concerned with parasitic forestry weeds like dwarf mistletoes (Viscaceae). Dwarf mistletoes are parasites of many conifer species; their impact on the North American forest industry is on the same scale as witchweeds in African agriculture (Baker and Knowles, 1992). The mistletoes preferentially parasitize young trees because the bark is thinner and easier to penetrate (Parker and Riches, 1993). Forestry practices have increased infestations because cutting and replanting mean there are more younger trees of the same age and vulnerability to infection, and fires are no longer allowed to burn – these once reduced dwarf mistletoe infestation. Managing parasitic weeds is difficult and often requires herbicides. Sometimes trap cropping is used to stimulate germination and growth of parasitic weeds so they can be managed in a crop that is not economically important before the weeds infest more important crops (Chittapur *et al.*, 2001). In these cases, the 'trap crop' is not usually a weed but crop species like flax (*Linum* spp.) not meant to be harvested for human use.

Mutualisms

Not all interactions between individuals produce negative effects. Mutualisms are interactions where both individuals benefit. Generally, a mutualist provides a service to its partner that the partner cannot provide for itself, and in return obtains a reward (Bronstein, 1994b). The types of benefits that mutualists gain include:

- nutrition (each organism supplies different essential nutrients to one another);
- protection (one individual protects the other); and
- transport (one mutualist gains mobility through the actions of the other).

Both organisms do not necessarily gain the same benefit. For example, one mutualist may provide nutrition while the other provides transport.

Types of mutualisms

Obligate mutualisms

In an obligate mutualism, both partners of the association require each other in order to survive. The most extreme examples of an obligate mutualism (often called a 'symbiosis') are lichens, which are associations between fungi and algae. The fungus forms the main body structure of the lichen and the alga provides carbohydrates. Because they no longer can live independently, neither the fungus nor the alga is considered to be a distinct species. Plants and their insect pollinators may be obligate mutualists, especially in tropical forests. For example, over 900 species of fig trees (*Ficus* spp.) are each pollinated by a separate species of aganoid wasp (Janzen, 1979), and similar obligate pollinators associations exist for *Yucca* spp. and some orchids. Weeds generally do not form obligate mutualisms with pollinators because most weeds are exotics that have not co-adapted with the available fauna.

Obligate mutualisms are beneficial to both mutualists but they can result in limited distribution or extinction. The extinction of one partner will almost certainly lead to the extinction of the other. If one mutualist has a wide potential distribution, its actual distribution could be much smaller if it is limited by its mutualistic relationship. Species that are part of an obligate mutualism generally make poor weeds unless the potential weed and its mutualist invade at the same time, and both species can withstand the environmental conditions of the new habitat. For example, of the 60 species of fig introduced into Florida, the three species that have become weedy only did so after their pollinator wasps were introduced (Nadel *et al.*, 1992; Richardson *et al.*, 2000). Similarly, two species of banyan figs (*Ficus* spp.) introduced to New Zealand did not set seed until their specific pollinating wasps arrived, apparently through long-distance dispersal, from Australia (Gardner and Early, 1996). This is not to claim that plants that are apparently obligate mutualists can never become weeds. It is possible for the plants to adapt to local fauna or vice-versa, hence the relationship between figs and their pollinator wasps may not be obligate in all cases (Richardson *et al.*, 2000).

Facultative mutualisms

In a facultative mutualism, both species can survive independently, but both benefit when they are found together. Facultative mutualisms are common. For example, mutualistic animals provide plants with 'services' like pollination (Chapter 4) and seed dispersal (Chapter 6). Mutualistic fungi and bacteria provide access to more or different sources of nutrients (Richardson *et al.*, 2000).

Pollen and seed dispersal of weeds is facilitated by the presence of native and non-native animal dispersers. In North America, weeds with simple flowers like ox-eye daisy (*Chrysanthemum leucanthemum*) and orange hawkweed (*Hieracium aurantiacum*) are visited by a wide range of insects that include 'domesticated' European honey bees (*Apis mellifera*) and native bumblebees (*Bombus* spp.) (Murphy and Aarssen, 1995). This even is true of weeds with more complex flowers like cow vetch (*Vicia cracca*). Native bumblebees had enough 'experience' with similar native flowers so that pollinating cow vetch was not a difficult task to master (e.g. Murphy and Aarssen, 1995).

Mycorrhizal associations are formed when a fungus infects plant roots. The fungus increases the effective 'root' surface area and therefore increases the supply of nutrients and water. In turn, the plant provides carbohydrates from photosynthesis to the fungus. The fungus can either penetrate the intercellular spaces (ectomycorrizhae) of roots or the cells themselves (endomycorrhizae). Mycorrhizae are so prevalent and important that most native flora and some weeds will grow only if certain fungal species are present in sufficient densities (Schroft, 1998; van der Heijden *et al.*, 1998; Jordan *et al.*, 2000; Dahlberg, 2001).

Mycorrhizal weeds may be specialists or generalists. When Monterey pines were introduced into New Zealand plantations, their spread was limited until spores from the right species of mycorrhizal fungi from plantation soil had accumulated (Richardson and Higgins, 1998). In contrast, other weeds like Russian thistle (*Salsola kali*) are generalists; since they can use most mycorrhizal fungal species, they can invade most habitats. While mycorrhizal weeds often have negative effects on ecosystems (as expected of any weed), they actually can have beneficial effects if the net result of having additional mycorrhizal species is to

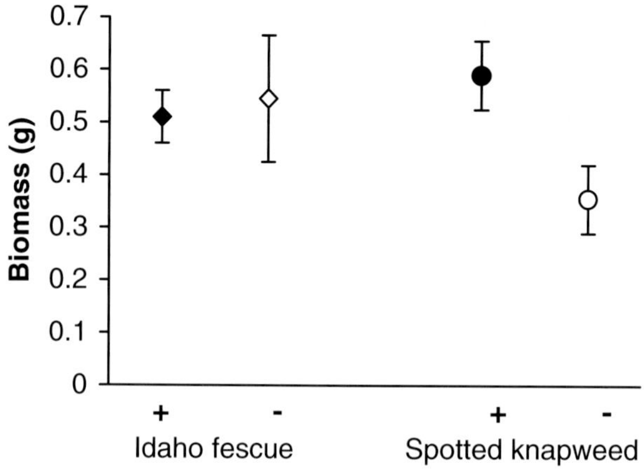

Fig. 9.3. Biomass (± 1 standard error) of Idaho fescue (*F. idahoensis*) and spotted knapweed (*C. maculosa*) when grown in competition with (+) and without (–) mycorrhizal fungi (redrawn from Marler *et al.*, 1999).

improve soil quality or attract organisms that attack weeds (Jordan *et al.*, 2000).

Many weeds are not mycorrhizal (Goodwin, 1992). This is why they are found in highly disturbed habitats, i.e. where the mycorrhizal 'system' has been disrupted by human activities. When mycorrhizal fungi are absent from soil, non-mycorrhizal species are favoured, e.g. most mustard weeds (Brassicaceae) and lambsquarters (*Chenopodium album*). However, non-mycorrhizal weeds may use an existing mycorrhizal network in a relatively undisturbed habitat to its advantage. For example in North American prairie, weedy spotted knapweed (*Centaurea maculosa*) is able to outcompete the native Idaho fescue (*Festuca idahoensis*), but only if mycorrhizal fungi are present (Marler *et al.*, 1999) (Fig. 9.3). Normally, neither species benefits individually from mycorrhizae as both are infected with the fungi but do not provide carbohydrates to them (hence the fungi constitute merely a passive infection). If, however, an extensive mycorrhizal network is formed among prairie natives, spotted knapweed will 'activate' its mycorrhiza and 'steal' nutrients from the normally dominant Idaho fescue via their common fungal 'pipeline'.

Plants often form associations with mutualistic bacteria that use ('fix') atmospheric nitrogen. Nitrogen in this form is normally unavailable to plants because they lack the enzymes needed to capture it. The plant gets another source of nitrogen while the bacteria receive carbohydrates and protection. These nitrogen-fixing bacteria are generally ubiquitous (Richardson *et al.*, 2000) and weedy nitrogen-fixing plants usually have no difficulty finding the required bacteria. e.g. gorse (*Ulex europeus*) and scotch broom (*Cytisus scoparius*) (Peterson and Prasad, 1998; Clements *et al.*, 2001). Nitrogen-fixing weeds may have a strong influence on their new habitat because they may change the nutrient dynamics. This is especially true in places like Hawaii where nitrogen-fixers are not a part of the native flora. For example, the fire tree (*Myrica faya*) enriches soil nitrogen and provides shade that improves germination and seedling growth of the native `ohi'a lehua (*Metrosideros polymorpha*). However, the increased amount of nitrogen in the soil also permitted the invasion of weeds formerly constrained by limited nitrogen, e.g. strawberry guava (*Psidium cattleianum*) (Wall and Moore, 1999). The same phenomenon can occur when nitrogen-fixing crops like clover (*Trifolium*) or lucerne (*Medicago*) are used; the increased soil nitrogen is of benefit to non-nitrogen-fixing crops (like maize) that will be planted the next year in rotation, but it also benefits weeds. Further,

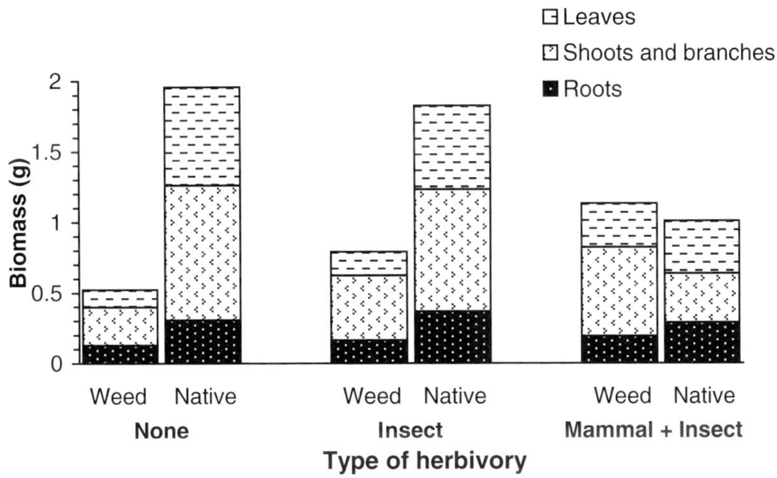

Fig. 9.4. Distribution of biomass to the roots, shoots and branches, and leaves of the introduced weed Japanese honeysuckle (*L. japonica*) and the native trumpet honeysuckle (*L. sempervirens*) when subject to no herbivory, insect herbivory, and mammal and insect herbivory (adapted and redrawn from data in Schierenbeck *et al.*, 1994).

if the nitrogen-fixing crops colonize non-agricultural areas like forests, then the weeds also will be able to colonize.

Complexity in the Real World: Interactions Between Ecological Processes

The net effect of interacting ecological processes is not straightforward. A good example is the interaction of competition and herbivory between the weedy exotic Japanese honeysuckle (*Lonicera japonica*) and the native trumpet honeysuckle (*L. sempervirens*) (Schierenbeck *et al.*, 1994). In the absence of herbivory, Japanese honeysuckle will be outcompeted by trumpet honeysuckle because it has a higher growth rate and accumulates more biomass (Fig. 9.4). However, being a native, trumpet honeysuckle is more vulnerable to herbivores and the resultant damage makes it less competitive than Japanese honeysuckle. Why does trumpet honeysuckle not adapt by developing strong defences against herbivores so it can be both impervious and a good competitor? The main reason is because there is a trade-off of limited resources allocated to competitive traits like growth, versus those allocated to defences against herbivores (Herms and Mattson, 1992).

A more complicated example of interactions between ecological processes is that of weedy tree-of-heaven (*Ailanthus altissima*) (Facelli, 1994). White oak (*Quercus alba*) leaf litter: (i) provides habitat for herbivores that attack tree-of-heaven saplings and adults, and (ii) delays the seedling emergence of tree-of-heaven. Separate experiments also show that competition from species like giant foxtail (*Setaria faberii*) reduces seedling biomass of the tree-of-heaven. We might expect that a combination of white oaks and species like giant foxtail would harm tree-of-heaven even further. However, when white oak litter and competition from species like giant foxtail occur simultaneously, the leaf litter has a greater effect on giant foxtail than it did on tree-of-heaven. Contrary to expectations from examining leaf litter and competition separately, a combination of both actually increases the ability of tree-of-heaven to survive.

Summary

In this chapter we discussed the basic processes of herbivory, parasitism and mutualism. Herbivory of shoots, leaves, roots and seeds is often non-fatal but since it is damaging, plants have adapted avoidance, tolerance and inducible defence mechanisms. Though non-fatal herbivory can alter the population dynamics of weeds (and other plants), the absence of herbivores helps to give weeds an advantage and the introduction of herbivorous biological control agents attempts to remove that advantage. The same is true for introduction of parasites for biological control of weeds, though many weeds are already parasitized. Weeds themselves can be parasites; this reinforces the idea that not all parasites are microscopic organisms like bacteria or fungi. Mutualisms are important in populations and communities, especially in terms of mycorrhizae and nitrogen-fixing bacteria. Most native plants cannot survive well without mutualists like mycorrhizae. Some weeds need mycorrhizae, but many weeds gain extra competitive advantages if mycorrhizae are diminished in numbers since native species will be at a disadvantage.

We conclude that while studying all of these processes separately was useful, the real world is a place where the processes act antagonistically or synergistically in complicated ways. A realistic ecological study would be one that examines the community – the hundreds of plant species (many of them weeds) and millions of other species that interact in a defined area – however arbitrary the definition of the community boundaries may be. As a result of all this complexity, it is necessary to discuss how this is addressed by studying what is called 'community structure and dynamics' in Chapters 10–14.

Questions

1. What strategies does the species you selected use to reduce the effects of herbivores? Explain them. Does the species you selected form mutualistic or parasitic association with another species? Does your species benefit or suffer from the association?
2. Herbivores have three immediate effects on plants: (i) the loss of photosynthetic tissue; (ii) the removal of meristems; and (iii) the loss of stored nutrients (Noy Meir, 1993). Explain how *each* of these can influence the growth and survival of a weed.
3. Research the relationship between a weedy plant and its biological control agents. Was the agent successful at controlling the weed? Explain why or why not.
4. Explain why the cactus moth borer follows meta-population dynamics in Australia where it was introduced to control the prickly pear cactus.
5. Explain how parasitic weeds might be controlled by the use of 'trap' crops.
6. What is the relationship between weeds and their mycorrhizal associations?

General References

Bronstein, J.L. (1994) Conditional outcomes in mutualistic interactions. *Trends in Ecology and Evolution* 9, 214–217.

Callaway, R.M. and Walker, L.R. (1997) Competition and facilitation: a synthetic approach to interaction in plant communities. *Ecology* 78, 1958–1965.

Kareiva, P.M. and Bertness, M.D. (1997) Re-examining the role of positive interactions in communities. *Ecology* 78, 1945–1951.

Richardson, D.M., Allsopp, N., D'Antonio, C.M., Milton, S.J. and Rejmánek, M. (2000) Plant invasion – the role of mutualisms. *Biological Review* 75, 65–93.

Literature Cited

Aarssen, L.W. (1995) Hypotheses for the evolution of apical dominance in plants: implications for the interpretation of overcompensation. *Oikos* 74, 149–156.

Agrawal, A.A. (2000) Overcompensation of plants in response to herbivory and the by-product benefits of mutualism. *Trends in Plant Science* 5, 309–313.

Andrews, T.S. (1995) Dispersal of seeds of giant *Sporobolus* spp after ingestion by grazing cattle. *Australian Journal of Experimental Agriculture* 35, 353–356.

Baker, F.A. and Knowles, K. (1992) Impact of dwarf mistletoes on jack pine forests in Manitoba. *Plant Disease* 76, 1256–1259.

Belsky, A.J., Carson, W.P., Jensen, C.L. and Fox, G.A. (1993) Overcompensation by plants – herbivore optimization or red-herring. *Evolutionary Ecology* 7, 109–121.

Blaney, C.S. and Kotanen, P.M. (2001) Post-dispersal losses to seed predators: an experimental comparison of native and exotic old field plants. *Canadian Journal of Botany* 79, 284–292.

Blossey, B., Casagrande, R., Tewksbury, L., Landis, D.A., Wiedenmann, R.N. and Ellis, D.R. (2001a) Nontarget feeding of leaf-beetles introduced to control purple loosestrife (*Lythrum salicaria* L.). *Natural Areas Journal* 21, 368–377.

Blossey, B., Skinner, L.C. and Taylor J. (2001b) Impact and management of purple loosestrife (Lythrum salicaria) in North America. *Biodiversity and Conservation* 10, 1787–1807.

Bronstein, J.L. (1994a) Conditional outcomes in mutualistic interactions. *Trends in Ecology and Evolution* 9, 214–217.

Bronstein, J.L. (1994b) Our current understanding of mutualism. *Quarterly Review of Biology* 69, 31–51.

Cates, R.G. (1975) The interface between slugs and wild ginger: some evolutionary aspects. *Ecology* 56, 391–400.

Chittapur, B.M., Hunshal, C. S. and Shenoy, H. (2001) Allelopathy in parasitic weed management: role of catch and trap crops. *Allelopathy Journal* 8, 147–159.

Clements, D.R., Peterson, D.J. and Prasad, R. (2001) The biology of Canadian weeds. 112. *Ulex europaeus* L. *Canadian Journal of Plant Science* 81, 325–337

Cory, J.S. and Myers, J.H. (2000) Direct and indirect ecological effects of biological control. *Trends in Ecology and Evolution* 15, 137–139.

Crawley, M.J. (1989) The relative importance of vertebrate and invertebrate herbivores in plant population dynamics. In: Bernays, E.A. (ed.) *Insect-Plant Interactions,* Vol. 1. CRC Press, Boca Raton, Florida, pp. 45–71.

Crawley, M.J. (1992) Seed predators and plant population dynamics. In: Fenner, M. (ed.) *Seeds: The Ecology of Regeneration in Plant Communities.* CAB International, Wallingford, pp. 157–191.

Cromar, H.E., Murphy, S.D. and Swanton, C.J. (1999) Influence of tillage and crop residue on post-dispersal predation of weed seeds. *Weed Science* 47, 184–194.

Cyr, H. and Pace, M.L. (1993) Magnitude and patterns of herbivory in aquatic and terrestrial ecosystems. *Nature* 361, 148–150.

Dahlberg, A. (2001) Community ecology of ectomycorrhizal fungi: an advancing interdisciplinary field. *New Phytologist* 150, 555–562.

Elliot, S.L., Sabelis, M.W., Janssen, A., van der Geest, L.P.S., Beerling, E.A.M. and Fransen, J. (2000) Can plants use entomopathogens as bodyguards? *Ecology Letters* 3, 228–235.

Facelli, J.M. (1994) Multiple indirect effects of plant litter affect the establishment of woody seedlings in old fields. *Ecology* 75, 1727–1735.

Foroni, J. and Nunez-Farfan, J. (2000) Evolutionary ecology of *Datura stramonium*: genetic variation and costs for tolerance to defoliation. *Evolution* 54, 789–797.

Gardner, R.O. and Early, J.W. (1996) The naturalisation of banyan figs (*Ficus* spp. Moraceae) and their pollinating wasps (Hymenoptera: Agaonidae) in New Zealand. *New Zealand Journal of Botany* 34, 103–110.

Goodwin, J. (1992) The role of mycorrhizal fungi in competitive interactions among native bunchgrasses and alien weeds: a review and synthesis. *Northwestern Science* 66, 251–260.

Herms, D.A. and Matton, W.J. (1992) The dilemma of plants – to grow or defend. *Quarterly Review of Biology* 67, 283–335.

Janzen, D.H. (1979) How to be a fig. *Annual Review of Ecology and Systematics* 10, 13–51.

Jennings, J.C., Apel-Birkhold, P.C., Bailey, B.A. and Anderson, J.D. (2000) Induction of ethylene biosynthesis and necrosis in weed leaves by a *Fusarium oxysporum* protein. *Weed Science* 48, 7–14.

Jordan, N.R., Zhang, J. and Huerd, S. (2000) Arbuscular-mycorrhizal fungi: potential roles in weed management. *Weed Research* 40, 397–410.
Kaltz, O. and Schmid, B. (1995) Plant venereal disease: a model for integrating genetics, ecology and epidemiology. *Trends in Ecology and Evolution* 10, 221–222.
Longman, K.A. and Jenik, J. (1974) *Tropical Rainforest and its Environment.* Longman, London.
Louda, S.M., Keeler, K.H. and Holt R.D. (1990) Herbivore influences on plant performance and competitive interactions. In: Grace, J.B. and Tilman, D. (eds) *Perspectives on Plant Competition.* Academic Press, San Diego, California, pp. 413–444.
Marler, M.J., Zabinski, C.A. and Callaway, R.M. (1999) Mycorrhizas indirectly enhance competitive effects of an invasive forb on a native bunchgrass. *Ecology* 80, 1180–1186.
Murphy, S.D. and Aarssen, L.W. (1995) *In vitro* allelopathic effects of pollen from three *Hieracium* species (Asteraceae) and pollen transfer to sympatric Fabaceae. *American Journal of Botany* 82, 37–45.
Musser, R.O., Hum-Musser, S.M., Eichenseer, H., Pfeiffer, M., Ervin, G., Murphy, J.B. and Felton, G.W. (2002) Herbivory: caterpillar saliva beats plant defences. *Nature* 416, 599–600.
Nadel, H., Frank, J.H. and Knight, R.J. (1992) Escapees and accomplices: the naturalization of exotic *Ficus* and their associated faunas in Florida. *Florida Naturalist* 75, 29–38.
Noy Meir, I. (1993) Compensating growth of grazed plants and its relevance to the use of rangelands. *Ecological Applications* 3, 32–34.
Olson, D. and Wallander, R. (1999) Carbon allocation in *Euphorbia esula* and neighbours after defoliation. *Canadian Journal of Botany* 77, 1641–1647.
Parker, C. and Riches, C.R. (1993) *Parasitic Weeds of the World: Biology and Control.* CAB International, Wallingford, UK.
Peterson, D.J. and Prasad, R. (1998) The biology of Canadian weeds. 109. *Cytisus scoparius* (L.) Link. *Canadian Journal of Plant Science* 78, 497–504.
Price, P.W., Bouton, C.E., Gross, P., McPheron, B.A., Thompson, J.N. and Weis, E.A. (1980) Interactions among three trophic levels: influence of plants on interactions between insect herbivores and natural enemies. *Annual Review of Ecology and Systematics* 11, 41–65.
Rambert, B., Hellqvist, S., Ekbom, B. and Banks, J.E. (2001) Assessment of trap crops for *Lygus* spp. in lettuce. *International Journal of Pest Management* 47, 273–276
Richardson, D.M. and Higgins, S.I. (1998) Pines as invaders in the southern hemisphere. In: Richardson, D.M. (ed.) *Ecology and Biogeography of Pinus.* Cambridge University Press, Cambridge, pp. 450–473.
Richardson, D.M., Allsopp, N., D'Antonio, C.M., Milton, S.J. and Rejmánek, M. (2000) Plant invasion – the role of mutualisms. *Biological Review* 75, 65–93.
Rosenheim, J.A., Kaya, H.K., Ehler, L.E., Marois, J.J. and Jaffee, B.A. (1995) Intraguild predation among biological-control agents: theory and evidence. *Biological Control* 5, 303–335.
Sabelis, M.W., van Baalen, M., Bakker, F.M., Bruin, J., Drukker, B., Egas, M., Janssen, A.R.M., Lesna, I.K., Pels, B., van Rijn, P.C.J. and Scutareanu, P. (1999) The evolution of direct and indirect plant defence against herbivorous arthropods. In: Olff, H., Brown, V.K. and Drent, R.H. (eds) *Herbivores: Between Plants and Predators.* Blackwell Scientific, Oxford, pp. 109–166.
Sandras, V.O. (1996) Cotton compensatory growth after loss of reproductive organs as affected by availability of resources and duration of recovery. *Oecologia* 106, 432–439.
Schierenbeck, K.A., Mack, R.N. and Sharitz, R.R. (1994) Effects of herbivory on growth and biomass allocation in native and introduced species of *Lonicera. Ecology* 75, 1661–1672.
Scholte, K. (2000). Growth and development of plants with potential for use as trap crops for potato cyst nematodes and their effects on the numbers of juveniles in cysts. *Annals of Applied Biology* 137, 31–42.
Schroft, G. (1998) A review of belowground interactions in agroforestry, focusing on mechanisms and management options. *Agroforestry Systems* 43, 5–34.
Shykoff, J.A. and Kaltz, O. (1998) Phenotypic changes in host plants diseased by *Microbotryum violaceum*: parasite manipulation, side effects, and trade-offs. *International Journal of Plant Sciences* 159, 236–243.
Silvertown, J. (1980) The evolutionary ecology of mast seedling in trees. *Biological Journal of the Linnaean Society* 14, 235–250.
Strauss, S.Y. and Agrawal, A.A. (1999) The ecology and evolution of plant tolerance to herbivory. *Trends in Ecology and Evolution* 14, 179–185.

van der Heijden, M.G.A., Klironomos, J.N., Ursic, M., Moutoglis, P., Streitwolf-Engel, R., Boller, T., Wiemeken, A. and Sanders, I.R. (1998) Mycorrhizal fungal diversity determines plant biodiversity, ecosystem variability, and productivity. *Nature* 396, 69–72.
Venecz, J.I. and Aarssen, L.W. (1998) Effects of shoot apex removal in *Lythrum salicaria* (Lythraceae): assessing the costs of reproduction and apical dominance. *Annales Botanici Fennici* 35, 101–111.
Wall, D.H. and Moore, J.C. (1999) Interactions underground. Soil biodiversity, mutualism, and ecosystem processes. *BioScience* 49, 109–117.
Zimdahl, R.C. (1999) *Fundamentals of Weed Science,* 2nd edn. Academic Press, San Diego, California.

10
Studying Populations and Their Interactions

Concepts

- It is not usually possible to sample an entire population or community, therefore, subsamples are commonly used to collect vegetation data in the field.
- Data should be considered in terms of its accuracy, precision and bias.
- Life tables and survivorship curves are used to interpret survival and mortality data within populations.
- Experiments can be done in controlled conditions in the greenhouse or growth chamber, or in natural conditions in the field. There are advantages and disadvantages to each of these.
- Experiments should be designed so that they answer the specific question that you are asking.

Introduction

In the past nine chapters we have presented a variety of experimental results to illustrate concepts we have discussed. We have described how data are collected in very general terms, because it was the results that we were concerned with rather than the methodology. The aim of this chapter is to describe commonly used methods and procedures to help you understand how scientific information is obtained. Scientific knowledge is built from the data we collect using well-designed studies and experiments. Poorly designed experiments can lead to erroneous conclusions. We do not discuss statistics as that is a subject unto itself (Underwood, 1997; Zar, 1999; Quinn and Keough, 2002); we merely want you to understand how ecological information is obtained. This chapter is divided into sections that address sampling population abundance, constructing survivorship curves and designing population interaction experiments.

Doing Good Science

What are the methods of discovery, and how do we lessen the impact of our personal biases on our interpretation of the results?

While we like to think that scientists come to conclusions based on the cold, hard facts, this is not always the case. Rarely are the results of an experiment so compelling that there is only one possible interpretation. Even long-accepted 'facts' are sometimes revisited and discredited. For example, the extinction of the tambalacoque tree (*Sideroxylon grandiflorum*) has been linked to the extinction of the dodo bird (*Raphus cucullatus*). It was said that the seeds of this tree had to pass through the gut of the dodo to be able to germinate. When the dodo became extinct, tambalacoque seeds were no longer able to germinate. It made common sense. New evidence, however, has shown that this well-accepted 'fact' is not true. In fact, tambalacoque seedlings have been observed, although rare. The decrease in this species seems to have been caused by the introduced monkeys that consumed the fruit before the seeds were ripe and therefore they were unable to germinate when dispersed (Witmer and Cheke, 1991). Thus, common sense does not necessarily make good science.

The scientific method

To help avoid the pitfalls of relying on common sense, ecologists use the 'scientific method'. It focuses on asking good questions, making hypotheses, designing experiments to test them and then using this new information to re-evaluate our understand-

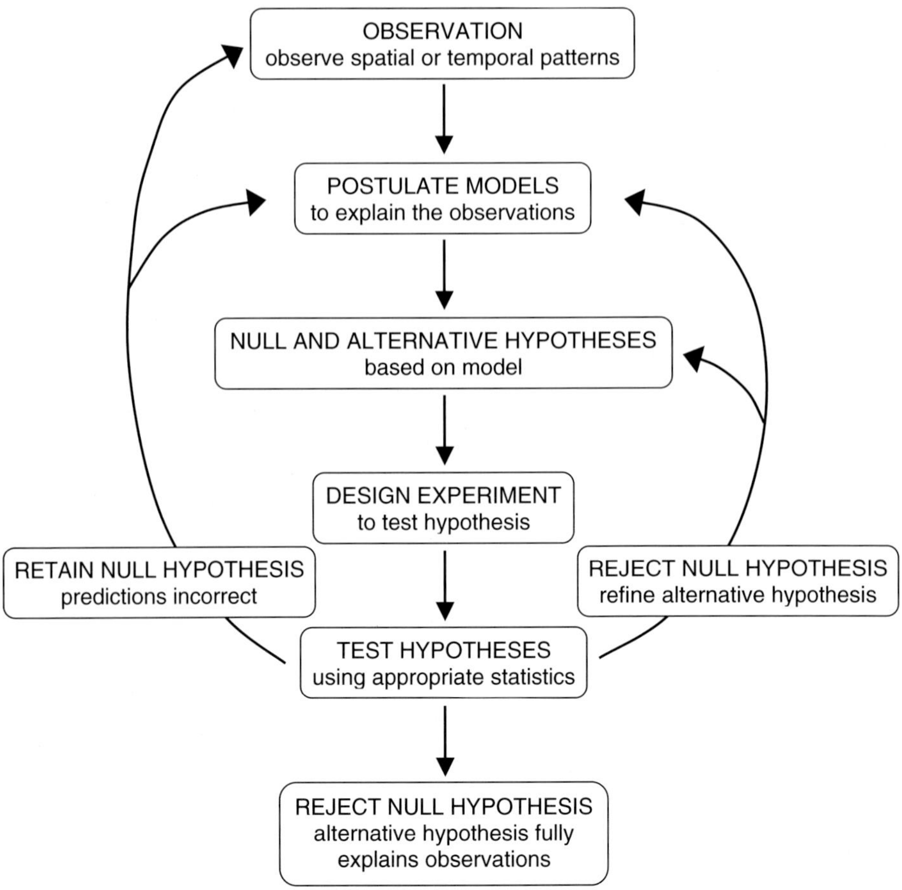

Fig. 10.1. The scientific method.

ing of the world (Fig. 10.1). The first step in the scientific method is to make observations of ecological patterns. This sounds trivial, but science based on inaccurate observations is useless. From observations, we then propose models to describe or explain our observations. For example, if we observe that a weed species tends to be found only in open habitats, we might hypothesize that light is an important factor determining the species growth or survival. Next, we create one or several more specific hypotheses. For example, we might come up with the following two hypotheses to explain our observations:

- H_1 – light has a direct positive effect on the growth of the weed and therefore it tends to be found in open habitats
- H_2 – herbaceous insects that eat the weed prefer to live in shady areas, therefore only seedlings in open sunny locations survive

We also state the 'null hypothesis'. This is a statement that there is no effect expected. For our example our null hypothesis (H_0) is:

- H_0 – light has no effect on the growth or survival of the weed.

For statistical reasons, we actually attempt to falsify our null hypothesis rather than 'prove' our hypothesis that there is an effect of light on growth or survival (Underwood, 1990). The reason this is done is similar to the legal ethic of being 'innocent until proven guilty'.

When we test hypotheses, there are two types of errors to be concerned with (Table 10.1). As type I and type II errors imply, the scientific method is not foolproof. Incorrect explanations about observations can result from mistakes at any point in the method. However, the scientific method does provide a common structure, and an opportunity to test whether what we perceive as fact is actually valid. Science is only as good as the investigator. So, how do we avoid bias and do good science? One way to do this is continually to reconsider what we believe to be true. Remember it was once common knowledge that the world was flat and that the sun revolved around the earth. These 'facts' once made sense and they were supported by the science of the day. Now, in light of our changing understanding of the universe, this seems ridiculous. Therefore, it is important to have an open and flexible mind.

Designing experiments

A hypothesis is tested using an experiment. There are numerous ways to design experiments, and many types of experiments can be used to test a single hypothesis. Experiments can be conducted either in environmentally controlled conditions such as in a greenhouse or a growth chamber, or in the 'field' where plants are grown and/or manipulated under natural or semi-natural conditions. The variables we are interested in are manipulated, and then we measure the response of the individual or population.

Table 10.1. Type I and type II errors in hypothesis testing.

Error		Significance
Type I	Null hypothesis is true, but rejected	You mistakenly believe that there is a significant effect when there is not
Type II	Null hypothesis is false, but not rejected	You mistakenly believe that there is no significant effect when there is

	Real effect	
Statistical result	Null hypothesis true	Null hypothesis false
Null hypothesis accepted	Correct	Type II error
Null hypothesis rejected	Type I error	Correct

To test our first hypothesis (H_1) from above, we might grow plants in growth chambers where light levels are manipulated, but all other variables (e.g. temperature, moisture) are kept the same. The benefit of a lab-based study is that the researcher can control most of the environmental conditions, and therefore it is possible to isolate exact causes. However, such controlled conditions are not necessarily realistic; plants grown in greenhouses and growth chambers are not subject to natural variation in abiot-

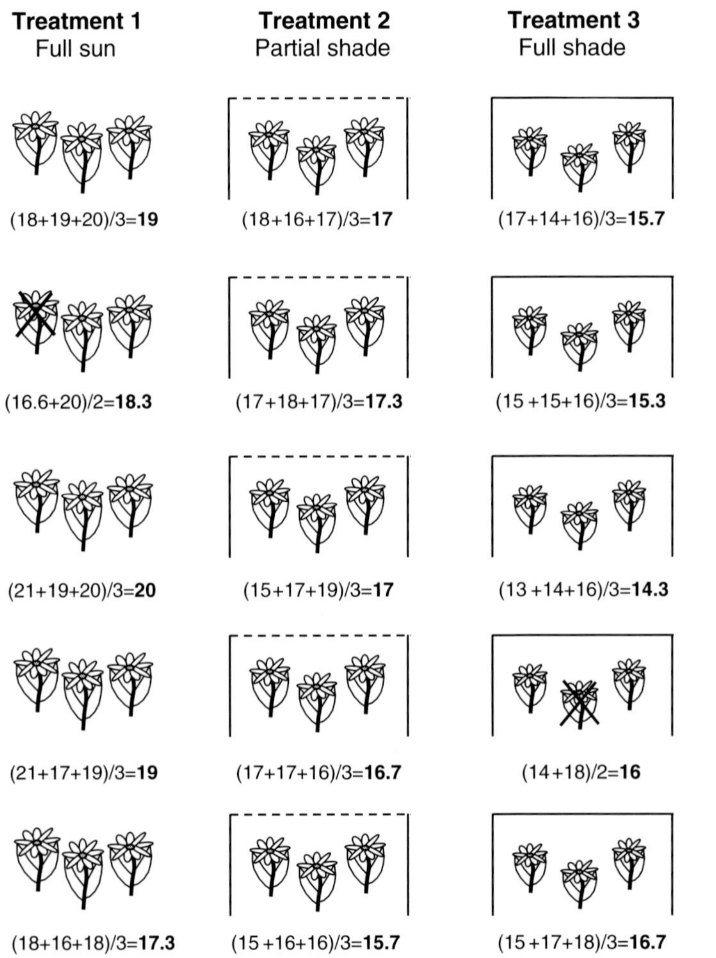

The mean height for each treatment is calculated using the mean height of each of the five replicates.

(19+18.3+20+19+17.3)/5	(17+17.3+17+16.7+15.7)/5	(15.7+15.3+14.3+16+16.7)/5
$mean_1$ = 18.7	**$mean_2$ = 16.7**	**$mean_3$ = 15.6**
SD_1 = 1.0	**SD_2 = 0.6**	**SD_3 = 0.9**

Fig. 10.2. An example of an experiment testing whether light affects plant growth (height). The design consists of three treatments (full sun, partial shade, full shade) with five replicates each. Each data point is the mean height of the three individuals in each replicate. In cases where a plant dies (shown by an 'X') the mean of the remaining individuals is taken. Shown are the means and standard deviations (SD) for each treatment.

ic conditions such as temperature fluctuations. Alternatively, we could find a natural population of our weed species or plant seeds or transplants, and then reduce ambient light by adding shade cloth over some individuals. In such field experiments, environmental conditions are not tightly controlled (e.g. a severe storm could flood your plants); however, it is a more natural set-up for the plants and may be more reflective of ecological reality.

The two key features of an experiment are treatments and replication. Treatments are the number of types of manipulations made. For example, our field experiment might have three treatments: full (ambient) light, partial shade and full shade (Fig. 10.2). The treatment with full light is called the control because there is no experimental manipulation; the other treatments are compared to it see if there is an effect.

Every treatment is replicated several times; for example, we might set up each treatment five times. Replication is done for practical reasons. It ensures that your experiment is not ruined simply because one plant dies for reasons unrelated to treatments; for example, if someone steps on your plant. There are also statistical reasons for replication; there will be natural variation within a population, and replication allows us to accommodate this. If we had used one replicate plant per treatment, and the one in full light happened to have genes that caused it to be short, then we would have concluded erroneously that plants in shade grow taller. We would be unlikely to make this mistake if we have a range of plants growing in each treatment. Therefore, replication increases our ability to detect differences due to experimental treatments rather than differences due to natural variation. The number of replicates required will increase as natural variation increases. Of course, this is an extremely simple experimental design. Most experiments are much more complicated because they incorporate numerous types of treatments and their interactions.

Accuracy, precision and bias

When we collect data, we have to make sure they are accurate, precise and unbiased (Fig. 10.3). Accurate data closely reflect the true value of the variable (e.g. height, biomass, density) you are estimating. Precision describes how close the values of replicated data are to each other, and therefore is a

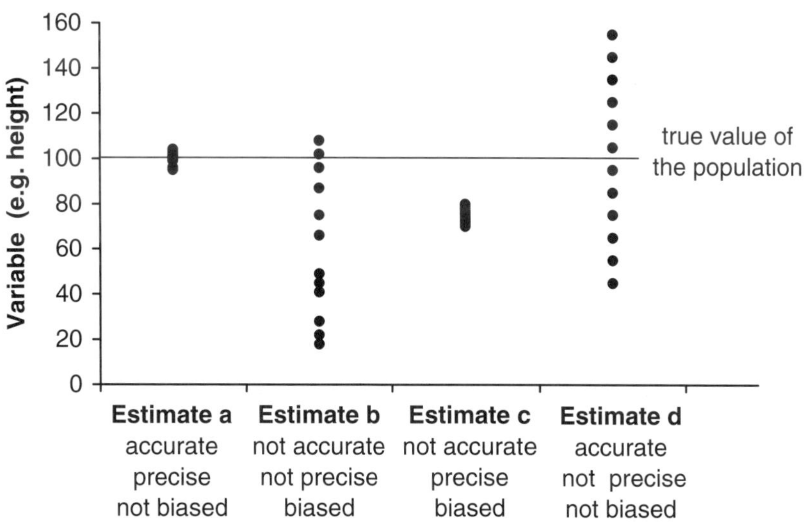

Fig. 10.3. The effect of accuracy, precision and bias on data. Four data sets with 12 samples each were collected from the same population.

measure of how reliable your measurements are. When replicated sampling values are consistently distorted in one direction, then the data are biased. Bias is a measure of how well the sampling procedure reflects the true value of the variable.

Sampling a population's abundance

Often in ecology we conduct studies to measure variables of a population (or community). These are not necessarily 'experiments' because we often are not manipulating variables or creating treatments – we are just measuring what is there. Before we sample a population, we must first decide:

- *who* the population members are;
- *what* an individual is;
- *where* the population boundaries are;
- *how* to count individuals.

In some cases these answers are straightforward, whereas in others they are more complex. To sample, we must be able to identify every member of a species and differentiate it from other similar species. This is difficult when individuals change in appearance with age (developmental plasticity) or with their environment (environmental plasticity). In some cases we can only identify plants to genera; for example, if we sample during a time when the plant is not flowering. Secondly, unlike mammals where every individual is a separate entity, plants can exist as clonal organisms, and therefore a researcher will have to decide whether to sample a ramet or genet (see Chapter 5). The third question of population boundaries was addressed in Chapter 2, where we also discussed measures of abundance.

The fourth decision of sampling methodology is the subject of the rest of this section. If we want to describe a population, we cannot count or measure every individual. Therefore, we measure some of the individuals and use that to represent the population as a whole. The samples we take should be both random and representative of the population we are sampling. Fortunately, randomly collected samples are usually representative (Underwood, 1997). One method to collect random samples is to generate random coordinates within a habitat and then sample at those points. The next step is to decide whether to sample using a plot or plotless sampling technique. Your decision will be based on the species and habitat type as well as on the type of data you wish to collect.

Plot sampling

Plot sampling often involves quadrats. These are physical sampling units that are placed over the vegetation and act as boundaries for the sample. The optimal number, size and

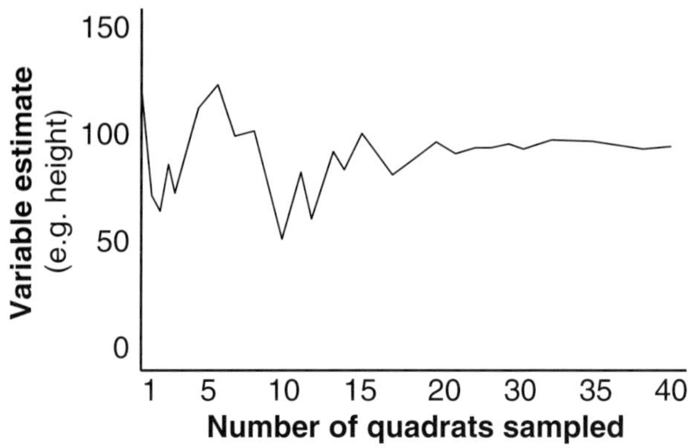

Fig. 10.4. Cumulative estimate (running mean) of a population variable estimate (e.g. height) as the number of quadrats sampled increases. Below 10 quadrats the estimates vary widely, but begin to level off after 15.

shape of quadrat will depend on the species being studied, the statistics to be carried out on the data, and the financial and physical resources of the researcher (Underwood, 1997; Zar, 1999; Quinn and Keough, 2002). Sampling many smaller quadrats is generally better than sampling fewer larger quadrats (Kershaw, 1973). This is because the accuracy of data typically increases as the number of quadrats increases. However, after a certain point increasing the number of quadrats will not improve the estimate of the variable, and will only increase the time and cost of the study. To determine whether enough quadrats have been sampled to get an accurate estimate of a variable, we can compare how the estimate of a variable (e.g. height) changes as the number of quadrats used increases (Fig. 10.4). The point where the curve levels out indicates minimum quadrat number. This 'running mean' approach is used commonly in ecology.

The size of the quadrat will depend primarily on the vegetation type. Understorey vegetation is often sampled with 1-m^2 quadrats, understorey trees and shrubs with 10-m^2 quadrats and canopy trees with 100-m^2 quadrats; however, these are just guidelines. The quadrat size is a compromise between the size of individual plants and

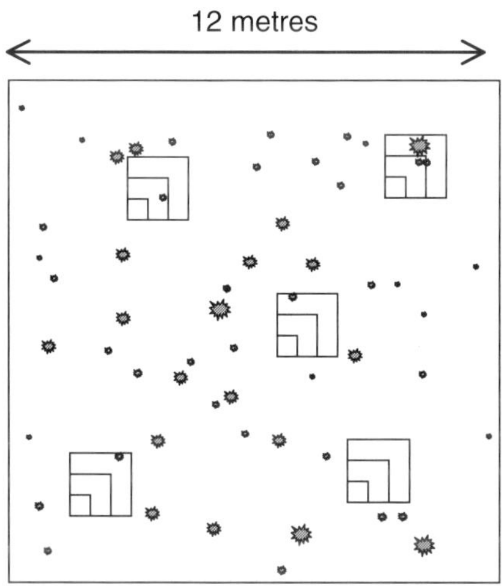

Quadrat size	**Frequency** {no. quadrats with species present / no. quadrats} x 100%	**Mean density** {(Σ no. present per quadrat) / no. quadrats} / quadrat area	**Mean cover** (Σ estimated % cover per quadrat) / no. quadrats
☐ 0.5 m x 0.5 m = 0.25 m^2	0/5 x 100 = 0%	{(0+0+0+0+0)/5} / 0.25 = 0/m^2	(0+0+0+0+0) / 5 = 0%
☐ 1 m x 1 m = 1 m^2	2/5 x 100 = 40%	{(1+1+0+0+0)/5} / 1 = 0.4/m^2	(1.5+1.5+0+0+0) / 5 = 0.6%
☐ 1.5 m x 1.5 m = 2.25 m^2	4/5 x 100 = 80%	{(1+3+1+1+0)/5} / 2.25 = 0.5/m^2	(1.5+13+1.5+1.5+0) / 5 = 3.5%
True value: 12 m x 12 m = 144 m^2	none	57 / 144 = 0.4/m^2	{(4x0.1) + (38x0.01) + (15x0.005) / 144} x 100% = 0.6%

Fig. 10.5. The effect of quadrat size on estimates of a population's density, cover and frequency. There is no frequency value for a 'true' population

density of common species. Various measures of abundance will be affected by quadrat size in different ways (Krebs, 1999). Frequency is more dependent on quadrat size than other measures of abundance: large quadrats will result in more species having 100% frequency, whereas in small quadrats many frequencies will be zero (Fig. 10.5). The smaller the quadrat, the more likely you are to 'miss' individuals.

In some cases, we need to adjust how we collect random samples to ensure that they are representative. For example, if there is an environmental gradient, we might want to change the sampling protocol. Transects are lines, often arranged in a rectangular or square grid, to help determine where to locate quadrats to test for changes along environmental gradients. The random placement of quadrats along transects means that an ecologist is likely to be able to detect patterns. For example, we might be interested in how a weed's distribution responds to changes in environmental variable like soil moisture. If quadrats were randomly placed throughout the site, some areas might be missed; for example, there may be no quadrats in wet soils. To accommodate environmental gradients, researchers can use a combination of regularly spaced quadrats along randomly placed transects (Fig. 10.6). The main advantage to this is that it allows ecologists to sample randomly (using transects) to reduce bias yet still maintain a systematic, representative sample (Zar, 1999).

The spatial arrangement of individuals within a population can affect estimates of abundance. Quadrat sampling assumes that individuals are randomly distributed and the environment is relatively homogeneous. In fact, this is rarely the case. Figure 10.7 shows estimates of the density, frequency and cover of three populations (calculated using randomly placed quadrats). Estimates are different even though the true values of the three populations are the same. Therefore, under some circumstances, mean density, frequency and cover may be of limited value because of sampling bias. There are ways to sample populations that are highly non-random, but this requires the use of advanced statistics (Cardina *et al.*, 1996; Dieleman and Mortensen, 1999; Gibson, 2002).

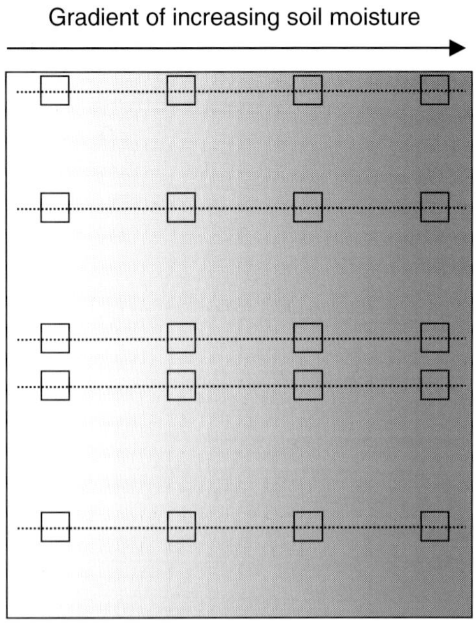

Fig. 10.6. Placement of regularly placed quadrats along randomly placed transects to accommodate an environmental gradient.

Plotless sampling

Sometimes it is not possible or appropriate to use plots for sampling. This is especially true for large plants. Several plotless methods have been developed to accommodate this (Brower *et al.*, 1998; Krebs, 1999). For example, in the point-quarter method, a random point is located (usually along a transect) and from this point four equal quadrants (not quadrat) are established (Fig. 10.8). In each quadrant, the tree (>10 cm in diameter) nearest to the point is identified, and its diameter at breast height (dbh) and the distance are recorded. The same process can be repeated for saplings (2.5–10 cm in diameter). This process is repeated at multiple points, and a variety of statistics can be calculated using these data (Table 10.2) (Engeman *et al.*, 1994). For example, Table 10.2 shows that the total tree density is 1663

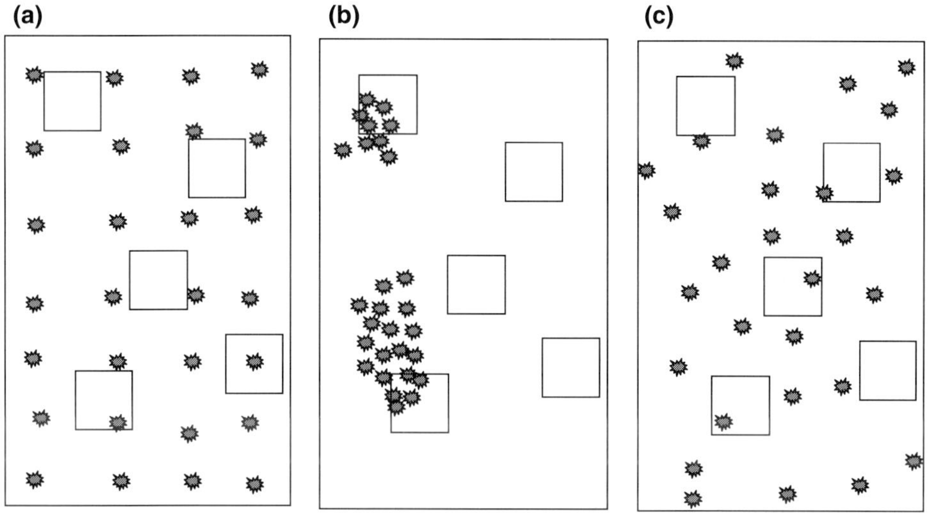

Distribution pattern	Frequency	Density	Cover
a) Regular pattern	40%	0.4 m^{-2}	3.2%
b) Clumped pattern	40%	2.0 m^{-2}	16.0%
c) Random pattern	60%	0.6 m^{-2}	4.6%
True value	-	0.7 m^{-2}	5.6%

Fig. 10.7. Influence of plant distribution on the estimation of a population's density, cover and frequency. Individuals within each population are arranged: (a) regularly, (b) in clumps, and (c) randomly. The quadrats are the same size and are randomly distributed within the population.

trees ha^{-1} and that species A makes up 25% of the trees.

Examples of How to Examine Ecological Phenomena: Survival, Competition and Herbivory

In earlier chapters of this text, we discussed population processes (such as survival), as well as interactions between populations (such as competition and herbivory). In this section we present some ways in which these phenomena are examined using ecological studies and experiments. There is an endless number of ways to examine ecological phenomena; here we present a few, to give you some insight into how ecology is 'done'.

Survival

In Chapter 3 we discussed how individuals within populations will vary with respect to their age, size and stage of development and presented survivorship curves as one way to display age-structured data. For example, we showed the age structure of spotted knapweed (*C. maculosa*) in Fig. 3.7, the survivorship curve of Drummond's phlox (*P. drummondii*) in Fig. 3.9, and the general idealized survivorship curves in Fig. 3.10. In this section we discuss how to collect survival data and construct survivorship tables and curves.

There are two ways to collect survivorship data:

- by looking at the age structure of a population at one point in time ('static life table');

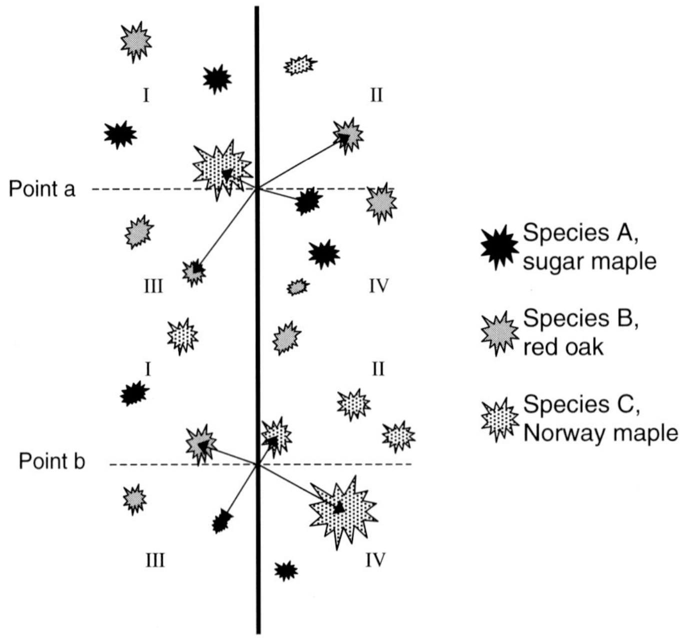

Data sheet

Point-quadrant	Species	Distance (m)	dbh (cm)	Basal area (πr^2 in m^2)
a-I	C Norway maple	3	100	0.7854
a-II	B Red oak	9	50	0.1963
a-III	B Red oak	9.5	35	0.0962
a-IV	A Sugar maple	4.6	32	0.0804
b-I	B Red oak	5.4	45	0.1590
b-II	C Norway maple	2.7	51	0.2043
b-III	A Sugar maple	6.1	22	0.0380
b-IV	C Norway maple	7.8	120	1.1310

Fig. 10.8. Methodology for the point-quarter technique. In each quadrant, the distance to the closest tree is measured, the tree is identified and its diameter at breast height (dbh) is recorded. The basal area is calculated as shown for each tree with r = radius.

- by following a cohort of individuals through time ('cohort life table').

Static life tables are used for organisms that can be aged. The researcher goes into the field and ages all individuals (or a random, representative subsample) of the population. Some long-lived plants such as trees can be aged and therefore it is possible to construct a static life table for them; however, most plants cannot be accurately aged. Static life tables are used more often in animals.

Table 10.2. Types of statistics calculated from data collected using the point-quarter technique.

Statistic	Equation	Explanation
Mean distance	MD = (sum of all distances) / (number of trees)	This is the mean distance from point to tree
Total tree density	TTD= 10,000 / (MD in m)^2	This puts tree density into units per hectare (10,000 m^2).
Species density	SD = [(number of trees of a species) / (total number of trees)] × TTD	
Species relative density	RSD = [(number of trees of a species) / (total number of trees)] × 100	
Species basal area	SBA = SD x MSBA	Diameter of each tree is converted to basal area using πr^2. Then, mean basal area (MSBA**)** for each species is calculated
Species relative basal area	RBA = (SBA / TBA) x 100	First, total basal area (TBA) is the sum of all individual basal areas.

Calculations based on data in Fig. 10.8.

Statistic	Calculations
Mean distance	= (3+9+9.5+4.6+5.4+2.7+6.1+7.8) / 8 = 48.1 / 8 = 6.01 m
Total tree density	= 10,000 / 6.01 = 1663 trees ha^{-1}
Total basal area	=0.7854+0.1963+0.0962+0.0804+0.1590+0.2043+0.038+1.131 = 2.691 m^2
Mean basal area	= 2.691 m^2 / 8 = 0.3364 m^2
Total basal area	= 0.3364 x 1663 = 559 m^2 ha^{-1}

	Species A Sugar maple	Species B Red oak	Species C Norway maple
Species density	= [2 / 8] × 1663 = 416 trees ha^{-1}	= [3 / 8] × 1663 = 624 tree ha^{-1}	Leave for students
Species relative density	= [2 / 8] x 100 = 25%	= [3 / 8] x 1663 = 37.5%	
Species mean basal area	= (0.0804+0.0380) / 2 = 0.0592 m^2	= (0.1963+0.0962+0.1590) / 3 = 0.1505 m^2	
Species basal area ha^{-1}	= 416 x 0.0592 = 24.6 m^2 ha^{-1}	= 624 x 0.1505 = 93.9 m^2 ha^{-1}	
Relative basal area	= (24.6 / 559) x 100% = 4%	= (93.9 / 559) x 100% = 17%	

A cohort is a group of individuals born within the same age class. Therefore, a plant cohort could be a group individuals that germinated the same decade, year, month or day – depending on our species and the level of detail we are interested in. To collect data for a cohort life table, a researcher marks individuals in a cohort and censuses them at regular intervals through time until they have all died. Cohort life tables are used more in plants because plants do not run away between sampling dates and are therefore easy to relocate; however, there are limitations for long-lived species.

The life table calculations are the same for data collected using a cohort or static approach. Life tables include data on the age class (n) and the number of individuals (n) alive at the start of an age class (n_x) (Table 10.3). To calculate survivorship (l) of each age class (x) (i.e. l_x), the number of individuals alive at the start of an age class (n_x) is divided by the number of individuals in the first age class (n_o). Therefore:

Table 10.3 Example of a life-table from a hypothetical population.

Age class x	Number alive at start of age class x, n_x	Proportion alive at start of age class x (survivorship) $l_x=n_x/n_o$	Number dying within age class x to $x+1$ $d_x=n_x-n_{x+1}$	Probability of death between age class x and $x+1$ $m_x=d_x/n_x$
0	1000	(1000/1000)=1	(1000–150)=850	(850/1000)=0.85
1	150	(150/1000)=0.15	(150–50)=100	(100/150)=0.67
2	50	(50/1000)=0.05	(50–20)=30	(30/50)=0.60
3	20	(20/1000)=0.02	(20–10)=10	(10/20)=0.50
4	10	(10/1000)=0.01	(10–5)=5	(5/10)=0.5
5	5	(5/1000)=0.005	(5–0)=5	(5/5)=1.0
6	0	(0/1000)=0	–	–

$$l_x = (n_x) / (n_o).$$

Survivorship data can be plotted to visualize them (Figs 3.9 and 10.9). A log scale is used because it turns constant mortality rate into a straight line, and it makes the data towards the end of the life cycle easier to interpret.

Age-specific mortality rate (m_x) is also included in a life table. This is calculated using the number of individuals dying within the specified age class ($d_x = n_x - n_{x+1}$) and dividing by the number within that age class:

$$m_x = (d_x) / (n_x).$$

These data are useful when a researcher is concerned with the mortality rate within age classes, rather than how many individuals in the population are surviving. From Table 10.3 we can see that 85% of individuals die within the first age class, and that as individuals age, their mortality rate decreases. The mortality rate of the final age class is always one (1) because, alas, everything must die sometime.

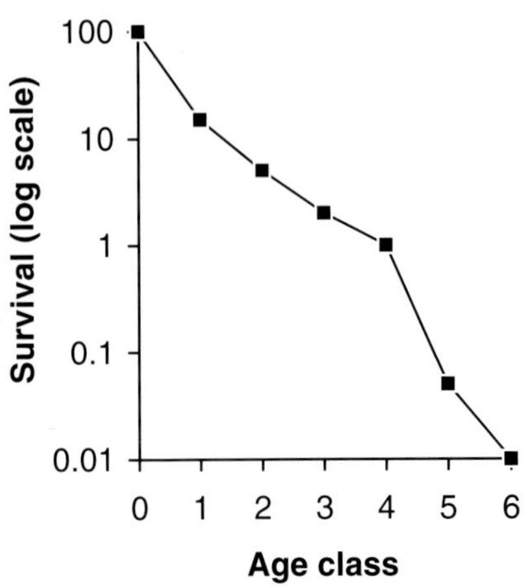

Fig. 10.9. Survivorship curve of hypothetical data in Table 10.3.

Competition

Our ability to understand competition is dependent on our ability to detect it experimentally. When designing competition experiments (or any experiment), it is important that the experiments you design address the question(s) you are asking (Connolly *et al*., 2001). The simplest type of competition experiment is to have three treatments: species A grown alone, species B grown alone, and species A and B grown in competition (Fig. 10.10). If individuals grow better alone than they do when in competition, then it is possible to conclude that competition has an effect on their growth rate. This type of experiment is of little practical value because it does not tell us anything about the type of competition (i.e. root vs. shoot) or how plants will react outside of the pot, in the wild.

Competition can be affected by the abiotic environment such as nutrient status. In Chapter 8 we discussed a study by Wilson and Tilman (1991). This is an example of an experiment designed to look at the effect of nutrient status (three levels of nitrogen) and type of competition (above vs. below ground) on three grasses (little bluestem,

Fig. 10.10. Basic experimental design to look at competition between two individuals of different species. Each species is grown separately (treatments 1 and 3) as well as together (treatment 2), and plant growth, biomass or other measurement is taken to compare them.

Schizachyrium scoparium, Kentucky bluegrass, *P. pratensis,* and quack grass, *Elytrigia repens*). This experiment has a much more complex design than the above experiment. While we do not know the exact set-up of the experiment, by reading the paper we can illustrate the general design. On an existing long-term study on the effect of nitrogen, the authors randomly selected eight 4 × 4 m plots each of low, medium and high nitrogen treatments (Fig. 10.11). Each plot was then subdivided into 45 subplots, and then 15 of each of three grass species were transplanted into the existing vegetation in the subplots. The 15 individuals of each species were then divided into three competition treatments: no competition (all neighbours removed), root competition (shoots of neighbours held back using nets), and root and shoot competition. Throughout the season, the researchers measured growth characteristics (tiller number, height) and then harvested the plants and measured biomass at the end of the season.

The experimental design allowed the authors to conclude that competition was important at all levels of nitrogen. In addition, below-ground competition dominated at low nitrogen, whereas both above- and below-ground competition were important at higher levels of nitrogen. The results of the Wilson and Tilman (1991) study have broad ecological application because they gives us information on the underlying mechanisms of competition. Only by using a complex experimental design were the authors able to look at the effect of an abiotic factor (nitrogen) on above- and below-ground competition of three grass species.

Density effects on competition

Plant density has a major influence on the outcome of competition. As the density of one species increases so will its effect on the other species. Low densities of a species may have no competitive effect on a target species, but as density increases, so does its competitive effect. For example, an agricultural weed may only influence crop yield if it is above a specific density. There are four general types of experiments that look at the effect of density on competition.

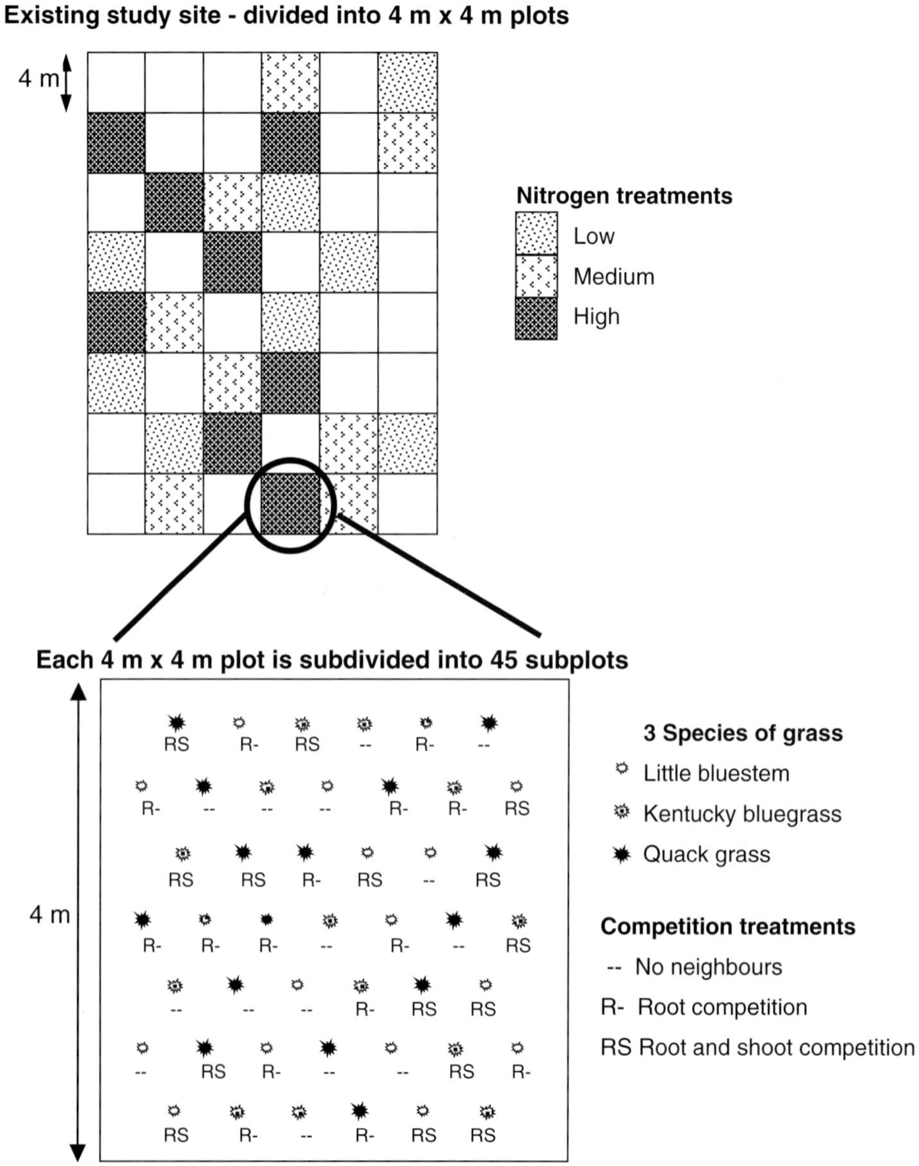

Fig. 10.11. Illustration of the experimental design used by Wilson and Tilman (1991). They examined the effect of three levels of nitrogen (high, medium and low) and three competition treatments (–, no neighbours, R-, root competition; RS, root and shoot competition) on three species of grass (little bluestem, Kentucky bluegrass, quack grass).

Four types of competition experiments

Density effects complicate the experimental design, as the selected density will determine, to some extent, competitive outcome. To circumvent this, four basic types of experimental designs with different density combinations are used (Fig. 10.12). Each experimental design has advantages and disadvantages (Silvertown and Lovett Doust,

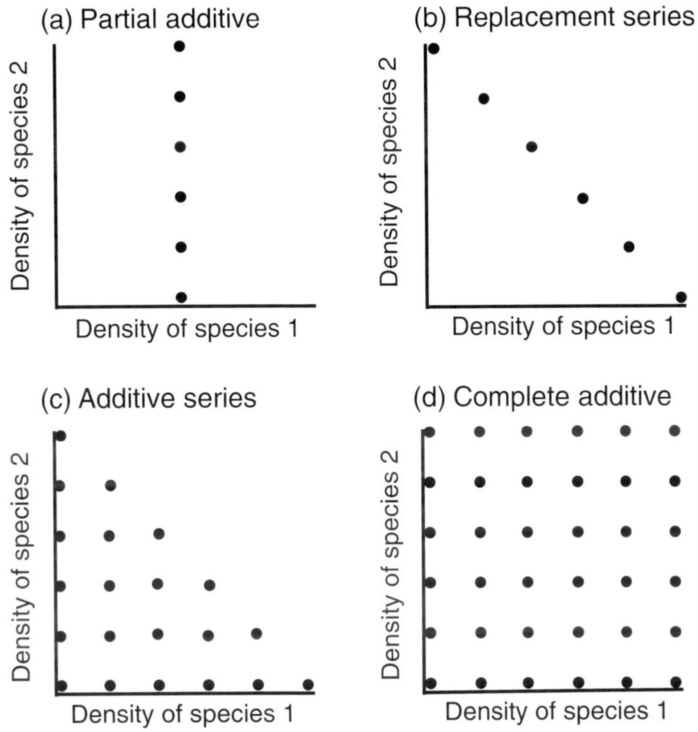

Fig. 10.12. Four types of competition experiments: (a) partial additive, (b) replacement series, (c) additive series, and (d) complete additive (redrawn from Silvertown and Lovett Doust, 1993).

1993) (Table 10.4). Obviously having more density combinations will give a better understanding of competitive effects, but with it comes an increased requirement of time and money. Numerous reviews have been written examining various aspects of competition experiments (Cousens, 1991; Goldberg and Barton, 1992; Gurevitch *et al.*, 1992; Goldberg and Scheiner, 1993; Hamilton, 1994; Freckleton and Watkinson, 2000; Joliffe, 2000; Connolly *et al.*, 2001). We will briefly examine each of the four types of experiments. The main difference among the designs is that they vary in the total density of individuals and in the proportion of each species present.

The partial additive design is the simplest type of competition experiment. Here, a 'target' species is kept at a constant density and grown in competition with a second species at a range of densities. This design is useful only when looking at the effect of increasing density on some component of plant growth. It is often used in agricultural experiments looking at weed density effects on crop yield (Buchanan *et al.*, 1980).

In the replacement series design, the total density of individuals is kept constant, while the relative densities of both species vary. Although this method was formerly in common use, researchers stopped using it because it was impossible to distinguish the effects of total density from relative density (Silvertown and Lovett Doust, 1993). Remember, in this design total density is kept constant; only the relative *proportion* of individuals changes.

The additive series and complete additive designs incorporate both proportion and density into the design. These experiments use a range of total densities and the relative proportion of each species also varies. The

Table 10.4. Advantages and disadvantages of four basic experimental designs used to examine the effect of density on plant competition (based on information in Silvertown and Lovett Doust, 1993).

	Advantages	Disadvantages
Partial additive	Good for applied situations where only concern is the effect of increased density of one species on another; for example, in agricultural systems; little expense, time	Limited application
Replacement series	Little expense, time	Cannot distinguish the separate effects that each species has on the other; uses relative density not absolute density
Additive series and complete additive	Density of both species varies independently therefore more useful; better understanding of competitive effects because all combinations of density used	Expense, and time consuming, but less so in additive series

difference between the additive series and complete additive is that the latter uses all combinations of densities. With a complete additive design, it is possible to construct a response surface (Fig. 10.13) (Firbank and Watkinson, 1985).

Landhäusser *et al.* (1996) used an additive series design to test whether three alternative species could be used to reduce competition by blue-joint grass (*Calamagrostis canadensis*) but allow white spruce (*Picea glauca*) to regenerate. They looked specifically at the effect of density and relative planting date of red and white clover (*Trifolium pratense* and *T. repens*) and fireweed (*Epilobium angustifolium*). They found that all three species reduced the growth and survival of blue-joint grass and

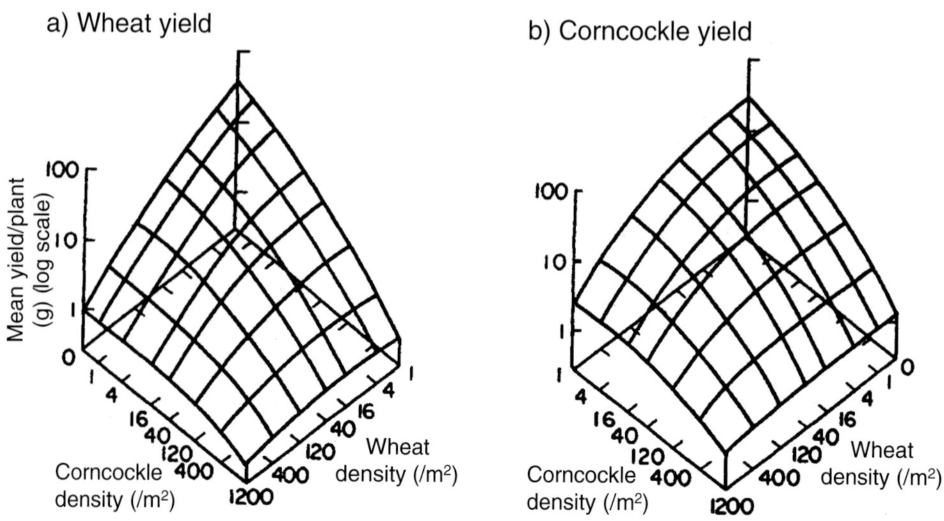

Fig. 10.13. Results of a competition experiment with a complete additive design. The effect of competition between wheat (*Triticum*) and corncockle (*Agrostemma githago*) was measured as the yield per plant. The effect of density on the yield of a) wheat, and b) corncockle is shown as a response surface diagram (Firbank and Watkinson 1985; with permission of Blackwell Science).

that the effect increased as the time between the planting of replacement species and blue-stem grass increased. The three alternative species reduced the growth of white spruce, but not to the extent that blue-joint grass did.

Using field experiments to examine competition

The above experiments are carried out in controlled situations where the environment is controlled and plants are grown in pots. The difficulty with pot experiments is that we are never sure that what we perceive to be competition will actually occur in real field situations. Greenhouse-type experiments can be used to generate predictions that can then be tested using field experiments (Freckleton and Watkinson, 2000)

Using both field and pot experiments, Berendse (1981, 1982, 1983) examined competition between narrow-leaved plantain (*Plantago lanceolata*) and sweet vernalgrass (*Anthoxanthum odoratum*). In the pot experiments, the entire soil column was colonized by roots of both species during the course of the competition experiment. However, since plantain usually has deep roots and vernalgrass shallow roots, a field experiment was done to allow for a more natural rooting profile. They used a replacement series with a constant density of 36 plants per 50 × 50 cm plot. Five plantain:vernalgrass ratios (36:0, 27:9, 18:18, 9:27, 0:36) were used. Two treatments were imposed: different rooting depths possible (50 cm depth), and different rooting depths prevented (20 cm depth). Plantain had a much higher shoot biomass when rooting depth was not constrained. Therefore, interpretation of experiments must be done remembering the constraints placed by the experimental design. In this case the pots produced an artefact that prevented the researchers from understanding root competition.

There are two types of field experiments used to examine competition. Either the density of neighbour plants around a target individual can be changed (neighbour manipulation) or else resources can be altered (increased or decreased) so that the intensity of competition is changed (remember, plants only compete when resources are limited).

Neighbour manipulation

In this approach, the density of neighbours around the target plant is either increased or decreased (Aarssen and Epp, 1990). Neighbours can be removed by a combination of clipping to ground level once or repeatedly, and/or applying herbicides. Sometimes roots are removed through hand digging. All these methods have the disadvantage of creating a disturbance around the target plant. If roots are left intact, then they may add nitrogen as they decompose, but digging loosens the soil and may increase or decrease root growth of the target. Trenching around the base is sometimes used to look at the effect of root competition vs. shoot competition. Increasing neighbour density is done by either planting seeds or individuals around the target plant. Again this creates a disturbance around the target plant, and the transplant may suffer from transplant shock.

Altering resource levels

Increasing resource levels can be done simply by adding nutrients to increase soil fertility. Wilson and Tilman used this technique in their previously described experiment. Alternatively, the levels of some nutrients can be lowered. For example, soil nitrogen can be decreased temporarily by adding a carbon source such as sugar and/or sawdust to the soil. This stimulates the growth of soil microbes, which accumulate the nitrogen in their bodies making it temporarily unavailable to plants (Morgan, 1994).

Alpert and Maron (2000) used such a methodology to test whether decreasing nitrogen (by using sawdust) would decrease the invasion of non-native species in a Californian grassland. They used patches of grassland where the native nitrogen-fixing bush lupin (*Lupinus arboreus*) had been killed with a herbicide (glyphosate). Patches of dead lupin tend to increase the invasion of non-native species because they release nitrogen when they die. Alpert and Maron (2000) used three treatments:

- control – dead lupin patch, no change in soil;
- tilled + sawdust – soil was tilled to incorporate sawdust into the soil; sawdust could not be left on surface as this created a mulch that affected other characteristics such as soil temperature and moisture;
- tilled only – to test for the effect of tillage alone.

They measured species presence and biomass for 2 years. They found that carbon addition (reduction in nitrogen) had no affect on the total above-ground biomass of native species (primarily forbs), but did decrease that of non-native species. This occurred because nitrogen reduction decreased the biomass of grasses (which were mostly non-native) while increasing the richness of broadleaf species (which were mainly native) (Table 10.5).

Herbivory

Herbivory can be tested using real herbivores or by simulating herbivory by manually cutting or removing plant tissue. The type of manipulation selected will depend on the type of plants and herbivores examined. For example, the effect of elephant raids on crop yield cannot be simulated easily in a greenhouse experiment. Tiffin and Inouye (2000) have reviewed the advantages and disadvantages of using natural vs. imposed herbivory.

Herbivore addition and removal

To test for the effect of a herbivore on vegetation, it is possible to either increase or decrease herbivore density. Herbivore removal is often easier but can necessitate killing the organisms. Herbivore additions are difficult, because organisms must be caught and then added and maintained in a new habitat. This may cause problems if animal behaviour varies with density or if organisms are territorial. Thus the biology of the animal must be understood before density changes are imposed. For example, if animals become more aggressive and eat less when in high density, you will not get

Table 10.5. Mean (SE) above-ground biomass and species richness of grasses and forbs with and without sawdust and tillage in a coastal Californian grassland 2 years after treatment (adapted from Alpert and Maron, 2000).

	Tillage + sawdust	Tillage	Control
Biomass (g dry above-ground biomass m^{-2})			
Grasses			
Native	1 (1)	18 (18)	0 (0)
Non-native	253 (40)	466 (95)	386 (56)
Broadleaves			
Native	44 (12)	23 (6)	23 (10)
Non-native	102 (26)	106 (17)	90 (16)
Species richness (number per 0.25 m^2)			
Grasses			
Native	1.8 (0.1)	2.0 (0)	1.6 (0.2)
Non-native	2.4 (0.2)	2.5 (0.3)	2.4 (0.3)
Broadleaves			
Native	4.4 (0.3)	3.4 (0.6)	3.0 (0.8)
Non-native	1.8 (0.3)	1.1 (0.3)	0.7 (0.3)

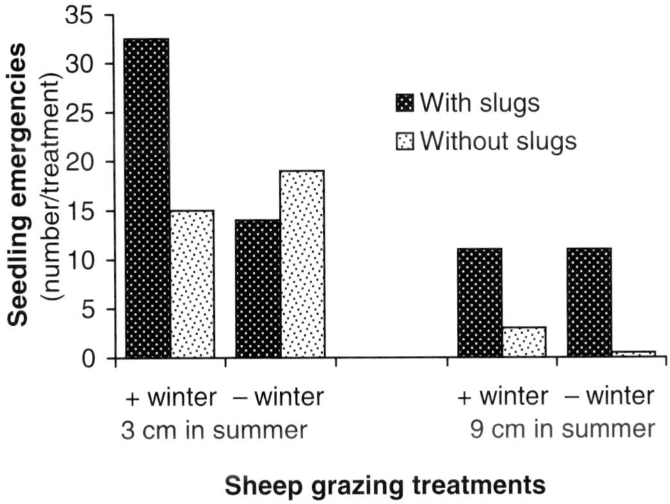

Fig. 10.14. Number of mouse-eared chickweed (*Cerastium fontanum*) seedlings emerging from plots with various combinations of herbivory. Sheep grazing in the summer was managed so that pastures were maintained at either 3- or 9-cm heights. In the winter pastures were either grazed (+) or not grazed (–). In addition, subplots were maintained either with or without slug herbivores (redrawn from Clear Hill and Silvertown, 1997).

the information you are looking for when you increase their density.

Clear Hill and Silvertown (1997) used herbivore addition and removal experiments to examine the interaction of slugs (dominant species grey field slug, *Deroceras reticulatum*) and sheep (*Ovis aries*) on the seedling establishment of several grassland species. Sheep herbivory was managed in two ways. First, some pastures were winter-grazed (+winter) while other were not (–winter). In addition, summer grazing was done to maintain grass height of 3 or 9 cm. Thus there are four sheep grazing treatments (Fig. 10.14). Within each sheep grazing treatment, slug density was manipulated by placing a metal ring (0.1 m^{-2}), 5 cm deep such that slugs could not cross. Then slugs were trapped in the plots to manipulate plot density so they were either absent or at density of 20 m^{-2}. Seeds of mouse-eared chickweed (*Cerastium fontanum*) were planted in each plot and their emergence counted.

The authors found that more seedlings emerged in sites that had been intensively grazed by sheep in the summer (to 3 cm). This probably occurred because intensive summer grazing created microsites suitable for seed germination. There was an interaction in the effect of sheep and slug grazing. The presence of slugs reduced seedling emergence in all treatments, except that with intense summer grazing but no winter grazing. In this treatment, slugs may have switched from eating chickweed seedlings to other food such as litter and new growth on established vegetation which was more available because they had not been removed by winter grazing sheep.

Clear Hill and Silvertown (1997) were able to do this study because they used organisms whose density was easy to manipulate. Slugs can be trapped and moved, and sheep can be enclosed or excluded using fencing. Experiments like this are much more complicated with more mobile organisms such as birds or mice. While they are easily excluded from treatments, it is harder to envision a way to increase their density but still maintain a natural habitat for them.

Seed predation experiments

A variety of types of seed predation experiments can be done. For example, a researcher could survey natural populations to assess levels of pre-dispersal seed predation, or alternatively, the researcher could set up experiments explicitly to test a hypothesis. Fenner and Lee (2001) conducted a survey of pre-dispersal seed predators in 13 species native to Britain and weeds in New Zealand. They collected flowering heads from 1000 individuals of each species in three locations in both countries. They dissected the flowering heads and noted whether predatory larvae were present and calculated the percentage infestation rate of each species (Table 10.6). Their results showed that infestation rate was higher in the native country (Britain) and that seed predators were almost absent in the invaded habitat (New Zealand). This is a survey approach and provides general information about the presence of pre-dispersal predators. Note that predators were not identified and their abundance per inflorescence was not counted.

Swanton *et al.* (1999) used an experimental approach to examine whether farming practices affected the rate of pre-dispersal seed predation of two pigweed species (*Amaranthus retroflexus* and *Amaranthus powellii*) when grown under a maize crop. They created different environments by varying the maize row width (37.5 and 75 cm) and maize density (75,000 and 100,000 ha^{-1}). They found that pre-dispersal seed predation was higher when maize was planted at low density but that row width had no effect (Table 10.6).

Cromar *et al.* (1999) looked at post-dispersal seed predation in a similar agricultural situation. They conducted two experiments which examined the effect of: (i) tillage (mouldboard, chisel and no-till) and (ii) crop residues (maize, soybean, wheat) on post-dispersal seed predation of lambsquarters (*Chenopodium album*) and barnyardgrass (*Echinochloa crusgalli*). In both experiments, seeds were placed in petri dishes which were buried flush with the soil surface and soil residue was placed over the top. These dishes had small mesh cages placed over them to exclude various seed predators. Mesh of 1.5 mm excluded all organisms and was used as a control to calculate losses due to effects such as wind. A mesh size of 7 mm excluded mammals and birds but allowed insects to enter. By comparing the retention rates of seeds under the two mesh sizes, seed loss due to predation by insects was estimated. They found that seed loss was lowest under chisel plough systems, and lowest under wheat and soybean litter (Table 10.7).

Table 10.6. Percentage infestation of inflorescences of 13 species in the aster family at three locations in Britain (native habitat) and New Zealand (adapted from data in Fenner and Lee, 2001).

Species	Britain				New Zealand			
	1	2	3	Mean	1	2	3	Mean
Yarrow, *Achillea millefolium*	0	0	0	0	0	0	0	0
English daisy, *Bellis perennis*	0	5.0	1.5	2.2	0	0	0	0
Canada thistle, *Cirsium arvense*	1.5	1.0	0	0.8	0	0	0	0
Bull thistle, *Cirsium vulgare*	2.0	28.0	4.0	11.3	0	0	0	0
Smooth hawk's-beard, *Crepis tectorum*	6.0	0	1.5	2.5	0	0	0	0
Mouse-ear hawkweed, *Hieracium pilosella*	0	0.5	0	0.2	0	0	0	0
Nipplewort, *Lapsana communis*	0	0	0	0	0	0	0	0
Oxeye daisy, *Leucanthemum vulgare*	35.5	30.5	12.0	26.0	0	0	0	0
Tansy ragwort, *Senecio jacobaea*	0	1.5	2.5	1.3	0	0	0	0
Common groundsel, *Senecio vulgaris*	0	0	0	0	0	0	0.5	0.2
Dandelion, *Taraxacum officinale*	0	3.0	3.0	2.0	0	0	0	0
Scentless mayweed, *Tripleurospermum inodorum*	23.5	48.0	27.5	33.0	0	0	0	0

Table 10.7 Percentage seed lost to post-dispersal seed predation in tillage and crop cover experiments by Cromar *et al.* (1999). Percentage predation rates (SE) are based on spring and autumn sampling periods averaged over 3 years. Within experiments, treatments followed by the same letter are not significantly different, according to Tukey's test.

	Tillage experiment			Crop cover experiment		
	No-till	Chisel plough	Mouldboard plough	Maize	Soybean	Wheat
Residue biomass (g dry wt m^{-2})	572 (15)	225 (37)	64 (21)	5.3 (41)	328 (40)	510 (66)
% predation	32% (2)[a]	24 (2)[b]	32 (2)[a]	31 (2)[b]	24 (2)[a]	21 (2)[b]

The effect of sampling

A final point we would like to make is that sampling itself can have an effect on what you are measuring. Touching a plant while making observations or measurements can induce a change in its growth, survival, resource allocation or many other variables. For example, Cahill *et al.* (2000) tested whether visiting a plant weekly and touching it (to simulate the act of taking measurements) would influence its survival or losses due to herbivory. They found that after 8 weeks, visitation affected leaf loss in two of the six species tested: visitation increased leaf loss in hemp dogbane (*Apocynum cannabinum*) but decreased it in sulphur cinquefoil (*Potentilla recta*). In addition, there was a decrease (but not significant) in survival in yellow toadflax (*Linaria vulgaris*). Leaf loss and survival of the other species (horsenettle, *Solarium carolinense*; Canada thistle, *Cirsium arvense*; Kentucky bluegrass, *Poa pratensis*) were not affected.

This phenomenon is sometimes called the observer effect. While we presented a herbivory example, the observer effect applies to all types of experiments where plants are repeatedly visited or measured. It should be considered when deciding what types of manipulations you will be doing during the course of an experiment.

Summary

Designing an experiment is the 'art of science' because it involves imagination and creativity. There is no one correct way to answer an ecological question. However, designing experiments also requires pragmatism. When an ecologist is interested in a specific ecological phenomenon, there is much work required to plan and design the experimental methodology. Lack of planning results in unusable data. There is a trade-off between the various types of controlled greenhouse-type experiments and field experiments. In the former, most environmental variables can be tightly controlled so it is possible to ask specific questions about ecological mechanisms. On the other hand, the observed mechanisms may not be important under real conditions. Often the best evidence comes when a variety of approaches are used to answer a question.

Questions

1. Using the weed you've selected, design a manipulative experiment to test one of the following:

- Can your weed out-compete another important weed in your area?
- How is your weed affected by a specific predator found in your region?

Clearly state the hypothesis you are testing, and the procedures you would follow. Describe the experimental site (e.g. forest, grassland, field) or the greenhouse set-up. Give your justification for choosing field

Continued

Questions, *continued.*

vs. lab experiment. Give specific experimental information such as the treatments used, number of replicates and quadrat size. Make sure that your experiment is reasonable, i.e. it could be carried out within a 1-year period. What resources (people, equipment, time and money) would you need to carry out this experiment?

2. Earlier in this chapter we proposed two hypotheses to explain why a certain weed tends to be found in open habitats. Design an experiment to test the second hypothesis, that is:

- H_2 – plant-eating insects that eat the weed prefer to live in shady areas, therefore only seedlings in open sunny locations survive

3. Four researchers sample a population's density by counting density in 10 1-m^2 quadrats. Which researcher(s) were the most accurate, precise and least biased. The true density of the population is 50 m^{-2}. What calculations do you need to do to make these decisions?

Researcher 1	Researcher 2	Researcher 3	Researcher 4
51	25	53	27
55	38	59	39
52	77	48	68
53	76	42	14
54	45	48	16
56	28	50	29
53	81	52	58
51	55	53	75
57	33	49	63
52	44	47	11

4. Explain why data may be precise but not accurate, or accurate but not precise. Can biased data be precise, but not accurate? Explain why.
5. Show how the values of frequency, density and cover were calculated in Fig. 10.7.
6. Complete the following life table. Construct two survivorship curves from the data: one with the *y*-axis on a log scale and the second with an arithmetic scale. Compare the shapes of the curves and explain why a log scale is more appropriate for displaying this type of data. On a separate graph, plot the mortality rate of the population (using arithmetic scale). Explain why it is unnecessary to plot mortality rate on a logarithmic scale. (Hint: it is a rate.)

Age class, x	Number alive at the start of age class x, n_x	Proportion alive at the *the* start of age class x (survivorship) $l_x=n_x/n_0$	Number dying within age class x to x+1 $d_x=n_x-n_{x+1}$	Probability of death between age class x and x+1 (age-specific mortality rate), $m_x=d_x/nx$
0	1000			
1	100			
2	50			
3	20			
4	12			
5	7			
6	3			
7	2			
8	0			

7. Design a pre-dispersal *or* post-dispersal seed predation experiment, using any weed species. Make sure you ask a focused ecological question and construct an appropriate hypothesis. Is your experiment lab- or field-based? Explain why.

General References

Brower, J.E., Zar, J.H. and von Ende, C. (1998) *Field and Laboratory Methods for Ecology*, 4th edn. W.C. Brown, Dubuque, Iowa.

Gibson, D.J. (2002) *Methods in Comparative Plant Population Ecology.* Oxford University Press, Oxford.

Underwood, A.J. (1997) *Experiments in Ecology.* Cambridge University Press, Cambridge.

Literature Cited

Aarssen, L.W. and Epp, G.A. (1990) Neighbour manipulations in natural vegetation: a review. *Journal of Vegetation Science* 1, 13–30.

Alpert, P. and Maron, J.L. (2000) Carbon addition as a countermeasure against biological invasion by plants. *Biological Invasions* 2, 33–40.

Berendse, F. (1981) Competition between plant populations with different rooting depths. II. Pot experiments. *Oecologia* 48, 334–341.

Berendse, F. (1982) Competition between plant populations with different rooting depths: III. Field experiments. *Oecologia* 48, 334–341.

Berendse, F. (1983) Interspecific competition and niche differentiation between *Plantago lanceolata* and *Anthoxanthum odoratum* in a natural hayfield. *Journal of Ecology* 71, 379–390.

Brower, J.E., Zar, J.H. and von Ende, C. (1998) *Field and Laboratory Methods for Ecology*, 4th edn. W.C. Brown, Dubuque, Iowa.

Buchanan, G.A., Crowley, R.H., Street, J.E. and McGuire, J.A. (1980) Competition of sicklepod (*Cassia obtusifolia*) and redroot pigweed (*Amaranthus retroflexus*) with cotton (*Gossypium hirsutum*). *Weed Science* 28, 258–262.

Cahill, J.F., Castelli, J.P. and Caspers, B.B. (2000) The herbivory uncertainty principle: visiting plants can alter herbivory. *Ecology* 82, 307–312.

Cardina, J.A., Sparrow, D.H. and McCoy, E.L. (1996) Spatial relationships between seedbank and seedling populations of common lambsquarters (*Chenopodium album*) and annual grasses. *Weed Science* 44, 298–308.

Clear Hill, B.H. and Silvertown, J. (1997) Higher order interaction between molluscs and sheep affecting seedling numbers in grassland. *Acta Œcologia* 18, 587–597.

Connolly, J., Wayne, P. and Bazzaz, F.A. (2001) Interspecific competition in plants: How well do current methods answer fundamental questions? *American Naturalist* 157, 107–126.

Cousens, R. (1991) Aspects of the design and interpretation of competition (interference) experiments. *Weed Technology* 5, 664–673.

Cromar, H.E., Murphy, S.D. and Swanton, C.J. (1999) Influence of tillage and crop residue on post-dispersal predation of weed seeds. *Weed Science* 47, 184–194.

Dieleman, J.A. and Mortensen, D.A. (1999) Characterizing the spatial pattern of *Abutilon theophrasti* seedling patches. *Weed Research* 39, 455–467.

Engeman, R.M., Sugihara, R.T., Pank, L.F. and Dusenberry, W.E. (1994) A comparison of plotless density estimates using Monte Carlo simulation. *Ecology* 75, 1769–1779.

Fenner, M. and Lee, W.G. (2001) Lack of pre-dispersal seed predators in introduced Asteraceae in New Zealand. *New Zealand Journal of Ecology* 25, 95–99.

Firbank, L.G. and Watkinson, A.R. (1985) On the analysis of competition within two-species mixtures of plants. *Journal of Applied Ecology* 22, 503–517.

Freckleton, R.P. and Watkinson, A.R. (2000) Designs for greenhouse studies of interactions between plants: an analytical perspective. *Journal of Ecology* 88, 386–391.

Gibson, D.J. (2002) *Methods in Comparative Plant Population Ecology.* Oxford University Press, Oxford.

Goldberg, D.E. and Barton, A.M. (1992) Patterns and consequences of interspecific competition in natural communities: a review of field experiments with plants. *American Naturalist* 139, 771–801.

Goldberg, D.E. and Scheiner, S.M. (1993) ANOVA and ANCOVA: field competition experiments. In: Scheiner, S.M. and Gurevitch, J. (eds) *Design and Analysis of Ecological Experiments.* Chapman and Hall, New York, pp. 60–93.

Gurevitch, J., Morrow, L.L., Wallace, A. and Walsh, J.S. (1992) A meta-analysis of field experiments on competition. *American Naturalist* 140, 539–572.

Hamilton, N.R.S. (1994) Replacement and additive designs for plant competition studies. *Journal of Applied Ecology* 31, 599–603.

Jolliffe, P.A. (2000) The replacement series. *Journal of Ecology* 88, 371–385.

Kershaw, K.A. (1973) *Quantitative and Dynamic Plant Ecology*, 2nd edn. Edward Arnold, London.

Krebs, C.J. (1999) *Ecological Methodology*, 2nd edn. Addison Westley Longman, Menlo Park, California.

Landhäusser, S.M., Stadt, K.J. and Lieffers, V.J. (1996) Screening for control of a forest weed: early competition between 3 replacement species and *Calamagrostis canadensis* or *Picea glauca*. *Journal of Applied Ecology* 33, 1517–1526.

Quinn, G.P. and Keough, M.J. (2002) *Experimental Design and Data Analysis for Biologists*. Cambridge University Press, Cambridge.

Silvertown, J.W. and Lovett Doust, J. (1993) *Introduction to Plant Population Biology*. Blackwell Scientific, Oxford.

Swanton, C.J., Griffiths, J.T., Cromar, H.E. and Booth, B.D. (1999) Pre- and post-dispersal weed seed predation and its implications to agriculture. In: *Proceedings of the 1999 Brighton Conference – Weeds*. British Crop Protection–Council, Farnham, UK, pp. 829–834.

Tiffin, P. and Inouye, B.D. (2000) Measuring tolerance to herbivory: accuracy and precision of estimates made using natural versus imposed damage. *Evolution* 54, 1024–1029.

Underwood, A.J. (1990) Experiments in ecology and management: their logics, functions and interpretations. *Australian Journal of Ecology* 15, 365–389.

Underwood, A.J. (1997) *Experiments in Ecology*. Cambridge, UK.

Wilson, S.D. and Tilman, D. (1991) Components of plant competition along an experimental gradient of nitrogen availability. *Ecology* 72, 1050–1065.

Witmer, M.C. and Cheke, A.S. (1991) The dodo and the tambalacoque tree: an obligate mutualism reconsidered. *Oikos* 61, 133– 138.

Zar, J.H. (1999) *Biostatistical Analysis*, 4th edn. Prentice Hall, Englewood Cliffs, New Jersey.

Part III

Community Ecology

11

Basic Community Concepts and Diversity

Concepts

- An ecological community consists of groups of species found together at the same time and space. Communities exist at any temporal or spatial scale. Species within a community may or may not be interdependent.
- When researching a community, the researcher must first decide where community boundaries are and what species to include. These are both fairly arbitrary decisions.
- Species diversity is a measure of the number of species present (richness) and their relative abundances (evenness).
- Diversity exists at different scales, e.g. from the genetic variation found in individuals to the diversity of species across biomes.
- Diversity can be measured using a variety of indices; the method chosen depends on the ecological information needed.
- Scientists have long debated about whether increased species diversity leads to a more stable ecosystem ('diversity–stability hypothesis').

Introduction

A community can be described as an assemblage of species or populations that occur in the same space and time (Begon *et al.*, 1990). Really, a community is a human construct: a group of species lumped together for our convenience, and not necessarily reflective of an ecological reality. That does not necessarily mean that communities are not an ecological reality. It is just not a precondition.

Early studies in community ecology primarily described community patterns and associations between species that were observed in nature. Later, community ecologists turned towards understanding the underlying processes responsible for the observed patterns. Nowadays, community ecologists focus more on testing theory that will allow us to explain and predict community changes caused by natural and anthropogenic forces. There is a dichotomy between describing observed patterns and understanding the processes that cause them. Pattern is the observed structure of vegetation, for example, the zonation of vegetation up a mountain, or weed composition in fields of different cropping systems. Processes are the mechanisms that create the observed pattern. These include species interactions, climate, disturbance and nutrient availability.

This chapter addresses aspects of community structure and diversity (the number and relative abundances of species present in a community). We discuss how to define and delineate communities, whether communities are integrated units, and then discuss patterns, causes and consequences of diversity. Chapter 12 then addresses community dynamics and how communities change over time, while Chapter 13 addresses how and why species invade communities and their effects on community structure and dynamics. As you will see, we are beginning to integrate the information you have learned in earlier chapters on populations and interactions as we move towards examining communities as a whole.

Defining Communities

We have said that a community is a group of populations of different species that occur in the same place at the same time. Although this appears to be a simple concept, much of the ecological literature is taken up with discussions on whether communities exist at all, and if they do, how will we recognize them (Clements, 1916, 1936; Gleason, 1917, 1926; Drake, 1990; Dale, 1994; Wilson, 1991, 1994)? While this appears to be a somewhat semantic argument, it does highlight the importance of considering what physical entity is being studied and what criteria are being used to define it.

We can delineate communities in a number of ways (Morin, 1999). We will present three ways: physically, taxonomically and statistically. Defining communities based on natural physical boundaries is simple for a community in a pond or on a cliff because they have distinct boundaries. However, these are the exception in natural systems. Boundaries in natural communities usually overlap one another. Managed systems tend to have more distinct boundaries but this is only because edges are imposed and maintained by human activities. Such physical boundaries are usually set based upon our perception of the community structure rather than on how the community actually functions. Thus, we view a forest, field or bog as communities whether or not we know how they function (Booth and Swanton, 2002). Even in agricultural fields, there is movement of soil, plants and animals across imposed field boundaries. Though we manage them as discrete units, there will be continuous exchange among fields. We must make reasonable decisions about community boundaries, but be cognizant that they are not 'real' entities and that these decisions may affect the interpretation of data.

An alternative way to physically delineate communities is to describe them based on climatic variables. This was done by Holdridge (1967), who classified the world into life zones (large-scale communities) based on annual precipitation, potential evapotranspiration (water loss as vapour from surfaces and stomata) and biotemperature (mean annual temperature above zero) (Fig. 11.1). For example, a habitat with low potential and high annual precipitation would be a rainforest in a tropical climate but would be a desert if annual precipitation were low. Whittaker (1975) combined the characteristics of the dominant plants with the physical environment to create vegetation biomes (Table 11.1). For example, he distinguished between temperate grasslands, shrubland, woodland, evergreen forest and deciduous forest. Numerous other schemes exist based on various combinations of biotic and abiotic factors (Hengeveld, 1990; Heywood, 1995). These types of classi fications are useful as a general descriptor of vegetation and allow us to make statements about large-scale community types.

The second way to define a community is based on taxonomic structure. We do this

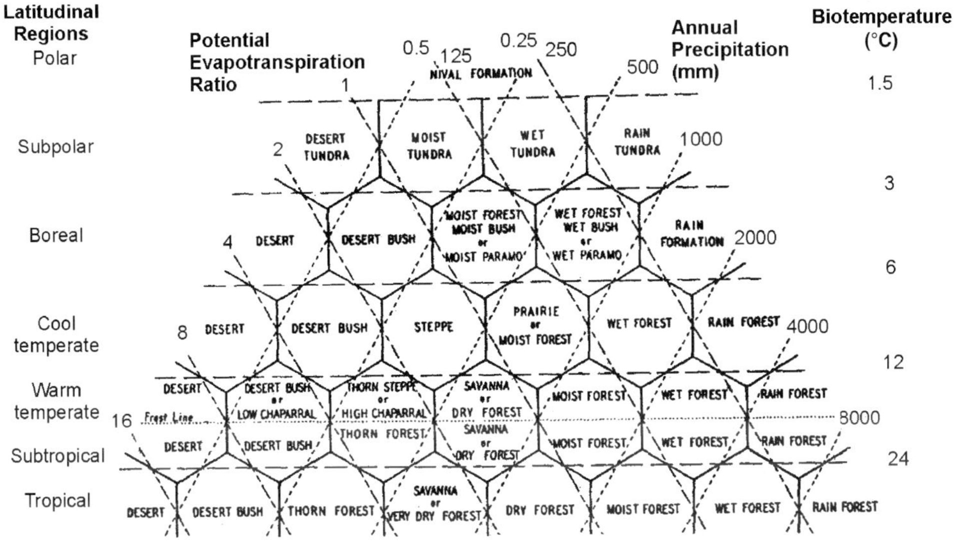

Fig. 11.1. Holdridge's Life Zone Classification System. Classification is based on annual precipitation, potential evapotranspiration (water loss through plant surfaces and stomata) and biotemperature (mean annual temperature above 0°C). (Holdridge, 1967; with permission of the Tropical Science Centre.)

Table 11.1. Major biomes of the world. (From Morin, 1999.)

1. Tropical rain forests
2. Tropical seasonal forests
3. Temperate rain forests
4. Temperate deciduous forests
5. Temperate evergreen forests
6. Taiga forests
7. Elfinwoods
8. Tropical broadleaf woodlands
9. Thornwoods
10. Temperate woodlands
11. Temperate shrublands
12. Savannas
13. Temperate grasslands
14. Alpine shrublands
15. Alpine grasslands
16. Tundras
17. Warm semidesert scrubs
18. Cool semideserts
19. Arctic-alpine semideserts
20. True deserts
21. Arctic-alpine deserts
22. Cool temperate bogs
23. Tropical freshwater swamp forests
24. Temperate freshwater swamp forests
25. Mangrove swamps
26. Salt marshes
27. Freshwater lentic communities (lakes and ponds)
28. Freshwater lotic communities (rivers and streams)
29. Marine rocky shores
30. Marine sandy beaches
31. Marine mudflats
32. Coral reefs
33. Marine surface pelagic
34. Marine deep pelagic
35. Continental shelf benthos
36. Deep-ocean benthos

Source: Whittaker (1975).

when we talk about a field of maize, a tallgrass prairie or a maple–beech forest. We may not know the exact species composition and their abundances, but we will know what species are likely to be present. We have an instinctive knowledge of how these communities differ from each other and could probably list their dominant plant and animal species and their important ecological processes.

The third method for defining communities is based on statistically detected associations among species. Methods used for this will be described in more detail in Chapter 14, but briefly, they involve examining a large data set of species abundances taken from multiple sites. Several types of statistical analysis sort this data into sites that have similar species composition. When data points are separated into distinct groups, then we can say that two (or more) community types are present (see Chapter 14 and Figs 14.6 and 14.7 for an explanation of these methods).

Community composition

Once we have defined the entity that we call a community, we must then decide what species we are going to include. Do we include *all* of the species present, only the plants or only a specific group of species? Most 'community' studies only consider part of the community. We talk about a 'plant community', 'bird community' or 'weed–crop community'. This reflects the taxonomic bias of individual researchers, but is also done for purely practical reasons. Communities are often intractable when we attempt to examine them as a whole because we cannot control all the variables (Drake *et al.*, 1996). This approach, of course, has limitations because the results may not be relevant to complex natural communities (Carpenter, 1996). With the exception of microcosm experiments, no studies that we know of examine the dynamics of the whole community. This will probably remain so out of necessity; however, in doing this we must remember that community dynamics may be caused by species or factors not included in the study.

When researchers ignore groups of organisms, ecological patterns might be missed, or alternatively observed patterns may not be explainable if they arise through interactions with excluded organisms (Booth and Swanton, 2002). For example, in a situation where soil-borne organisms control the community structure of plants (Jordan *et al.*, 2000), interactions between the plant species may be incorrectly used to explain a pattern if soil-borne organisms are omitted from the study. Mycorrhizal fungi, for example, can influence the competitive outcome in a tall-grass prairie (Smith *et al.*, 1999) and their interaction with vegetation should be considered as part of community dynamics.

Often the importance of a species will not be obvious from its size or abundance. A 'keystone species' has a disproportionate effect on community function relative to its biomass (Paine, 1966, 1969). Keystones are not necessarily the most abundant or largest species – it is their effect that determines their importance (Power *et al.*, 1996). For example, kangaroo rats (*Dipodomys* spp.) are keystone species in the Sonoran and Chihuan deserts because they preferentially feed on large-seeded plant species (Brown and Munger, 1985; Brown and Heske, 1990). Parasitic mistletoe can be a keystone species because of the large group of animals associated with it (Watson, 2001). Alternatively, a dominant tree species may not be a keystone. A weed could become a keystone if it alters nutrient cycles, soil properties or provides food for invasive animals. For example, when the fire tree (*Myrica faya*) invaded Hawaii, it changed the nitrogen dynamics, which in turn influenced which other species could survive (Chapter 13) (Walker and Vitousek, 1991).

Are Communities Integrated Units?

One of the earliest debates in ecology centred on whether a plant community is like an organism composed of interdependent species or whether a community is simply a group of species with similar environmental requirements. The two major scientists involved in the debate were Clements (1916, 1936), who proposed the organismic or holistic view, and Gleason (1917, 1926), who proposed the continuum or individualistic view. According to Clements, a community was greater than the sum of the individual species and would have 'emergent' properties unforeseen based on species alone. Under this view, each community type would have a specific and predictable species composition. Gleason, on the other hand, saw communities as random collections of co-occurring species. When species did have similar distributions, Gleason saw this as coincidence rather than interdependence.

One way to observe whether species are *in*dependent or *inter*dependent is to graph their abundances across an environmental gradient. The gradient could be as simple as increasing soil moisture, or as complex as an altitudinal gradient up a mountain where many environmental factors change. If communities are not tight associations of interacting species, then species' distributions will overlap and there will be no discrete boundaries between them (Fig. 11.2a). In this case, defining a community is difficult because there are no obvious species groupings. If species do occur in close association, then their distributions along a gradient will be similar and species' boundaries will coincide (Fig. 11.2b). The area of transition between communities is called an ecotone. Ecotones usually have many species because members of both communities will be present, albeit in low abundance. In Fig. 11.2b, there are three communities with two ecotones shown.

Does the experimental evidence support

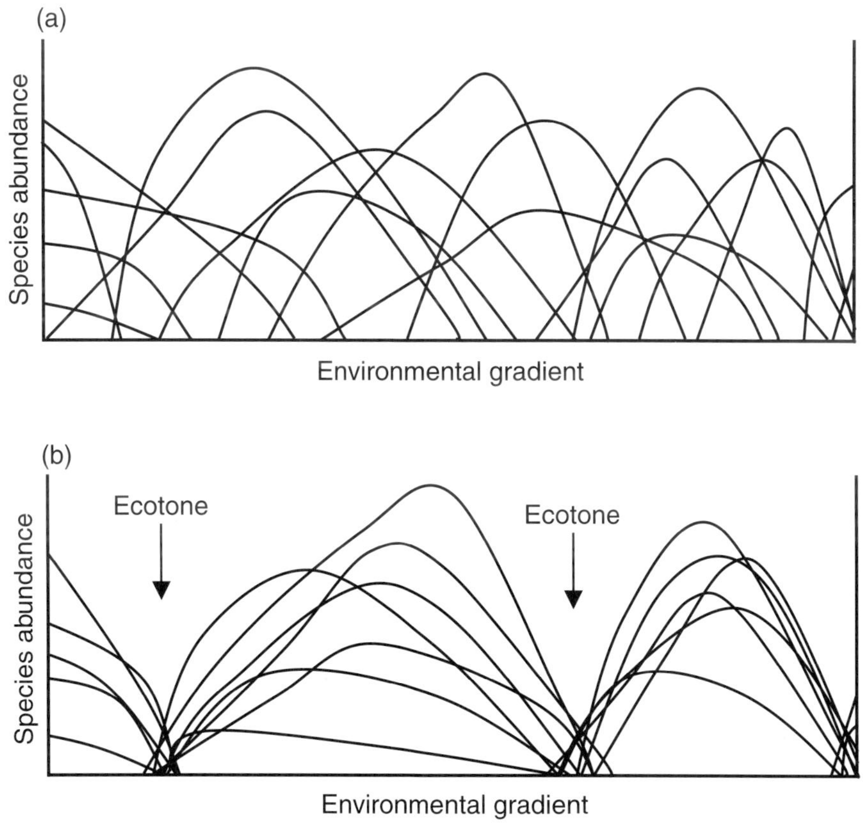

Fig. 11.2. Theoretical distributions of species abundances over an environmental gradient according to: (a) Gleason's individualistic concept and (b) Clements' continuum concept (based on Whittaker, 1975).

Clements or Gleason? Well, neither view will adequately describe *all* vegetation patterns. There is a general consensus that both views contribute to our understanding of community structure. Gleason's model is closer to current ideas, and most plant communities seem to follow Gleason's individualistic model. Certainly, as Gleason suggested, every species will have unique sets of environmental requirements or tolerances and will therefore have a unique distribution. However, species interactions can change how and where a plant will live (Chapters 8 and 9), and therefore environmental tolerances alone do not determine distribution.

As Clements suggested, some species are interdependent; we have seen this in our discussion of mutualisms (Chapter 9). A keystone species may have strong interactions with many species and therefore the distribution of the keystone will determine the distribution of other species (Power *et al.*, 1996). Finally, in situations where gradients are strong, for example between a lake and a forest, between the north and south slopes of a hill, or where the physical environment changes abruptly, then there also will be abrupt changes in community composition and therefore the Clements model will apply. Really, both models were developed from similar evidence (Booth and Larson, 1999), but were interpreted at different scales. Clements looked at large-scale vegetation patterns, whereas Gleason was more concerned with individual species patterns.

Matters of scale

We think of ecological communities as existing at scales that we can observe (e.g. we watch a forest or field over decades), but communities exist at many temporal and spatial scales. A leaf is the substrate for a community of mites, bacteria and fungi. It would be inappropriate to look at this community of microorganisms on a spatial scale of kilometres or on a temporal scale of decades, but these scales might be appropriate for forest studies. Community ecologists are beginning to recognize the importance of scale when designing and interpreting ecological experiments (Levin, 1986; Allen and Hoekstra, 1990, 1991; Menge and Olsen, 1990; Hoekstra *et al.*, 1991).

We can understand communities better if we consider them at multiple scales. The effect of scale means that a community may be responding to local (e.g. succession), regional (e.g. climate) or global changes (e.g. plate tectonics). Thus, subsequent changes in community structure may emerge from a micro-scale (1 m^2) to a mega-scale (>10^{12} m^2) (Delcourt *et al.*, 1983; Davis, 1987). Factors that influence a community's structure will function at many scales. Smaller-scale processes such as species interactions (e.g. competition or predation) and responses to abiotic factors (nutrient levels) will determine local community patterns. These local processes will be nested within large-scale environmental or climatic conditions that will either directly control species distribution or indirectly influence the small-scale processes (Díaz *et al.*, 1998; Menge and Olson, 1990; Woodward and Diament, 1991). The spatial patterns we observe in a community are the result of species responding to these multiple scales.

Patterns of community structure will emerge at a many scales. We might be interested in community-level properties such as the number of species (species richness), or we may be interested in changes in species composition. However, focusing on only one of these can distort our view of the community (Levin, 1986) because a pattern may emerge in one but not the other. For example, the number of species (richness) in a community may remain constant over time whereas the species make-up changes (Brown *et al.*, 2001).

Community Attributes

In Chapter 2, we discussed the various attributes of populations that can be measured: distribution, abundance and demography. Communities, too, have specific types of attributes used to characterize and compare them. These attributes are based on features of a community and do not describe individuals or populations. They include species composition, physiognomy and diversity (Barbour *et al.*, 1999).

The most basic way to describe a community is to list all the species present. However, as we have seen, it is not usually possible to list *all* species, therefore this option is often not possible. Instead, we may list the dominant species. A more general approach is to describe the general appearance of a community (physiognomy). Physiognomy includes such variables as:

- vertical structure of the vegetation (e.g. canopy, shrub layer, understorey);
- spacing of individuals, (e.g. random vs. clumped, sparse vs. dense);
- life forms of the dominant species (tree, shrub, herb).

We can go a long way towards understanding vegetation physiognomy by describing a few simple features. For example, we could describe the physiognomy of a tropical rainforest by dividing it into five stratified layers of vegetation: ground-level vegetation, the shrub and sapling layer, and three canopy layers (lower, mid-crown, emergent) (Fig. 11.3). Without knowing any species names, we could still understand the basics of how this community functions. Finally, we could use a measure or index of diversity to describe the variety of organisms in a community. The remainder of the chapter will discuss diversity in detail.

Diversity

Diversity describes the wide variety of organisms found in the world. It encompasses ways to quantify how many groups (e.g. species) are in a given community and their relative abundances. We usually think of diversity in terms of species, but we could also consider other types such as genetic or plant family diversity. In the last decade, diversity (popularly known as biodiversity) has become a political issue, and preserving diversity is seen as a good thing.

Figure 11.3. Physiognomy of a tropical rainforest showing five layers of vegetation: emergent canopy, main canopy, lower canopy, shrub and sapling layer, and ground layer. (Smith and Smith 2001; Copyright © 2001 by Benjamin Cummings. Reprinted by permission of Pearson Education Inc.)

If you go for a walk in the summer and look at the vegetation around you, you can get different impressions of diversity, depending on where you are walking and how closely you look. For example, casual observation of a lawn usually leaves the impression that it is a monoculture. However, if you ask homeowners, they will probably complain about the many weeds in their lawns. The same is true if you pass a farm field or forest plantation. Farmers and plantation managers may want to grow a crop monoculture, but there are likely to be many weeds as well. Conversely, we usually expect 'natural' areas to be teeming with different kinds of organisms. However, if you walk along a pathway in a small woodland almost anywhere in the world and look closely, the vegetation is comprised of only a few species and a lot of these are weeds. Some of these weeds are pleasing to the eye and many people do not mind them being there. We do not expect woodlands to look like monocultures, so a diversity of weeds does not look out of place to the untrained eye. If we compare managed and natural areas, we might find that each contains the same number of species, and many species are weeds. Only in managed areas might many people worry and try to do something about it. This perception is important because it colours our view of when and where we will accept the existence of diversity.

Basic components of diversity: richness and evenness

Diversity can be quantified simply by counting the number of species present and comparing their relative abundances. Species diversity in its simplest form is the number of species present in an area or in a community (species richness). You could calculate the diversity of your backyard by counting the number of species there. It is a crude method of measurement, but it does give a good idea of how many types of 'things' there are in a community. However, even this simple measure of diversity can be complicated when we consider sample effects. Two researchers calculating richness of a community will come to different conclusions based on the area they sampled. A researcher who samples a small area will obtain a lower estimate of richness than someone who samples extensively.

The second component of species diversity is 'species evenness'. Evenness compares the abundance of each species in a community, and tells you whether there are many rare species and a few common ones or if most species are equally common. Evenness is more informative than species richness, because it indicates whether the community is dominated by one or a few species or whether most species are represented by approximately equal numbers of individuals. For example, Table 11.2 summarized the species density data from four fields. The species richness is the same for all fields (four species in each); however, evenness differs among fields. Fields 1 and 3 have the same evenness, as all species are equally represented. Field 2 is dominated by downy brome (*Bromus inermis*) and field 4 is dominated by viper's bugloss (*Echium vulgare*) and therefore abundance is uneven. That different species dominate in two fields gives us a hint about the characteristics

Table 11.2. Density of species found in four hypothetical fields.

	Density (number of individuals m^{-2})			
Species	Field 1	Field 2	Field 3	Field 4
Downy brome	92	101	61	25
Canada thistle	103	13	63	12
Wild carrot	104	15	65	60
Blueweed	97	11	60	100

of the fields that species richness would miss. Downy brome prefers field margins or abandoned farm fields; blueweed is more typical of gravelly areas. Of course, sampling is normally more complex, but this illustrates why species richness fails to give more than a rough illustration of community structure.

Rank-abundance curves

Species evenness reflects the relative abundance of species in a community. When discussing species diversity, it is common to rank species from the most abundant to the least common. On the *x*-axis, species are ranked from high to low abundance, and on the *y*-axis abundance is plotted on a logarithmic scale. Figure 11.4 shows the rank–abundance curves of the four field communities. Most communities, however, have more than four species and thus curves are more complex. There are four theoretical manifestations of rank–abundance curves: geometric series, log series, log-normal and broken stick model (Fig. 11.5).

In a geometric series, each number is a constant multiple of the number immediately preceding it (1, 3, 9, 27, 81... is an example; every number is a multiple of 3). Biologically, this occurs when the success of the dominant species is overwhelming. This type of community has low evenness with one (or a few) dominant species and is typical of harsh, resource-limited environments such as deserts, arctic tundra and recently scoured volcanic flowplains. A log series is similar to the geometric series except that it does not decrease as rapidly. Biologically, this is similar to the geometric series, except that the dominant species does not tie up as many resources. These series are usually found in human-disturbed areas such as parks, farms and forest plantations. A community with a log-nor-

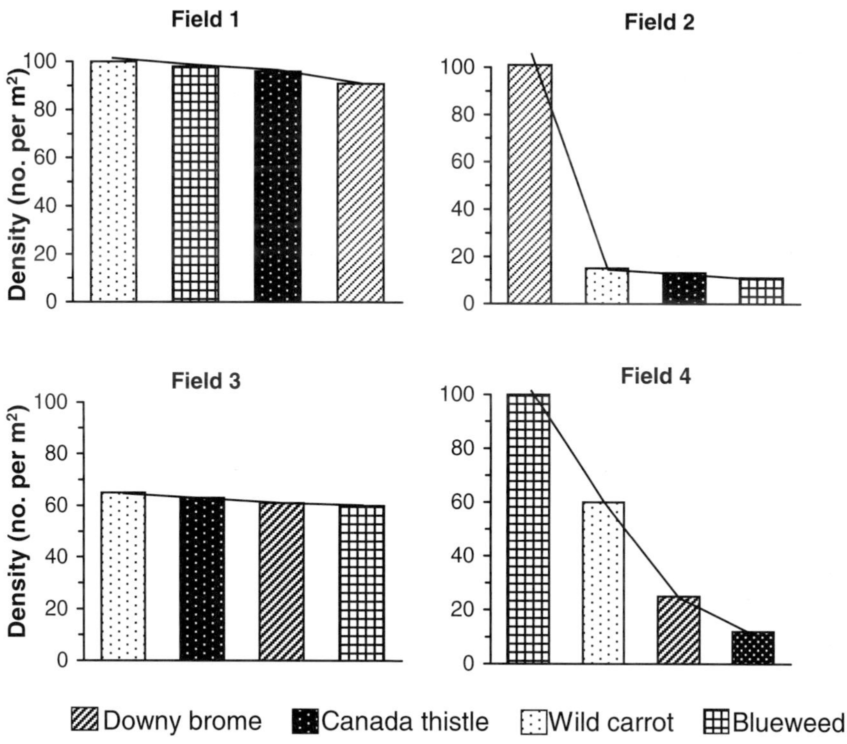

Fig. 11.4. Rank–abundance curves of the four field communities in Table 11.2.

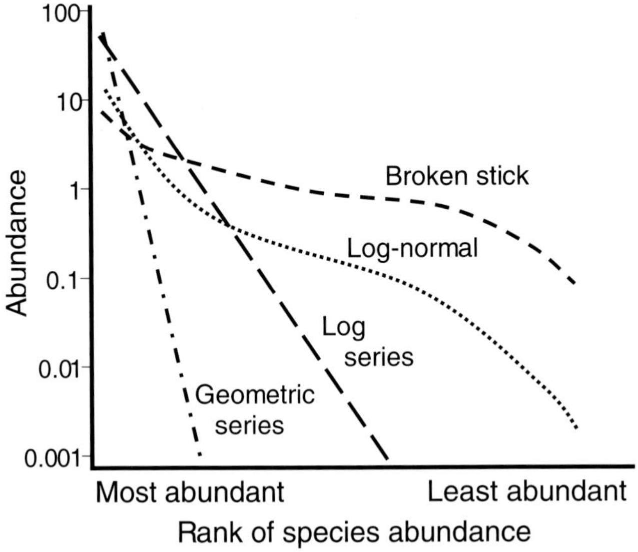

Fig. 11.5. Four models of species–abundance relationships – broken stick, log-normal, log series and geometric.

mal distribution has only a few common species, many intermediately abundant species and only a few rare species (Preston, 1962). This curve tends to represent communities that are relatively undisturbed (late successional meadows, for example). In the broken stick model (MacArthur, 1957) the limiting resource is shared evenly among all species (there is no competition) and this causes high evenness in the community. This model is rather unusual but occasionally occurs where there is eutrophication (pollution, for example) and one resource that would otherwise be limiting to a group of similar organisms becomes excessively abundant.

Types of Diversity and Their Measurement

Diversity can be defined and examined at different spatial scales (Franklin, 1993; Angermeier, 1994). At smaller scales (e.g. individuals, populations, species), genetic diversity is a focus because it is the basis of natural selection and responds to it. At larger scales (e.g. community, ecosystem, landscapes, biomes), diversity becomes important as a *possible* indicator of whether communities (for example) differ from one another or whether an ecosystem (for example) is functioning well or not. Diversity is often classified into categories based on scale (Whittaker, 1975; McNaughton, 1983; Brockway, 1998; Loreau, 2000). The two main categories of diversity are the total species diversity in an area (inventory diversity) and the species diversity along an environmental gradient (differentiation diversity). They are subdivided as follows.

- Total species diversity in an area ('inventory diversity'):
  - α- (alpha-) diversity: the total species diversity within one defined habitat or community (e.g. an area the size of a farm field, forest plantation, woodlot or wetland);
  - γ- (gamma-) diversity: the total species diversity within an ecological landscape (e.g. an area the size of a watershed);

- ε- (epsilon-) diversity: the total species diversity within a larger ecological landscape (e.g. an area the size of a biome).

- Species diversity along an environmental gradient ('differentiation diversity'):
  - β- (beta-) diversity: the comparison of diversity between habitats, communities and/or along an environmental gradient (e.g. changes in diversity along a 1-km long moisture gradient);
  - δ- (delta-) diversity: the comparison of diversity between landscapes (usually about the size of watersheds).

These categories are rather arbitrary. For example, we defined γ-diversity as the inventory (total) species diversity within a landscape roughly the size of a watershed. However, some define it more like differentiation diversity; others equate γ- and δ-diversity because both are concerned with landscapes about the size of watersheds, thereby focusing on spatial scale rather than considering whether inventory and differentiation diversity should be separated (Brockway, 1998; Goetz, 2000).

Scientists have developed ways to measure and interpret biodiversity, again mainly at the scales of species. One reason why it is important to do this is that the relative increase or decrease in species diversity might be a good indicator of the complex ecological structure and function that exists, but would otherwise be difficult to measure (Lovejoy, 1994).

Measuring α-diversity

Measuring species richness

Species richness is a common measurement used to represent species diversity. A simple, typical approach is to randomly sample individuals within a defined area and identify them to species. One common procedure (called Margalef's diversity index, D_{Mg}) used is relatively easy to understand:

- count the total number of species found in the community (symbolized by a letter S);
- count the total number of individuals sampled (symbolized by a letter N);
- calculate species richness with the formula $D_{Mg} = S–1/\ln N$.

D_{Mg} is an index used only as a rough and quick estimate of the species diversity in a community. This is because D_{Mg} is sensitive to the total area sampled, the number of individuals sampled and whether all or most of the species that exist were represented in the sample.

Measuring species abundance

Species abundance can be measured using relative and proportional methods. The 'relative' methods simply rank species in order of their abundance and provide a broad comparison of whether communities are composed of one dominant species, a few dominant species or species that are relatively equal in abundance (Magurran, 1988) (Fig. 11.5). These approaches tend to be less popular because they are not quantitative and are less relevant to weed ecology. More broadly, there are dozens of methods of measuring diversity, all with varying advantages and disadvantages and popularity (Table 11.3; see Magurran, 1988; Cousins, 1991; Schlesinger *et al.*, 1994; Conroy and Noon, 1996; Yorks and Dabydeen, 1998; Stiling, 1999; Wilson *et al.*, 1999, for longer discussions).

'Proportional' species abundance methods measure and incorporate both evenness and species richness. In Chapter 14, you will be provided with some examples of calculations; in the current chapter, we provide the theoretical basis for these methods. A common one used in weed ecology is called the Shannon–Wiener diversity index (note that this is sometimes incorrectly called the Shannon-*Weaver* index) (Magurran, 1988; Magnussen and Boyle, 1995). The Shannon–Wiener diversity index is accurate only if sampling is able to detect all species actually present in an ecosystem and if indi-

Table 11.3. Comparing the scale, sophistication, and biases and limitations of diversity measures discussed in this chapter. This table is modified from the approach used by Magurran (1988; see Table 4.5 on page 79 of this reference).

Name	Scale	Level of sophistication	Bias and limitations
Margalef's index	α-diversity	Measures only gross species richness	Very sensitive to number and thoroughness of sample
Shannon–Wiener diversity index and evenness measure	α-diversity	Measures species richness and evenness and compares diversity between two different sampling areas within a given habitat	Moderately sensitive to sample size but sampling must meet criteria of randomness and completeness (all possible species represented)
Simpson's dominance index	α-diversity	Measures dominance of one species vs. other	Low sensitivity to sample size but says little about species richness
Whittaker's statistic	β-diversity	Measures the rate of turnover of species along a gradient	Low sensitivity to sample size and independent of α-diversity; implicitly assumes gradients are linear and continuous (many are not
Jaccard's coefficient of similarity	β-diversity	Measures the similarity of species composition between two communities	Moderately sensitive to sample completeness and does account for differences in numbers of individuals of different species
Steinhaus coefficient of similarity	β-diversity	Measures the similarity of species composition between two communities; accounts for differences in numbers of individuals of different species and relative success of species common between communities	Moderately sensitive to sample completeness

viduals were sampled randomly from populations too large to ever sample completely. The Shannon–Wiener diversity index requires data on the number of individuals of each weed species found in the sample, the total number of individuals (of all species) found in the sample, and the proportion of this total that each species represents.

Another common proportional abundance method is based on 'dominance'. Dominance measurements preferentially increase the weight or importance of the abundances of the most common species. They do not provide a good assessment of species richness but are useful when examining evenness. A common dominance index in weed ecology and, especially, in forest ecology is Simpson's dominance index because it quickly shows whether a community is characterized by only a few common species. Simpson's dominance index is based on the probability that any two individuals sampled will turn out to be the same species. The more a community is dominated by any one species, the more likely it is that you will keep sampling the same species over and over and the less diverse the community is. The advantage of Simpson's dominance index is that it is simpler to calculate than the Shannon–Wiener index and requires only that you sample and estimate the number of individuals in any given species and the total number of individuals in a site.

Some simple ways to measure β-diversity

Why the Whittaker method is useful

Recall that we said that the distinction between α- and β-diversity (or any other scales) can be arbitrary. However, the advantage of measuring β-diversity is that it is explicitly meant to compare different areas, communities or landscapes, and this normally implies that you are examining the effect of some type of putative environmental difference on species composition. The key characters desirable in measuring β-diversity are:

- ability to account for changes in the species composition along gradients in a community (e.g. as soil moisture, temperature or nitrogen concentrations change across an area);
- additivity (i.e. values of β-diversity remain constant no matter where you sample, e.g. whether you sample along an entire gradient or just from two ends of the gradient);
- independence from measurements of α-diversity (i.e. you can compare different areas regardless of whether the α-diversity is high or low);
- independence from sample size (i.e. you can make comparisons even if the sample size is small).

When all of these factors are considered, one of the best methods is called Whittaker's method or statistic. It requires you to measure the total number of species in the sample and the mean species richness per sample.

Coefficients of similarity

Another common way to calculate β-diversity is to use coefficients of similarity such as the Jaccard, Sørenson and Steinhaus indices. These are better for comparing distinct habitats, rather than ones connected by some type of environmental gradient. They are based on comparing the number of species that are common or unique to two communities; Steinhaus' coefficient also accounts for the numerical (quantitative) differences in species found in each community. Since the indices are calculated in different ways, they may give different answers, even though they are measuring the same basic aspects of diversity. The reason is related to their relative sophistication. Steinhaus' coefficient more accurately accounts for the differences in numbers of individuals. It also shows when a species succeeds in one habitat but not another.

The idea that the coefficients of similarity can give different answers illustrates a pitfall in using diversity indices: numbers tell you nothing unless you understand their actual assumptions and meaning in ecological terms. Magurran (1988) emphasized this when she compared the utility of different measures and the assumptions of each (Table 11.3).

Global Patterns of Species Diversity

If we use species richness as a measure of diversity and compare this among communities around the world, there are observable global patterns. In general, diversity is much higher in the tropical regions than in temperate or polar regions, e.g. the species richness of woody plants decreases from the equator to the poles (Currie and Paquin, 1987) (Fig. 11.6). This pattern is not exact, however. For example, at the same latitude, there are more tree species in eastern North America than in the west. A number of mechanisms to explain these patterns have been proposed – these tend to be classified as either large- or small-scale mechanisms (Fraser and Currie, 1996; O'Brian, 1998; Whittaker *et al.*, 2001).

At large scales, climate and historical factors are important determinants of species richness. Climatic effects can be quantified by examining 'water-energy dynamics' (O'Brian, 1998). Energy (heat and light) and water are necessary for photosynthesis; thus photosynthesis will increase up to a point as energy and water increase. Available energy can be quantified using

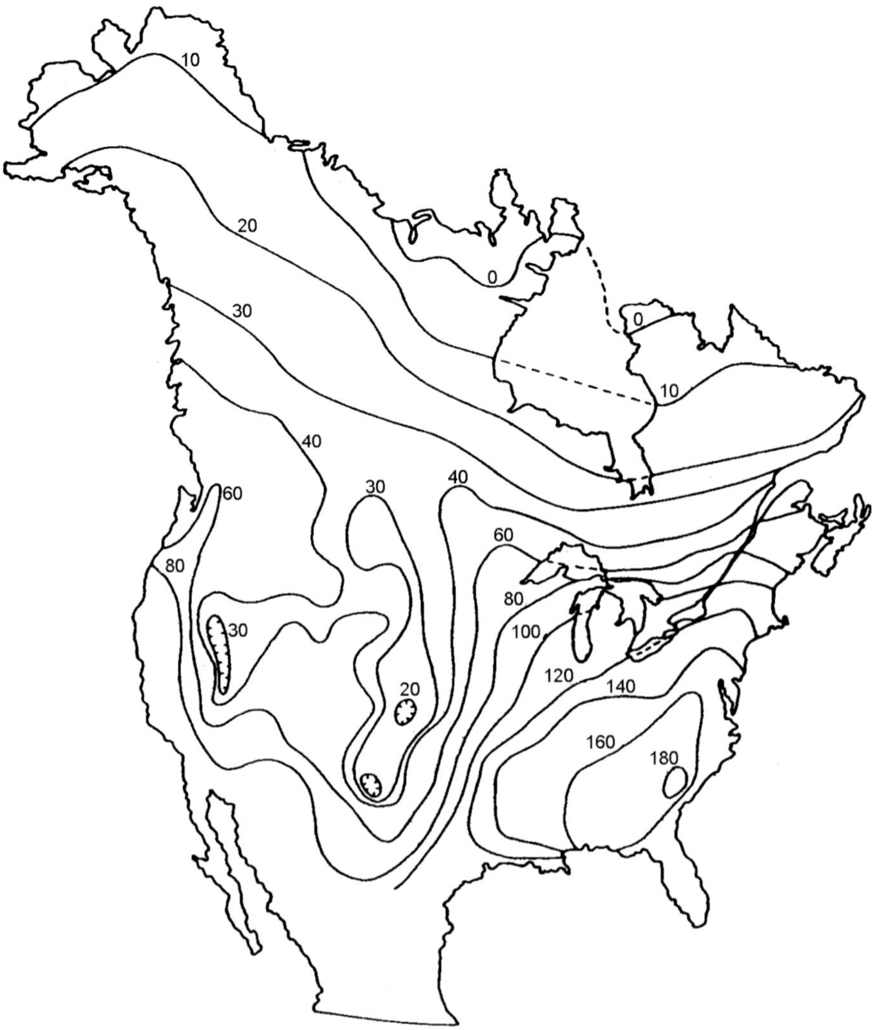

Fig. 11.6. Pattern of species richness for trees in North America. (Currie and Paquin 1987; with permission of *Nature* and the author.)

temperature or potential evapotranspiration (PET); water availability can be quantified using annual precipitation. O'Brian *et al.* (1998) successfully described species, genus and family richness of woody plants in southern Africa using PET and annual rainfall. Historical factors include the time since glaciation, speciation rate and dispersal effects. Communities closer to the equator have had much more time since glaciation and therefore have had more time for evolution to occur and new species to arise.

The intermediate disturbance hypothesis

On smaller scales, local factors moderate general richness patterns. For example, topography, species interactions and disturbance will affect local richness. One important way to explain species diversity is the 'intermediate disturbance hypothesis' (Fig. 11.7). The key aspects to this hypothesis are defining 'disturbance', explaining what disturbance means for species diversity and determining when disturbance is 'interme-

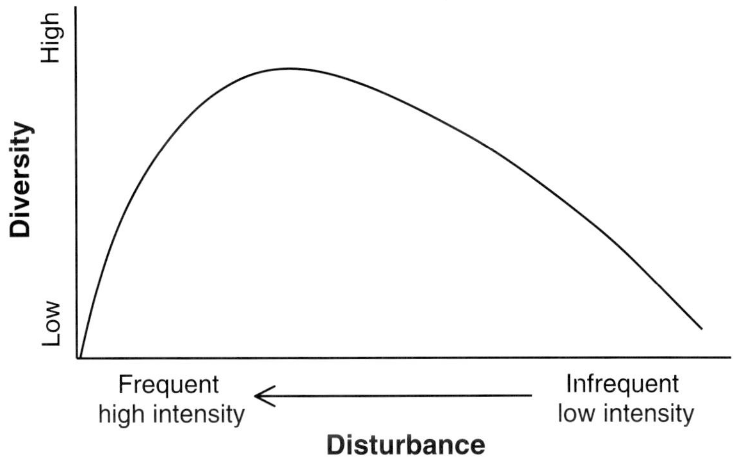

Fig. 11.7. Intermediate disturbance hypothesis (based on Connell, 1978).

diate'. Disturbance is any event that disrupts the existing structure or function of the environment. Disturbance can vary in temporal scale, i.e. it can be periodic (e.g. someone steps on a weed, a tree is blown over, a forest fire erupts, a hurricane occurs) or continual (e.g. grazing, burial, yearly high tides, ice scouring). Disturbances also can vary in 'intensity'. Ecologists tend to use 'intensity' to describe a physical characteristic of the disturbance, e.g. a forest fire that burns at high temperatures, destroys all vegetation and mineralizes almost all the nutrients, or herbivores that eat the meristems of plants so they cannot regrow. Ecologists also could examine disturbances on the basis of spatial scale (e.g. stepping on a weed is small scale and a hurricane is large scale). However, spatial scale can be incorporated into the definition of intensity (e.g. a hurricane is more intense because it is more violent and affects many individuals).

A habitat's structure and function results from many frequencies and intensities of disturbance. For example, grasslands may have areas of high- and low-intensity grazing, gopher holes, and piles of animal faeces with locally high concentrations of nitrogen. The various frequencies and intensities of these disturbances help to open and allow colonization of gaps of different sizes at different times. This is part of an ecological process called 'patch' or 'gap' dynamics that creates a mosaic of vegetation types with different individuals and, in general, species responding in different ways to the disturbance regimes (Grubb, 1985). Thus, while disturbance harms or destroys some individuals (Grime, 1979; van der Maarel, 1993), it also releases resources and allows new individuals to colonize; weeds, unfortunately, are well adapted to almost any disturbance.

The intermediate disturbance hypothesis relates diversity to the frequency and intensity of disturbance in an ecosystem (Grime, 1973; Horn, 1975; Connell, 1978; Petratis *et al.*, 1989; Wilkinson, 1999; Buckling *et al.*, 2000; Whittaker *et al.*, 2001). At one extreme, high frequencies and intensities of disturbance change selection pressures so rapidly that the number of species that survive will be small (most become extinct). As the frequency and/or intensity of disturbance declines, diversity will increase because: (i) there is sufficient disturbance to open a niche for new species, and (ii) there is enough time between major disturbances to allow more species to colonize successfully. It is at these 'intermediate' levels of disturbance that diversity is highest. When the frequency of disturbance is very low, the relative constancy of selection pressures means a few well-adapted species can dominate and competitively exclude most existing species and would-be colonizers. As a result, few new species establish themselves and diversity is low.

A major problem with the intermediate disturbance hypothesis is related to defining exactly what time periods and states 'high', 'intermediate' and 'low' frequencies of disturbance represent. For example, does 'high disturbance' mean a 1000 ha forest fire that burns at 300°C every year or does it mean a 10 ha forest fire that burns at 100°C every 10 years? What if there was a mix of these conditions, e.g. a 1000 ha forest fire that burns at 100°C every 50 years – is this high, intermediate or low disturbance? There are no standard criteria to define frequencies and intensities, hence there is disagreement over how, when and if the intermediate disturbance hypothesis explains diversity patterns. A good approach is to establish the historical pattern of disturbance (if possible) and then relate the current disturbance frequency and intensity to this. Even

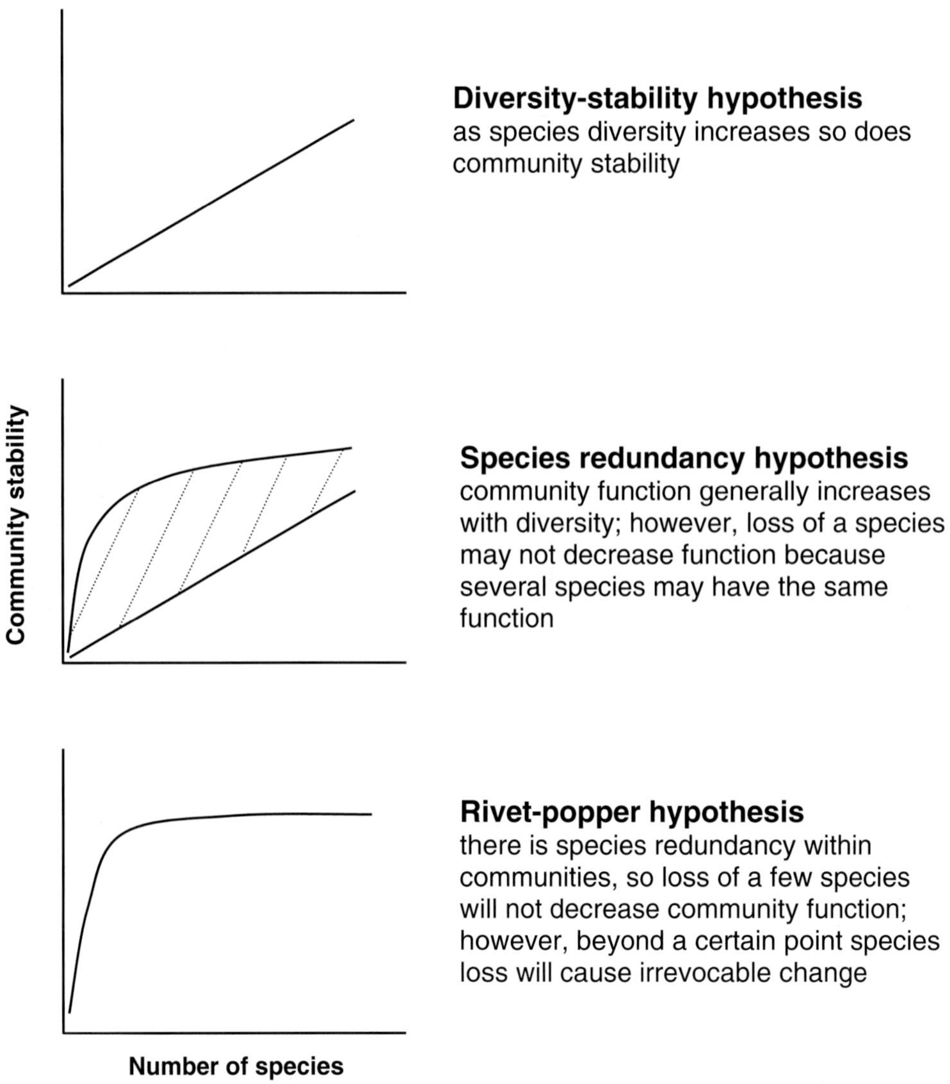

Fig. 11.8. Hypotheses about the relationship between diversity and community function (redrawn based on Johnson *et al.*, 1996).

if this is done, however, the historical pattern will vary across the planet, e.g. our discussion of why tropical ecosystems have relatively high diversity included the idea that they have had less recent disturbance from glaciation. The general lesson from this is to ensure that you understand the context of your study rather than trying to apply too general a concept like 'intermediate disturbance'. The basic theory is sound; its application requires careful analysis within the particular situation you are studying.

Is Species Diversity Important to Ecosystem Function?

Even if we can recognize patterns of species diversity and their general causes, this still does not test the fundamental question of whether species diversity is merely an outcome of natural selection or whether it has important ecological roles that will influence natural selection. In the last 50 years, people have debated whether the ecological role of diversity is to make ecosystems less vulnerable to destruction, i.e. whether species diversity allows more ecosystem stability.

Hypotheses used to explain the functional role of diversity

Diversity-stability hypothesis

The species diversity-stability hypothesis (whether species diversity allows more ecosystem stability) originally focused on the idea that complex trophic structures resulted in more stable communities (MacArthur, 1955) (Fig. 11.8). Elton (1958) refined this further when he hypothesized that human-disturbed communities, where many species had been 'extirpated' (made extinct in certain locations), were prone to pest outbreaks and unpredictable fluctuations in populations. May (1974) asked what is, in hindsight, an obvious question: does diversity confer ecosystem stability or does ecosystem stability allow for increased diversity? He used mathematical models to test how the number and type of linkages within food webs related to stability. May found that the relationship of food web stability to diversity was more of an illusion because it all really depended on a relatively constant environment (not unlike what would be found when there were intermediate level disturbances). A constant environment was the limiting factor and therefore stability causes diversity. This reversal of the diversity-stability hypothesis caused scientists to test the hypothesis further and offer alternatives or refinements.

Species redundancy hypothesis

One question raised by May's work is whether species are so similar (redundant) that elimination of most of them would not affect ecosystem function (see reviews in Schulze and Mooney, 1993). This is the 'species redundancy hypothesis'. It implies that most species are interchangeable and that conservation of most species is not necessary in terms of ecosystem function. However, Walker (1992) explained that redundancy is better interpreted as follows. First, we might classify species in definable groups based on species that have the same function in an ecosystem; some of these species and functions are extremely important to the total ecosystem function – recall our earlier discussion about what we called 'keystones'. Then, we need to determine if there are enough species to expand their roles ('take over the job') when another species in the same functional group becomes extinct. Finally, we need to examine whether any of the functional groups has been or is likely to be eliminated entirely. If the keystone groups exist and are not threatened, then species diversity may be less important (Walker, 1992; Schulze and Mooney, 1993; Vitousek and Hooper, 1993; Martinez, 1996). There is, however, a further caveat to the species redundancy hypothesis. It probably applies only if extirpations or a few extinctions are occurring – as opposed to mass extinctions on a planet-wide scale.

Rivet-popper hypothesis

Unfortunately, extinctions can occur on a planet-wide scale and this appears to be what humans are doing now. Ehrlich and Ehrlich (1981) recognized this problem and proposed the 'rivet-popper hypothesis'. Using the analogy of species as rivets on an airplane wing, the Ehrlichs argued that there is functional and structural redundancy built into a system like an airplane wing or an ecosystem (more rivets than needed, more species than needed). The problem is that the loss of one rivet or one species contributes to a cumulative effect. By itself, a loss of one rivet or one species may not cause a sudden change. The loss of several rivets or species also may not cause any changes. However, at some unpredictable point, the losses of many rivets or many species will cause an irrevocable change because their cumulative effect is to undermine the structure and function of the wing or ecosystem. Effectively, the Ehrlichs propose that ecosystem function is compromised only as species loss increases but the number required to cause problems is unknown. Therefore, it is not wise to extirpate any species because we do not know how many other species have been lost before and if the loss of just one more species might be the one that pushes the ecosystem over the threshold and into oblivion.

Bumper hypothesis (changes in critical systems)

Schulze and Mooney (1993, p. 497) noted that a species that is not now a keystone species eventually may become one. The analogy used was automobile parts. A bumper is usually not a critical system – unless you happen to get into a low-speed accident that otherwise would cause serious damage to life and property. In the same way, a species might indeed be redundant or less important today – but what happens if the environment changes? Some people believe the environment is constant but, at scales from years to aeons, environments inevitably do change, so this hypothesis probably applies to many situations.

Empirical tests of hypotheses about diversity and ecosystem function

The hypotheses discussed above are all plausible but it has been difficult to determine which, if any, are good explanations for the continued existence and role of species diversity. The primary controversy has been whether diversity does indeed create a better, more stable ecosystem or whether ecosystems can survive with just a few species. The reason it has been difficult to resolve this controversy is that it is difficult to test all these hypotheses. It is challenging (to say the least) to manipulate entire ecosystems or find appropriate *post hoc* tests using existing ecosystems. In either case, the studies would, and do, take a long time to complete.

In the absence of much evidence, the debate over diversity has become heated, as it is apparent human disturbance is causing rapid and massive extinctions and extirpations (Ehrlich and Ehrlich, 1981; Wilson, 1985). The debate at first centred on whether these extinctions were really so rapid and massive but then shifted focus to whether diversity has any ecological meaning (remember that few tests of the hypotheses have been completed). Exacerbating this is the use of so many different definitions of words associated with diversity. Consequently, trying to compare hypotheses and tests about diversity is like trying to read an instruction manual that keeps switching languages at random (King and Pimm, 1983; Pimm, 1991; Drake *et al.*, 1996). Further, few studies have actually asked exactly the same questions, hence comparing studies is difficult (McNaughton, 1988).

What tests and evidence exist?

Although there has been a recent flurry of tests, the empirical evidence to date has not resolved the diversity issue. McNaughton (1985) and Frank and McNaughton (1991) provided some of the first empirical evidence that diversity and stability were linked. However, the controversy exploded again when Tilman and Downing (1994) and Naeem *et al.* (1994, 1995) used field and controlled environment studies, respectively, explicitly to test the diversity-stability hypothesis. In these studies, species diversity was determined to be an important factor in ecosystem function and stability, but not consistently. For example, greater drought resistance in prairie ecosystems occurred because species diversity was correlated with increased biomass (Tilman and Downing, 1994). Primary productivity rose with increased species diversity and this appeared to indicate increased ecosystem stability (Naeem *et al.*, 1994). However, Naeem *et al.* (1994) found contradictory results, e.g. important ecosystem processes like nutrient retention were not affected by species diversity (Berendse, 1993).

Contradictory results and ideas on diversity's role in ecosystems

You will find that *many* papers have been written about diversity's role in ecosystems (Clements *et al.*, 1994; Naeem *et al.*, 1994, 1999; Johnson *et al.*, 1996; Tilman *et al.*, 1996, 1997; Huston, 1997; McGrady-Steed *et al.*, 1997; Naeem and Li, 1997; Tilman, 1997; Wardle *et al.*, 1997; Chapin *et al.*, 1998; Hooper and Vitousek, 1998; Lundberg and Frodin, 1998; Naeem, 1998). The general conclusion of these papers has been that diversity and stability may be related but there are enough problems with experimental designs and semantics that conclusive evidence about the exact nature of the relationship between diversity and stability is still lacking. Some suggest that different diversity hypotheses apply to different types of ecosystems and ecosystem processes (Lawton, 1994; Naeem *et al.*, 1995; Symstad *et al.*, 1998), thereby explaining contradictory results. Another explanation for contradictory results may be as simple as the fact that diversity is too broad a concept to use as a good indicator, but it continues to have political and illustrative appeal and power for that very reason (Angermeier and Karr, 1994; Schläpfer, 1999). Fortunately, the controversy over diversity-stability has not impeded assessing the risks and damages caused by weeds and other invasives and how humans are creating these problems (Lundberg and Frodin, 1998; Dukes and Mooney, 1999; Levine and D'Antonio, 1999; McKinney and Lockwood, 1999; Mack *et al.*, 2000). It has, however, complicated the argument over the mechanisms and consequences of plant invasions (Chapter 13).

The Political Aspects of Diversity

Because there are a lot of hypotheses and analogies but few examples of evidence, diversity has become as much a political as a scientific controversy (Ehrlich and Wilson, 1991; Pitelka, 1993; McFadden and Parker, 1994; Tisdell, 1994). The debate has both scientific and political aspects that are intertwined. The political issue is usually based on the problem that conservation of species diversity can impinge on economic wellbeing and vice versa (Franklin, 1993). In many cases, scientists are asked to defend and define species diversity in terms of benefits to humans (e.g. medicines, aesthetics) rather than actually answer the scientific questions about species diversity and stability (Kunin and Lawton, 1996). In the absence of much evidence, scientific ideas about species diversity can be used to support either side of the political issue. Some argue that since natural selection inevitably eliminates species, human-caused extirpation or extinction is akin to natural selection. Scientists who support the idea that diversity creates stability are not likely to agree with this interpretation, since human-caused extinction outpaces expected extinctions by several orders of magnitude. Most scientists prefer to project themselves as following a version of the medical Hippocratic oath: first do no harm or, in other words, do not knowingly eliminate species. Scientists probably would prefer to see the diversity-stability debate confined to the domain of its origin: testing scientific hypotheses. The problem is that the political controversy is outpacing the ability of science to keep up.

Summary

In Chapters 2–7 we talked about characteristics of individual plant populations. Then, in Chapters 8 and 9 we discussed how populations interact, and in Chapter 10 presented some methods used to study populations and their interactions. In this chapter, we built on these basic concepts to introduce communities. We presented various ways of defining and describing communities, and introduced ideas of how communities change over time.

Diversity is a broad concept that captures the idea that life on earth exhibits a lot of variation. This variation exists at many scales, from genetic through ecosystem. Different measurements attempt to quantify diversity and thereby provide insight into differences between ecological communities. However, our measures all have assumptions that may not always be met and they also should never be used without considering the type (scale) of question being asked. More generally, the entire concept of diversity is problematic because it is difficult to use such a broadly defined term to ask quantitative and qualitative questions like:

- Does more species diversity help ecosystems to function better and last longer?
- Does loss of species diversity by human (or other) disturbance threaten the existence of ecosystems?

For weed ecology, the relevance of these questions may appear indirect but they are actually no less important than the general debate we described in this chapter. For example, if there are many weed species, this may increase diversity – but is this beneficial? The answer might be yes if the increased diversity of weeds is a result of environmental conditions favouring low numbers of each weed species, because they compete with one another or are eaten or parasitized. If this situation prevents any one weed species from outcompeting natives, crops or trees in plantations, then increased species diversity (based on more weed species) is beneficial. In general, however, weeds respond to most disturbances and their ultimate impact is usually to reduce species diversity, at least in non-managed habitats. This may affect ecosystem function and certainly reduces the diversity of native species. Our general conclusion is that diversity does matter in that the evidence supports the idea that ecosystems with more species and more genetic variation are less likely to succumb to a sudden environmental change. The question really is 'how much' diversity is needed for an ecosystem to function over the long term, and this has yet to be answered.

Questions

1. What type of communities does your selected weed species tend to be found in? Is your weed species a keystone species? Has it influenced the biodiversity and/or ecosystem function of any habitats in which it is found? If so, is there an explanation for how it exerts its influence?
2. Do communities really exist?
3. Why does diversity exist?
4. How would a study designed to measure α-diversity differ from one designed to measure β-diversity?
5. Why is it so difficult to determine the role of diversity in ecosystems?
6. If weeds were to increase species diversity in a habitat, would this be interpreted as a benefit to the ecosystem? How would your interpretation differ using the various hypotheses about diversity's role in ecosystems?

General References

Gaston, K. (1996) *Biodiversity: a Biology of Numbers and Difference*. Blackwell Science, Cambridge, Massachusetts.

Power, M.E., Tilman, D., Estes, J.A., Menge, B.A., Bond, W.J., Mills, L.S., Daily, G., Castilla, J.C., Lubchenco, J. and Paine, R.T. (1996) Challenges in the quest for keystones. *BioScience* 46, 609–620.

Walker, B., Kinzig, A. and Langridge, J. (1999) Plant attribute diversity, resilience, and ecosystem function: the nature and significance of dominant and minor species. *Ecosystems* 2, 95-113.

Whittaker, R.J., Willis, K.J. and Field, R. (2001) Scale and species richness: towards a general, hierarchical theory of species diversity. *Journal of Biogeography* 28, 453–470.

Literature Cited

Allen, T.F.H. and Hoekstra, T.W. (1990) The confusion between scale-defined levels and conventional levels of organization in ecology. *Journal of Vegetation Science* 1, 5-12.

Allen, T.F.H. and Hoekstra, T.W. (1991) Role of heterogeneity in scaling of ecological systems under analysis. In: Kolasa, J. and Pickett, S.T.A. (eds) *Ecological Heterogeneity*. Springer-Verlag, New York, pp. 47–68.

Angermeier, P.L. (1994) Does biodiversity include artificial diversity? *Conservation Biology* 8, 600–602.

Angermeier, P.L. and Karr, J.R. (1994) Biological integrity versus biological diversity as policy directives. *BioScience* 44, 690–697.

Barbour, M.G., Burk, J.H., Pitts, W.D., Gilliam, F.S. and Schwartz, M.W. (1999) *Terrestrial Plant Ecology*, 3rd edn. Benjamin/Cummings, Menlo Park, California.

Begon, M., Harper, J.L. and Townsend, C.R. (1990) *Ecology: Individuals, Populations and Communities*. Blackwell Scientific, Oxford.

Berendse, F. (1993) Ecosystem stability, competition and nutrient cycling. In: Schulze, E.D. and Mooney, H.A. (eds) *Biodiversity and Ecosystem Function*. Springer-Verlag, Berlin, pp. 409–431.

Booth, B.D. and Larson, D.W. (1999) Impact of language, history, and choice of system on the study of assembly rules. In: Weiher, E. and Keddy, P. (eds) *Ecological Assembly Rules: Perspectives, Advances, Retreats*. Cambridge University Press, Cambridge, pp. 206–229.

Booth, B.D. and Swanton, C.J. (2002) Assembly theory applied to weed communities. *Weed Science* 50, 2–13.

Brockway, D.G. (1998) Forest plant diversity at local and landscape scales in the Cascade Mountains of southwestern Washington. *Forest Ecology and Management* 109, 323–341.

Brown, J.H. and Heske, E.J. (1990) Control of desert grassland transition by a keystone rodent guild. *Science* 250, 1705–1707.

Brown, J.H. and Munger, J.C. (1985) Experimental manipulation of a desert rodent community – food addition and species removal. *Ecology* 66, 1545–1563.

Brown, J.H., Ernest, S.K.M., Parody, J.M. and Haskell, J.P. (2001) Regulation of diversity: maintenance of species richness in changing environments. *Oecologia* 126, 321–332.

Buckling A., Kassen, R., Bell, G. and Rainey, P.B. (2000) Disturbance and diversity in experimental microcosms. *Nature* 408, 961–964.

Carpenter, S.R. (1996) Microcosm experiments have little relevance for community and ecosystem ecology. *Ecology* 77, 677–680.

Chapin, F.S. III, Sala, O.E., Burke, I.C., Grime, J.P., Hooper, D.U., Laurenroth, W.K., Lombard, A., Mooney, H.A., Mosier, A.R., Naeem, S., Pacala, S.W., Roy, J., Steffen, W.L. and Tilman, D. (1998). Ecosystem consequences of changing biodiversity – experimental evidence and a research agenda for the future. *Bioscience* 48, 45-52.

Clements, D.R., Weise, S.F. and Swanton, C.J. (1994) Integrated weed management and weed species diversity. *Phytoprotection* 75, 1–18.

Clements, F.E. (1916) *Plant Succession*. Publication Number 242. The Carnegie Institution, Washington, DC.

Clements, F.E. (1936) The nature and structure of the climax. *Journal of Ecology* 22, 9–68.

Connell, J. H. (1978) Diversity in tropical rainforests and coral reefs. *Science* 199, 1302–1310.

Conroy, M.J. and Noon, B.R. (1996) Mapping of species richness for conservation of biological diversity: conceptual and methodological issues. *Ecological Applications* 6, 763–773.

Cousins, S.H. (1991) Species diversity measurement: choosing the right index. *Trends in Ecology and Evolution* 6, 190–192.

Currie, D.J. and Paquin, V. (1987) Large-scale biogeographical patterns of species richness of trees. *Nature* 329, 326–327.

Dale, M.B. (1994) Do ecological communities exist? *Journal of Vegetation Science* 5, 285-286.

Davis, M.B. (1987) Invasions of forest communities during the Holocene: beech and hemlock in the Great Lakes region. In: Gray, A.J., Crawley, M.J. and Edwards, P.J. (eds) *Colonization, Succession and Stability*. Blackwell Scientific, Oxford, pp. 373–393.

Delcourt, H.R., Delcourt, P.A. and Webb, T. III (1983) Dynamic plant ecology: the spectrum of vegetation change in space and time. *Quaternary Science Reviews* 1, 153–175.

Díaz, S., Cabido, M. and Casanoves, F. (1998) Plant functional traits and environmental filters on a regional scale. *Journal of Vegetation Science* 9, 113–122.

Drake, J.A. (1990) Communities as assembled structures: do rules govern pattern? *Trends in Ecology and Evolution* 5, 159–164.

Drake, J.A., Hewitt, C.L., Huxel, G.R. and Kolasa, J. (1996) Diversity and higher levels of organization. In: Gaston, K. (ed.) *Biodiversity: a Biology of Numbers and Difference*. Blackwell Science, Cambridge, Massachusetts, pp. 149–168.

Drake, J.A., Huxel, G.R. and Hewitt, C.L. (1996) Microcosms as models for generating and testing community theory. *Ecology* 77, 670–677.

Dukes, J.S. and Mooney, H.A. (1999) Does global change increase the success of biological invaders? *Trends in Ecology and Evolution* 14, 135–139.

Ehrlich, P.R. and Wilson, E.O. (1991) Biodiversity studies: science and policy. *Science* 253, 758–762.

Ehrlich. P.R. and Ehrlich, A.H. (1981) *Extinction: The Causes and Consequences of the Disappearance of Species*. Random House, New York.

Elton, C.S. (1958) *The Ecology of Invasions by Animals and Plants*. Chapman and Hall, London.

Frank, D.A. and McNaughton, S.J. (1991) Stability increases with diversity in plant communities: empirical evidence from the 1988 Yellowstone drought. *Oikos* 62, 360–362.

Franklin, J.F. (1993) Preserving biodiversity: species, ecosystems or landscapes? *Ecological Applications* 3, 202–205.

Fraser, R.H. and Currie, D.J. (1996) The species richness-energy hypothesis in a system where historical factors are thought to prevail: coral reefs. *American Naturalist* 148,138–159.

Gleason, H.A. (1917) The structure and development of the plant association. *Bulletin of the Torrey Botany Club* 44, 463–481.

Gleason, H.A. (1926) The individualistic concept of the plant association. *Bulletin of the Torrey Botany Club* 53, 7–26.

Goetz, D. (2000) Biodiversity of landscape sections: assessing and analyzing gamma-diversity by means of vegetation complexes. *Phytocoenologia* 30, 1–12.

Grime, J.P. (1973) Competitive exclusion in herbaceous vegetation. *Nature* 242, 344–347.

Grime, J.P. (1979) *Plant Strategies and Vegetation Processes*. John Wiley & Sons, New York.

Grubb, P.J. (1985) Plant populations and vegetation in relation to habitat, disturbance and competition: problems of generalization. In: White, J. (ed.) *The Population Structure of Vegetation*. Junk, Dordrecht, pp. 595–621.

Hengeveld, R. (1990) *Dynamic Biogeography*. Cambridge University Press, Cambridge.

Heywood, V.H. (ed.) (1995) *Global Biodiversity Assessment*. United Nations Environment Programme, Cambridge University Press, Cambridge.

Hoekstra, T.W., Allen, T.F.H and Flather, C.H. (1991) Implicit scaling in ecological research. *BioScience* 41, 148–154.

Holdridge, L. (1967) *Life Zone Ecology*. Tropical Science Centre, Costa Rica.

Hooper, D.U. and Vitousek. P.M. (1998) Effects of plant composition and diversity on nutrient cycling. *Ecological Monographs* 68, 121–148.

Horn, H.S. (1975) Markovian properties of forest succession. In: Cody, M.L. and Diamond, J.M. (eds) *Ecology and Evolution of Communities*. Belknap Press, Cambridge, Massachusetts, pp. 196–211.

Huston, M.A. (1997) Hidden treatments in ecological experiments: re-evaluating the ecosystem function of biodiversity. *Oecologia* 110, 449–460.

Johnson, K.H., Vogt, K.A., Clark, H.J., Schmitz, O.J. and Vogt, D.J. (1996) Biodiversity and the productivity and stability of ecosystems. *Trends in Ecology and Evolution* 11, 272–277.

Jordan, N.R., Zhang, J. and Huerd, S. (2000) Arbuscular-mycorrhizal fungi: potential roles in weed management. *Weed Research* 40, 397–410.

King, A.W. and Pimm, S.L. (1983) Complexity, diversity, and stability: a reconciliation of theoretical and empirical results. *American Naturalist* 122, 229–239.

Kunin, W.E. and Lawton, J.H. (1996) Does biodiversity matter? Evaluating the case for conserving species. In: Gaston, K. (ed.) *Biodiversity: a Biology of Numbers and Difference.* Blackwell Science, Cambridge, Massachusetts, pp. 283–308

Lawton, J.H. (1994) What do species do in ecosystems? *Oikos* 71, 367–374.

Levin, S.A. (1986) Pattern, scale, and variability: an ecological perspective. In: Hastings, A. (ed.) *Community Ecology.* Springer-Verlag, Berlin, pp. 1–13.

Levine, J.M. and D'Antonio, C.M. (1999) Elton revisited: a review of evidence linking diversity and invasibility. *Oikos* 87, 15–26.

Loreau, M. (2000) Are communities saturated? On the relationship between alpha, beta and gamma diversity. *Ecology Letters* 3, 73–76.

Lovejoy, T.E. (1994) The quantification of biodiversity: an esoteric quest or a vital component of sustainable development? *Philosophical Transactions of the Royal Society of London, Series B* 345, 81–87.

Lundberg, P. and Frodin, P. (1998) Ecosystem resilience and productivity: are predictions possible? *Oikos* 81, 603–606.

MacArthur, R. (1955) Fluctuations of animal populations and a measure of community stability. *Ecology* 36, 533–536.

MacArthur, R.H. (1957) On the relative abundance of bird species. *Proceedings of the National Academy of Sciences of the USA* 43, 293–295.

Mack, R.N., Simberloff, D., Lonsdale, W.M., Evans, H., Clout, M. and Bazzaz, F.A. (2000) Biotic invasions: causes, epidemiology, global consequences, and control. *Ecological Applications* 10, 689–710.

Magnussen, S. and Boyle, T.J.B. (1995) Estimating sample size for inference about the Shannon-Weaver and the Simpson indices of species diversity. *Forest Ecology and Management* 78, 71–84.

Magurran, A.E. (1988) *Ecological Diversity and its Measurement.* Helm, London.

Martinez, N.D. (1996) Defining and measuring functional aspects of biodiversity. In: Gaston, K.J. (ed.) *Biodiversity: a Science of Numbers and Difference.* Blackwell Scientific, Oxford, pp. 114–148.

May, R.M. (1974) *Stability and Complexity in Model Ecosystems,* 2nd Edition. Princeton University Press, Princeton, New Jersey.

McFadden, M.W. and Parker, K. (1994) Human values and biological diversity: are we wasting our time? *The Canadian Entomologist* 126, 471–474.

McGrady-Steed, J., Harris, P.M. and Morin, P.J. (1997) Biodiversity regulates ecosystem predictability. *Nature* 390, 162–195.

McKinney, M.L. and Lockwood, J.L. (1999) Biotic homogenization: a few winners replacing many losers in the next mass extinction. *Trends in Ecology and Evolution* 14, 450–453.

McNaughton, S.J. (1983) Serengeti grassland ecology: the role of composite environmental factors and contingency in community organization. *Ecological Monographs* 53, 291–320.

McNaughton, S.J. (1985) Ecology of a grazing ecosystem: the Serengeti. *Ecological Monographs* 55, 259–294.

McNaughton, S.J. (1988) Diversity and stability. *Nature* 333, 204–205.

Menge, B.A. and Olson, A.M. (1990) Role of scale and environmental factors in regulation of community structure. *Trends in Ecology and Evolution* 5, 52–57.

Morin, P.J. (1999) *Community Ecology.* Blackwell Science, Malden, Massachusetts.

Naeem, S. (1998) Species redundancy and ecosystem reliability. *Conservation Biology* 12, 39–45.

Naeem, S. and Li, S. (1997) Biodiversity enhances ecosystem reliability. *Nature* 390, 507–509.

Naeem, S., Thompson, L.J., Lawler, S.P., Lawton, J.H. and Woodfin, R.M. (1994) Declining biodiversity can alter the performance of ecosystems. *Nature* 368, 734–738.

Naeem, S., Thompson, L.J., Lawler, S.P., Lawton, J.H. and Woodfin, R.M. (1995) Empirical evidence that declining species diversity may alter the performance of terrestrial ecosystems. *Philosophical Transactions of the Royal Society of London, Series B* 347, 249–262.

Naeem, S., Tjossem, S.F., Byers, D., Bristow, C. and Li, S.B. (1999) Plant neighbourhood diversity and production. *Ecoscience* 6, 355–365.

O'Brian, E.M. (1998) Water–energy dynamics, climate, and prediction of woody plant species richness: an interim general model. *Journal of Biogeography* 25, 379–398.
O'Brian, E.M., Whittaker, R.J. and Field, R. (1998) Climate and woody plant diversity in southern Africa: relationships at species genus and family levels. *Ecography* 21, 495–509.
Paine, R.T. (1966) Food web complexity and species diversity. *American Naturalist* 100, 65–75.
Paine R.T. (1969) A note on trophic complexity and community stability. *American Naturalist* 103, 91–93.
Petratis, P.S., Latham, R.E. and Niesenbaum, R.A. (1989) The maintenance of species diversity by disturbance. *Quarterly Review of Biology* 64, 393–418.
Pimm, S.L. (1991) *The Balance of Nature? Ecological Issues in the Conservation of Species and Communities.* University of Chicago Press, Chicago, Illinois.
Pitelka, L.F. (1993) Biodiversity and policy decisions. In: Schulze, E.D. and Mooney, H.A. (eds) *Biodiversity and Ecosystem Function.* Springer-Verlag, Berlin, pp. 481–493.
Power, M.E., Tilman, D., Estes, J.A., Menge, B.A., Bond, W.J., Mills, L.S., Daily, G., Castilla, J.C., Lubchenco, J. and Paine, R.T. (1996) Challenges in the quest for keystones. *BioScience* 46, 609–620.
Preston, F.W. (1962) The canonical distribution of commonness and rarity: parts 1 and 2. *Ecology* 43, 185–215, 421–432.
Schlesinger, R.C., Funk, D.T., Roth, P.L. and Myers, C.C. (1994) Assessing changes in biological diversity over time. *Natural Areas Journal* 14, 235–240.
Schläpfer, F. (1999) Expert estimates about effects of biodiversity on ecosystem processes and services. *Oikos* 84, 346–352.
Schulze, E.D. and Mooney, H.A. (1993) Ecosystem function of biodiversity: a summary. In: Schulze, E.D. and Mooney, H.A. (eds) *Biodiversity and Ecosystem Function.* Springer-Verlag, Berlin, pp. 497–510.
Smith, M.D., Hartnett, D.C. and Wilson, G.W.T. (1999) Interacting influence of mycorrhizal symbiosis and competition on plant diversity in tallgrass prairie. *Oecologia* 121, 574–582.
Smith, R.L. and Smith, T.M. (2001) *Ecology and Field Biology*, 6th edn. Benjamin Cummings, San Francisco, California.
Stiling, P.D. (1999) *Ecology: Theories and Applications.* Prentice Hall, Upper Saddle River, New Jersey.
Symstad, A.J., Tilman, D., Willson, J. and Knops, J.M.H. (1998) Species loss and ecosystem functioning: effects of species identity and community composition. *Oikos* 81, 389–397.
Tilman, D. (1997) Distinguishing between the effects of species diversity and species competition. *Oikos* 80, 185.
Tilman, D. and Downing, J.A. (1994) Biodiversity and stability in grasslands. *Nature* 367, 363–365.
Tilman, D., Wedin, D. and Knops, J. (1996) Productivity and sustainability influenced by biodiversity in grassland ecosystems. *Nature* 379, 718–720.
Tilman, D., Knops, J., Wedin, D., Reich, P., Ritchie, M. and Siemann, E. (1997) The influence of functional diversity and composition on ecosystem processes. *Science* 177, 1300–1302.
Tisdell, C. (1994) Conservation, protected areas and the global economic system: how debt, trade, exchange rates, inflation, and macroeconomic policy affect biological diversity. *Biodiversity and Conservation* 3, 419–436.
van de Maarel, E. (1993) Some remarks on disturbance and its relations to diversity and stability. *Journal of Vegetation Science* 4, 733–736.
Vitousek, P.M. and Hooper, D.U. (1993) Biological diversity and terrestrial ecosystem biogeochemistry. In: Schulze, E.-D. and Mooney, H.A. (eds) *Biodiversity and Ecosystem Function.* Springer-Verlag, Berlin, pp. 3–14.
Walker, B.H. (1992) Biodiversity and ecological redundancy. *Conservation Biology* 6, 18–23.
Walker, L.R. and Vitousek, P.M. (1991) An invader alters germination and growth of a native dominant tree in Hawaii. *Ecology* 72, 1449–1455.
Wardle, D.A., Bonner, K.I. and Nicholson, K.S. (1997) Biodiversity and plant litter: experimental evidence which does not support the view that enhanced species richness improves ecosystem function. *Oikos* 79, 247–258.
Watson, D.M. (2001) Mistletoe – a keystone resource in forests and woodlands worldwide. *Annual Review of Ecology and Systematics* 32, 219–249.
Whittaker, R.J. (1975) *Communities and Ecosystems,* 2nd edn. MacMillan and Company, New York.
Whittaker, R.J., Willis, K.J. and Field, R. (2001) Scale and species richness: towards a general, hierarchical theory of species diversity. *Journal of Biogeography* 28, 453–470.

Wilkinson, D.M. (1999) The disturbing history of intermediate disturbance. *Oikos* 84, 145–147.
Wilson, E.O. (1985) The biological diversity crisis. *BioScience* 35, 700–706.
Wilson, J.B. (1991) Does vegetation science exist? *Journal of Vegetation Science* 2, 189–190.
Wilson, J.B. (1994) Who makes the assembly rules? *Journal of Vegetation Science* 5, 275–278.
Wilson, J.B., Steel, J.B., King, W.M. and Gitay, H. (1999) The effect of spatial scale on evenness. *Journal of Vegetation Science* 10, 463–468.
Woodward, F.I. and Diament, A.D. (1991) Functional approaches to predicting the ecological effects of global change. *Functional Ecology* 5, 202–212.
Yorks, T.E. and Dabydeen, S. (1998) Modification of the Whittaker Sampling Technique to assess plant diversity in forested areas. *Natural Areas Journal* 18, 185–189.

12
Community Dynamics: Succession and Assembly

Concepts

- Communities are dynamic over time; their composition changes over days, years or centuries.
- Primary succession occurs on newly created land, whereas secondary succession occurs after a disturbance removes only part of the plant biomass and substrate.
- Some processes that influence the type and rate of succession are facilitation, tolerance and inhibition. Succession is directed by site availability, species availability and species performance.
- Communities are assembled over time. Membership of a community will be determined by the interaction of species traits and the biotic and abiotic filters of the habitat.
- Species present in a community (actual species pool) must be members of the total, habitat, geographical and ecological species pools.

Introduction

In Chapter 11, we focused on how communities 'look' at one point in time; that is community structure as related mainly to diversity. The composition of any community will fluctuate daily, seasonally, annually, and will change over decades, centuries and millennia (Fig. 12.1). Over a year, communities change visibly as a result of phenology and short-term weather patterns. Even when we try to maintain a stable community, such as in a perennial garden, composition is dynamic. Year-to-year species composition changes in response to environmental factors. A wet year may favour some species, increasing their abundance over another species. Even the most fastidious of gardeners will have different species abundances from year-to-year and some species will be inadvertently lost or introduced over time. Weed communities in agricultural systems will change over time even when the cropping and tillage systems are consistent. Such short-term fluctuations in community composition may or may not result in long-term compositional changes.

Over the long term (centuries to millennia), communities change as a result of selection pressures. Over intermediate time

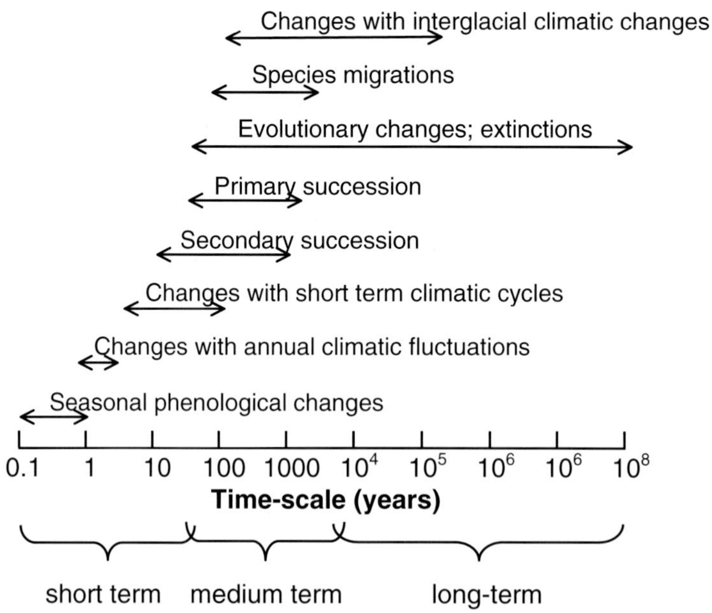

Fig. 12.1. Vegetation change over different time scales (redrawn from Miles, 1987).

scales (decades to centuries), there are changes in vegetation that we call succession. Succession is the directional change in community composition and is different from seasonal or random fluctuations in vegetation. This intermediate scale of vegetational change is the focus of this chapter. In some cases we can observe a community change over our life span. For example, we can watch an abandoned field proceed from a community of annual and perennial weeds during our childhood, to one with scattered shrubs in our adulthood, and possibly to the initial stages of a forest in our old age. In other cases, the development of vegetation and its change over time is almost imperceptible; for example, succession following the retreat of a glacier is very slow (Fastie, 1995).

Community Dynamics

The classical ecological theory assumed that natural communities existed in a state of 'dynamic equilibrium'. Communities were said to develop over time into a specific stable community type (the climax community). Equilibrium communities were thought to be controlled primarily by competition, and species co-existence was thought to be dependent on niche differentiation and resource partitioning. Implicit in this was the idea that following a disturbance, a community would return to its original state (Perrings and Walker, 1995). These ideas, however, do not fit with current empirical evidence (Pickett *et al.*, 1992; Holling *et al.*, 1995). For example, we now know that following a disturbance, different types of communities can develop and therefore there is no such thing as a single climax community (Walker, 1981; McCune and Allen, 1985; Dublin *et al.*, 1990). Abrams *et al.* (1985), for example, showed that clear-cut jack pine (*Pinus banksiana*) stands can develop into either sedge meadows, hardwood and shrub communities, or return to a jack pine community depending on the season of cutting and whether the site was burned by natural or controlled burns (Fig. 12.2).

The theory of communities reaching equilibrium has now been replaced by the

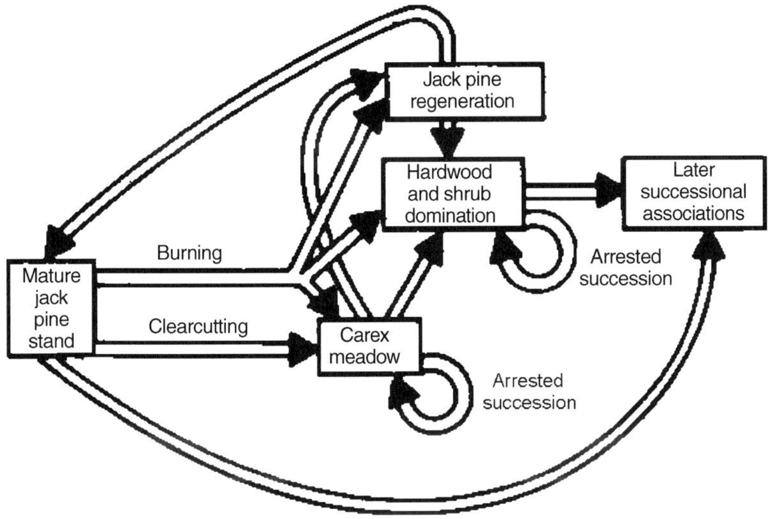

Fig. 12.2. Multiple successional pathways following burning or clearcutting of a mature jack pine forest (Abrams *et al.*, 1985; Copyright 1985, with permission from Elsevier Science).

non-equilibrium concepts that focus on community-level processes and changes over time rather than on any single climax community state (DeAngelis and Waterhouse, 1987). According to Pickett *et al.* (1994), the 'balance [equilibrium] of nature' is a cultural myth or metaphor rather than a scientific concept. Non-equilibrium concepts recognize that while some communities may be at equilibrium at some scales, this is not necessarily the 'normal' situation. Pickett *et al.* (1992) suggested that we use the 'flux of nature' as our metaphor rather than considering nature to be in balance.

Once we accept that communities are not necessarily at equilibrium we can begin to describe how they change over time. The term *stability* is often used to describe how communities resist change in response to disturbance or stress, but it is a vague term and has been defined in many ways (Lěps *et al.*, 1982; Pimm, 1991). Community stability is broken into three components: persistence, resistance and resilience (Table 12.1). Thus, a community's stability is a function of how long it remains the same (persistence), whether it resists stress or disturbance (resistance), and its ability to return to its original state following a disturbance (resilience). The combination of these three components determines stability; however, a community very strong in one component is not necessarily strong in the others. For

Table 12.1. Terms associated with descriptions of community stability.

Term	Definition	Source
Persistence	'The ability of a community to remain relatively unchanged over time'	Barbour *et al.* (1999)
Resistance	'The ability of a community to remain unchanged during a period of stress'	Barbour *et al.* (1999)
Resilience	'The ability of a community to return to its original state following stress or disturbance'	Barbour *et al.* (1999)
Elasticity	'The speed at which the system returns to its former state following a perturbation'	Putman (1994)

example, a community may be persistent but not resistant or resilient, and therefore will be susceptible to disturbance. A community may be highly resilient under a certain magnitude of disturbance, but once a threshold is passed, a transition will occur and an alternative state will be reached with either the same or new species composition. Such changes in community structure and function can be irreversible and quite abrupt (Perrings and Walker, 1995).

History and Development of Successional Theory

The term succession was originally used by Thoreau in 1860 to describe changes in forest trees (McIntosh, 1999). It remained largely unused until Cowles (1899) studied primary succession on the sand dunes of Lake Michigan near Chicago. He described how they developed through various associations to a forest climax. Cowles (1901) recognized that succession is 'not a straight-line process. Its stages may be slow or rapid, direct or tortuous and often they are retrogressive.'

Clements (1916, 1936), a contemporary of Cowles, was a more forceful individual and therefore his rather dogmatic writing on succession overshadowed those of Cowles (McIntosh, 1999). Clements described in detail what we now think of as the classical succession paradigm. He described it as a directional, progressive, orderly change in vegetation that would ultimately converge to a stable, predictable climax community. Clements believed that the vegetation in each successional pathway altered the environment and ameliorated it for later invading species. Thus, early species facilitated the invasion of later species. From this, Egler (1954) later developed the idea of 'relay floristics' where early species prepared the environment for later ones. Thus, Clements' view of succession proposed that autogenic (internal) processes controlled the development of the community climax. Clements believed that communities were more than the sum of their species and that they had properties of a superorganism (Chapter 11).

Two criticisms arose about Clements' ideas. First, Clements invoked climate as the sole determinant of community composition and neglected other factors. As we have seen in earlier chapters, other biotic and abiotic factors can be important community determinants. Second, Clements did not recognize the possibility of multiple successional pathways, an idea already hinted at by Cowles. Nevertheless, Clements' work is important for his observations of community dynamics and his recognition of the importance of disturbance to the process of succession (Walker, 1999).

In response to Clements' work, Gleason (1917, 1926) noted that successional events were *not* predictable and that succession proceeded *independently* following a disturbance. He believed that communities were simply collections of species with similar environmental tolerances. One criticism of Gleason's work was that he ignored the importance of species interactions in determining community composition, instead focussing almost entirely on abiotic processes (Tansely, 1935).

Gleason's views were greeted sceptically at first, and not taken seriously until decades later in the 1950s. By then ecologists had begun to recognize the work of Watt (1947) and others, who said that communities were a mosaic of patches at different successional stages. While overall community structure might remain constant, individual patches were dynamic. The differences between Clements and Gleason are not as extreme as they first appear. Clements looked at overall change in species composition over a pathway, whereas Gleason looked at smaller-scale changes in individual patches. Therefore, Clements observed temporal changes, and Gleason, spatial patterns.

Patterns of Primary and Secondary Succession

There are two main types of succession: primary and secondary. Primary succession occurs on newly created land (e.g. following a volcanic eruption), whereas secondary

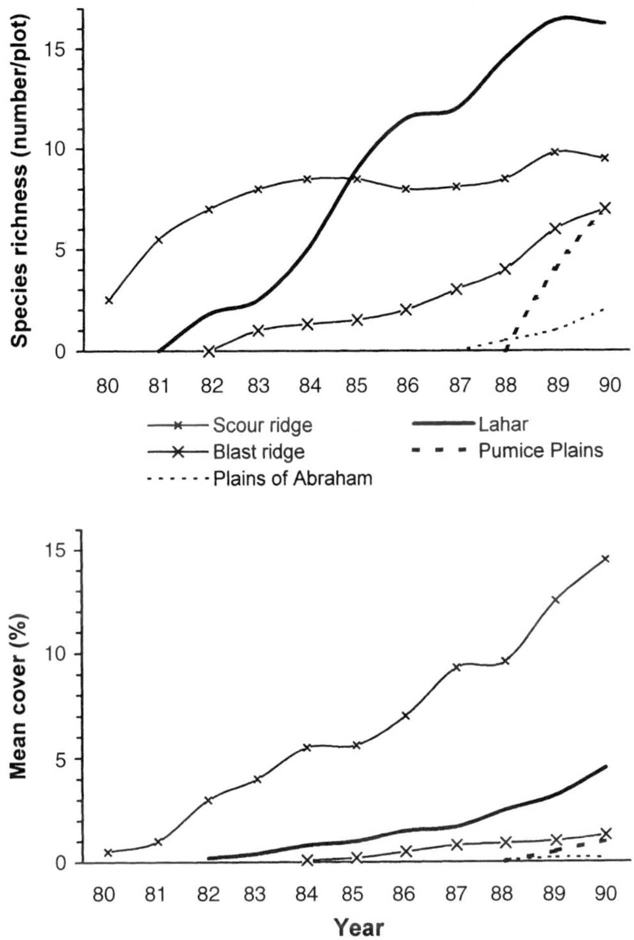

Fig. 12.3. Change in: (a) species richness (number in 250-m^2 plots), and (b) percentage cover during primary succession at five sites on Mount St Helens (redrawn from del Moral and Bliss, 1993).

succession occurs after a disturbance removes only part of the plant biomass and substrate (e.g. following a forest fire).

Primary succession

Primary succession occurs on newly created land where no plants have grown previously or where there is no effective seed bank on site. Habitats undergoing primary succession are usually environmentally harsh because there is no vegetation to ameliorate the abiotic environment and there tends to be few nutrients and little water. Primary succession is a slow process because there is no soil structure or remnants of vegetation to aid regeneration.

del Moral and Bliss (1993) compared the rate of primary succession in five habitat types following the eruption of Mount St Helens in 1980 (Fig. 12.3). They found that the type of soil substrate and vegetation that remained influenced the rate and types of successional pathway. Species richness (α-diversity) increased faster when some vegetation remained or was nearby (on the scour and lahar). After 10 years, the mean percentage cover was low at all sites except on the scour habitat where some original vege-

Table 12.2. Five dominant species in the major study areas of del Moral and Bliss (1993). Species are ranked according to percentage cover. The Scoured Ridge omitted and intact late successional vegetation site added for comparison (from del Moral and Bliss, 1993).

	Sites				
Species	Secondary succession, up to 15 cm of ash (tephra) deposited on vegetation	Lahar, mud flow leaving barren substrate	Blasted ridge, lava, ash and residual soil remaining	Pumice plains, heterogeneous pumice desert	Abraham pumice, pumice of 2–3 cm diameter, desert pavement
Thin bentgrass, *Agrostis diegoensis*	1				5
Dwarf lupin, *Lupinus lepidus*	2	2	1		
Spreading phlox, *Phlox diffusa*	3				
Newberry's fleeceflower, *Polygonum newberryi*	4	1			
Pink mountain-heather, *Phyllodoce empetriformis*	5				
Cardwell's penstemon, *Penstemon cardwellii*		3			
Common catsear, *Hypochaeris radicata*		4	4	4	3
Alpine buckwheat, *Eriogonum pyrolifolium*		5			
Western pearly everlasting, *Anaphalis margaritaceum*			2	1	1
Fireweed, *Epilobium angustifolium*			3	2	2
White hawkweed, *Hieracium albiflorum*			5	3	4
Rusty saxifrage, *Saxifraga ferruginea*				5	

tation and soil remained. Even the lahars, which showed increasing richness over time still had only 5% cover. Species composition also differed among the habitat types (Table 12.2). Species composition in three habitats (ridge, pumice and plains) was similar, but these differed substantially from the relatively unaffected tephra habitat. Communities established on mud flows (lahar) were intermediate in species composition reflecting its proximity to propagules. Even though the lahar and tephra vegetation had similar species composition, their structure varied considerably as the mean percentage cover for these sites in 1990 was 4% and 47%, respectively.

del Moral and Bliss (1993) noted that the characteristics of early invading species may determine how succession proceeds because once a species is established, its own seed production will outnumber propagules of other species that immigrate from other communities. Furthermore, early invaders tended to be located in the more favourable microhabitats, thus improving their own survival. Finally, environmental conditions during the establishment phase may favour some species over others, further directing the successional pathway.

Secondary succession

Secondary succession is the most commonly observed type of succession. It is usually initiated by a natural or human-caused disturbance; for example, fire, hurricanes and flooding initiate secondary succession. Cessation of human activities can also initiate secondary succession, such as when an agricultural field is abandoned. In all cases, there are existing vegetation or seed banks available to aid regeneration.

The rate at which secondary succession proceeds is dependent on the type of soil substrate remaining and on whether established vegetation is nearby and can provide propagules for regeneration. The rate is also dependent on the type of agent initiating succession. For example, a hurricane, clear-cut forest or small brush fire can initiate secondary succession, but the vegetation will develop at different rates.

Old field succession is probably the most studied type of secondary succession. It follows the cessation of farming activity such as ploughing and herbicide usage. A classic example of secondary succession is described in Brown and Southwood (1987) who characterized the succession of old fields on sandy soils at Silwood Park in southern Britain. When ploughing stopped, annuals were the first to colonize. Biennials briefly dominated in the second year, followed by an increase in perennials. These changes were reflected in the rapid increase in α-diversity over the first year followed by a gradual decline as perennials and then shrubs and trees dominated (Fig. 12.4).

Patterns and Processes of Succession

Patterns of succession

We have looked at specific examples of primary and secondary succession, but what generalizations can we make about communities undergoing succession, and what ecological processes are important over the course of a successional pathway? As communities develop, the types of plants growing there will change, often in a predictable manner. There are some general patterns as succession proceeds. First, plant cover, biomass and species richness tend to increase over time, at least to a point after which they may level off or decrease. The most obvious

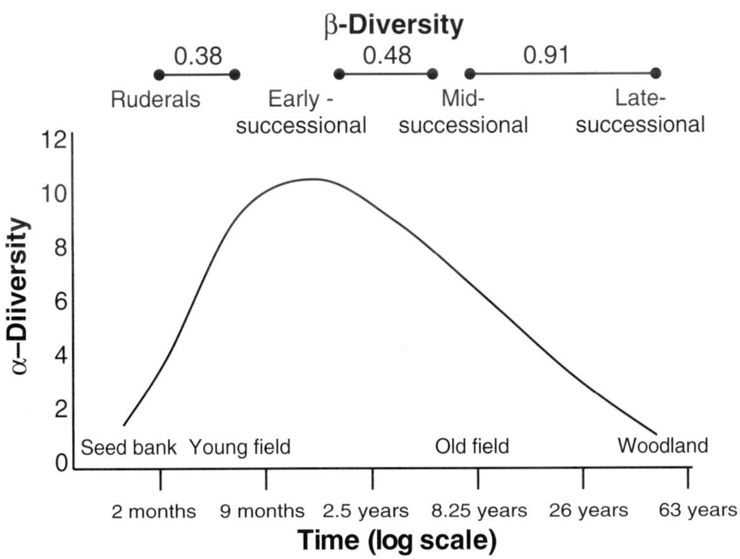

Fig. 12.4. Changes in α-diversity of green plants along a secondary successional pathway. Differences between stages in terms of β-diversity are shown along the top (redrawn from Brown and Southwood, 1987).

Table 12.3. Comparison of physiological and life history traits and population dynamics of plants from early and late stages of successional (from Pianka, 1970; Bazzaz, 1979; Huston and Smith, 1987).

Trait	Early succession	Late succession
Seed and dispersal		
Seed size and number	Many, small seeds	Few, large seeds
Dispersal distance	Long	Short
Dispersal mechanisms	Wind, birds, bats	Gravity, mammals
Seed viability	Long	Short
Size at maturity	Small	Large
Maximum life span	Short (often < 1 year)	Long (usually > 1 year)
Timing and frequency of reproductive events	Early, often monocarpic	Late, usually polycarpic
Growth rate	Fast	Slow
Structural strength	Low	High
Survivorship	Often Deevy Type III	Often Deevy Type I and II
Population size	Often variable over time	Fairly constant over time
Resource acquisition rate	High	Often low
Recovery from nutrient stress	Fast	Slow
Root to shoot ratio	Low	High
Photosynthetic rate	High	Low
Photosynthetic rate at low light	Low	High
Respiration rate	High	Low

change is the increase in canopy height from small annuals and herbaceous perennials to shrubs and trees during many successional pathways in temperate and tropical environments.

Huston and Smith (1987) summarized plant traits associated with early and late successional species (Table 12.3). These traits have been equated with *r*- and K-selected species (see also Chapter 3); however, Grubb (1987) cautioned against this. The high intrinsic rate of population increase (*r*) associated with early successional species can be achieved both by *r*- and K-selected species. A high *r* results from high fecundity in *r*-selected species and from low mortality in K-selected species (Grubb, 1987). During early stages of succession there is a high rate of species change and replacement, but as the community ages, species turnover rate declines. This is due primarily to the rapid growth rate and short life span of early successional species compared to late successional species.

Grime (1977) illustrated patterns of succession on his C-S-R triangle (Fig. 12.5a). In early secondary succession, species are ruderals (R) because there is abundant light and nutrients. The successional pathway favours competitive species (C) and then long-lived stress-tolerators (S) as nutrients and light become limiting. The direction of the pathway will vary with the level and consistency of potential productivity (nutrient status). Higher potential productivity will lead to species that are competitive in the intermediate stages. Potential productivity rarely stays constant throughout a pathway. Nutrients can also be added or removed through human actions. Using this model, successional pathways of specific habitats can also be mapped; for example, in a fertile forest, the pathway favours competitive species for approximately 25 years, and then favours toward stress-tolerators as the biomass continues to accumulate (Fig. 12.5b). Conversely, in an annually harvested meadow, the pathway begins with competitive species and the ruderals as biomass decreases.

This way of illustrating succession is unlikely to fit all situations. Ecke and Rydin (2000) examined the C-S-R model and found that primary succession on uplifted sea coast meadows did not follow the expected trend.

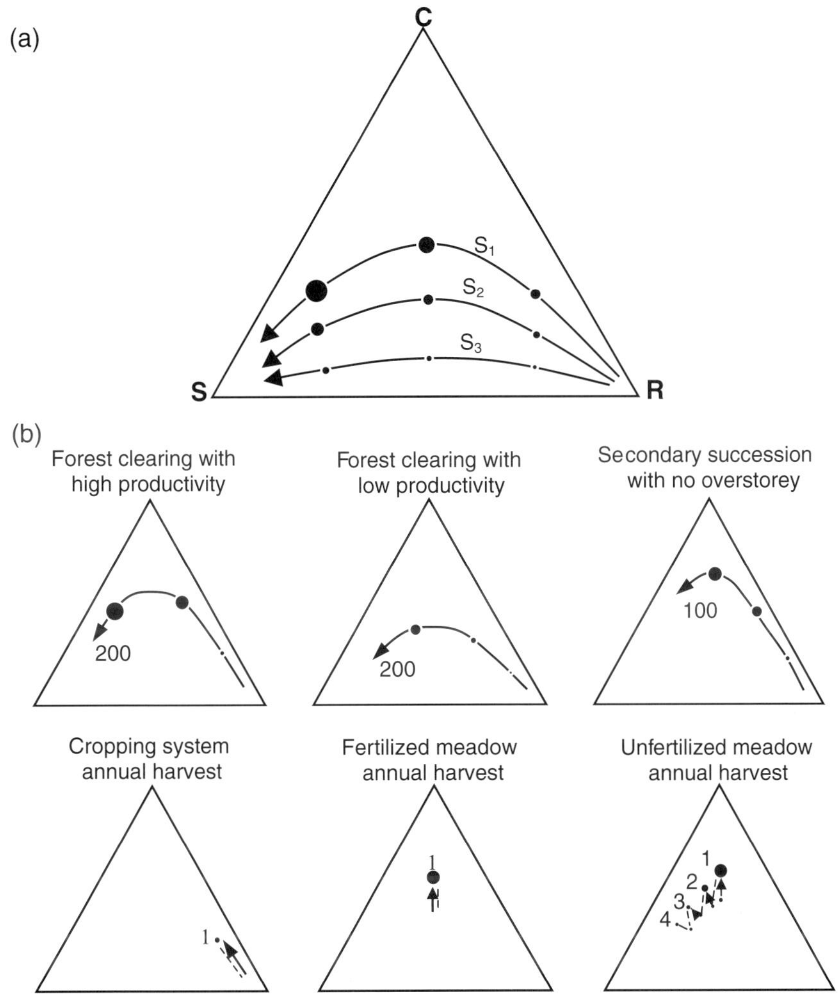

Fig. 12.5. Succession represented on Grime's C-S-R triangle showing: (a) general successional patterns under conditions of high (S_1) medium (S_2) and low (S_3) potential productivity, and (b) successional pathways of specific habitat types. Size of circles indicates the amount of plant biomasss present at that stage of succession. Numbers indicate the approximate time (years) in the successional pathway (redrawn from Grime 1977, 1987).

Here they found that ruderals were not able to colonize the disturbed sites, but that species tolerant of disturbance were dominant in early succession. This is a strategy not considered in the C-S-R model (Steneck and Dethier, 1994; Ecke and Rydin, 2000). Thus we can gain general insight using the C-S-R model but it may not be applicable in all situations.

Processes of succession: facilitation, inhibition and tolerance

In 1954, Egler proposed two models to explain succession. The 'initial floristics model' described how most species were present at the initial stages of succession but that the later successional species became more prominent over time as a result of longer lifespan, slower growth rate and larg-

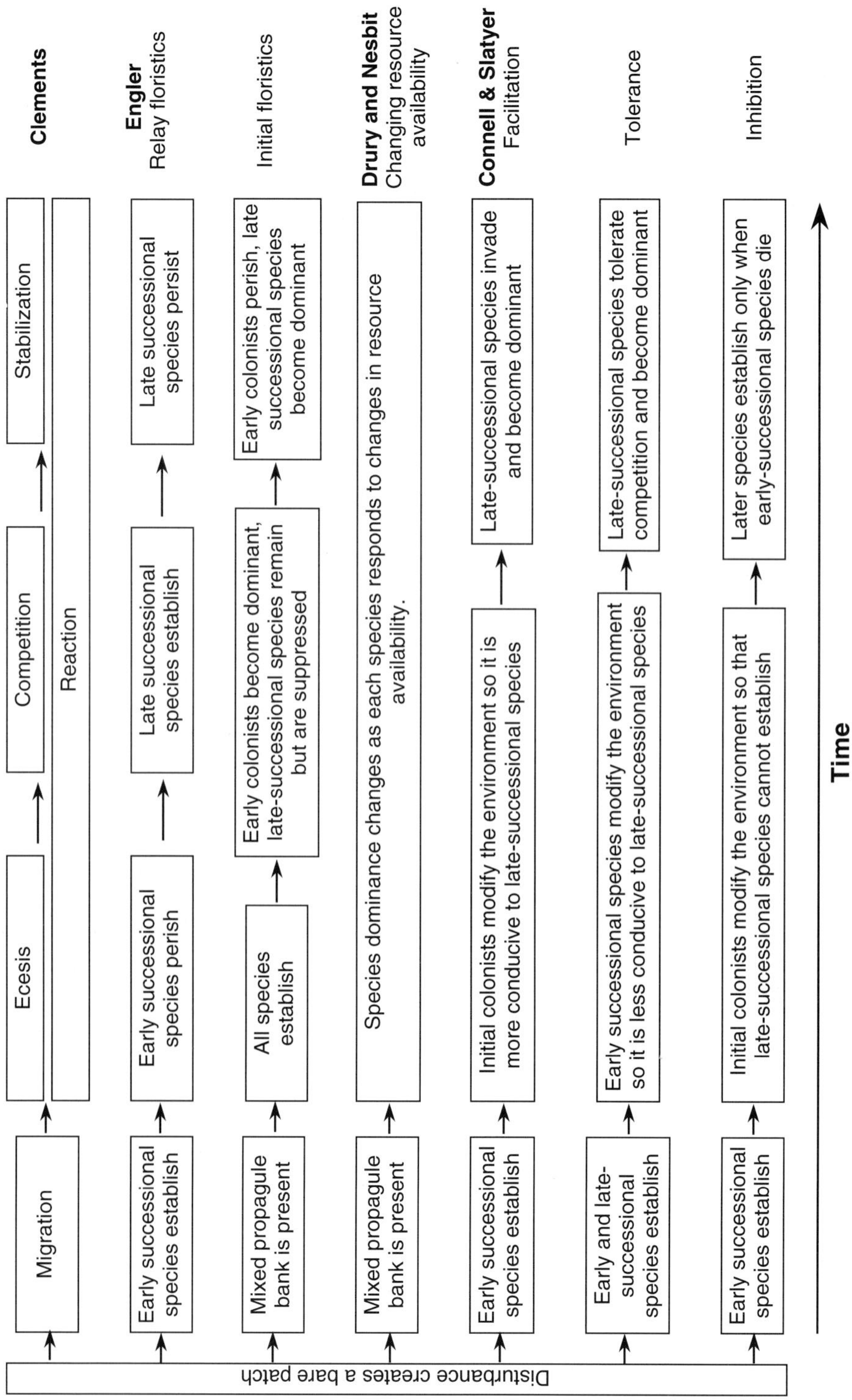

Fig. 12.6. Models of succession (Luken, 1990).

er size at maturity (Fig. 12.6). Drury and Nisbet (1973) expanded on this idea and tested whether physical stress and competition might also be important processes in succession. Second, Egler's 'relay floristics model' proposed that species prepare the environment for later appearing species, making it inhospitable to themselves.

In 1977, Connell and Slatyer further developed some of these ideas and presented three models to explain the mechanisms of succession: facilitation, inhibition and tolerance (Fig. 12.6). These models are still used as a basis to describe the processes underlying succession. These processes interact and the development of most pathways is a combination of all three. Furthermore, the relative importance of the three processes will change over time.

Facilitation

This model develops Clements' original ideas that early species facilitated the invasion of later ones. For example, they may ameliorate temperatures by providing shade, or may increase soil nitrogen by fixing atmospheric nitrogen. In this way, late invading species cannot appear until after the environment is 'improved' by earlier species. There are many examples of facilitation in natural communities; however, these occur mostly in primary succession where early environments are severe (Walker and Chapin, 1987). For example, after Mount St Helens erupted, nitrogen-fixing lupins (*Lupinus lepidus*) were among the first species to colonize. When they died, other species could then take advantage of the released nitrogen (del Moral and Bliss, 1993).

Early accounts of primary succession often reported the importance of lichens, moss and algae in facilitating the establishment of pioneer vascular plants (Griggs, 1933; Booth, 1941) because they were thought to build soil and create microhabitats for vascular plants. In fact, this is not necessarily the case. On Mount St Helens, for example, lichens, moss and algae did not establish in many habitats because of the dry summers and porous rock, and therefore they did not play a role in primary succession (del Moral and Bliss, 1993). In other situations, moss and lichens may actually inhibit the succession of vascular plants on volcanic soils (Frezen *et al.*, 1988).

Inhibition

Inhibition occurs when existing plants prevent or inhibit the establishment of subsequent species. This is caused by a combination of physical, chemical or biotic means. Usually, the early plant becomes established in a site and pre-empts biological space. There are numerous examples of inhibition in succession. In fact the lupins, mentioned in the above section on facilitation, inhibited establishment of other individuals while they were alive. It was only after their death that lupins facilitated the establishment of others.

The inhibitory effect of a species can last long after the individual has died. Litter, for example, can have an inhibitory effect that favours some species over others. In an old field succession, for example, the litter of giant foxtail (*Setaria faberii*) is a physical and chemical deterrent to annual fleabane (*Erigeron annuis*) (Facelli and Facelli, 1993). When the litter is removed, annual fleabane is able to establish and then reduces the growth of giant foxtail. The process of inhibition tends to favour early colonists over late colonists, thus dispersal ability and getting established first are very important.

Tolerance

Connell and Slatyer (1977) viewed facilitation and inhibition as extremes on a continuum, and placed tolerance in the centre. Under this model, early species have no effect on subsequent ones. The timing and success of each species' colonization is determined by its tolerance to environmental conditions; for example, late successional, long-lived species are more likely to tolerate shade than early successional species.

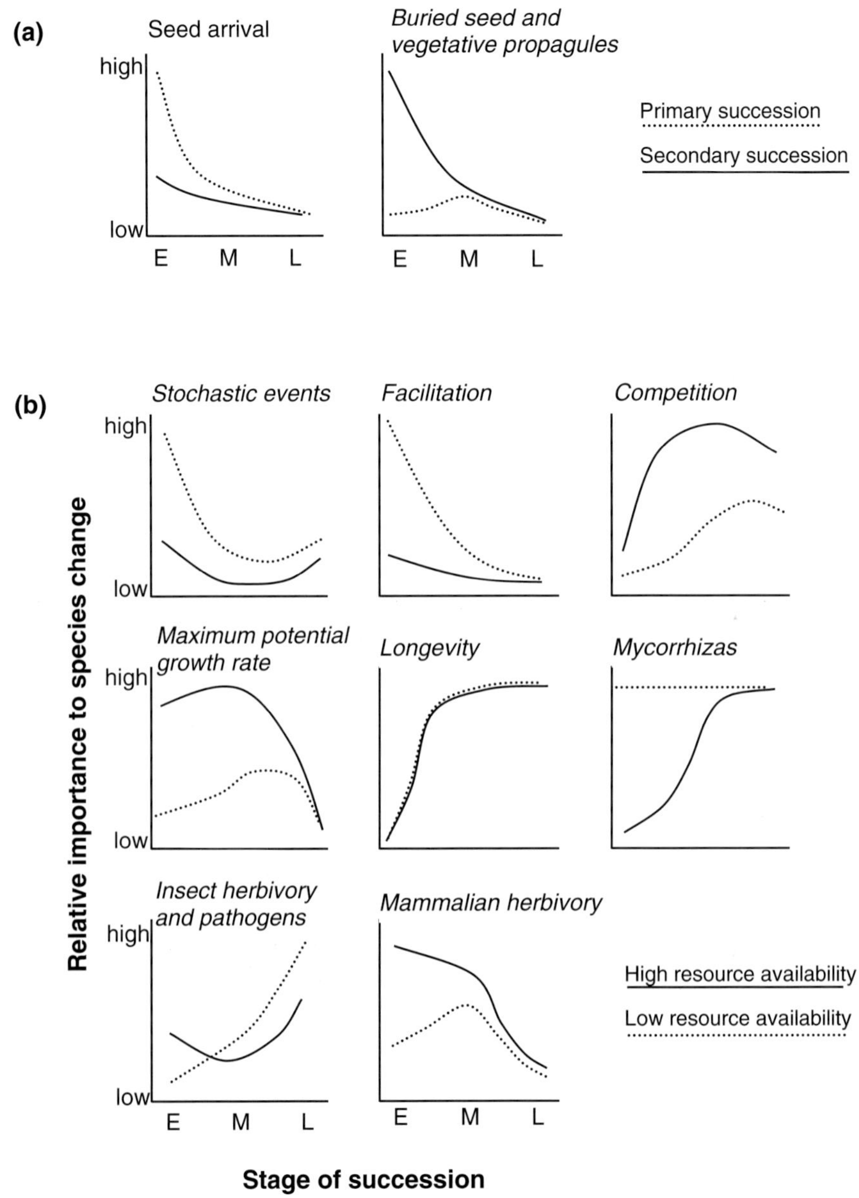

Fig. 12.7. Influence of the: (a) type of succession, and (b) level of resource availability on successional process over the course of early (E), middle (M), and late (L) succession (redrawn from Walker and Chapin, 1987).

Effect of environmental severity on successional processes

The types of processes that are important in determining species change will change with the stage of succession (early, mid- and late), the types of succession (primary vs. secondary) (Fig. 12.7a) and the level of available resources (water and nutrients) (Fig. 12.7b) (Walker and Chapin, 1987). For example, seed arrival is highly important in the early stages of primary succession because

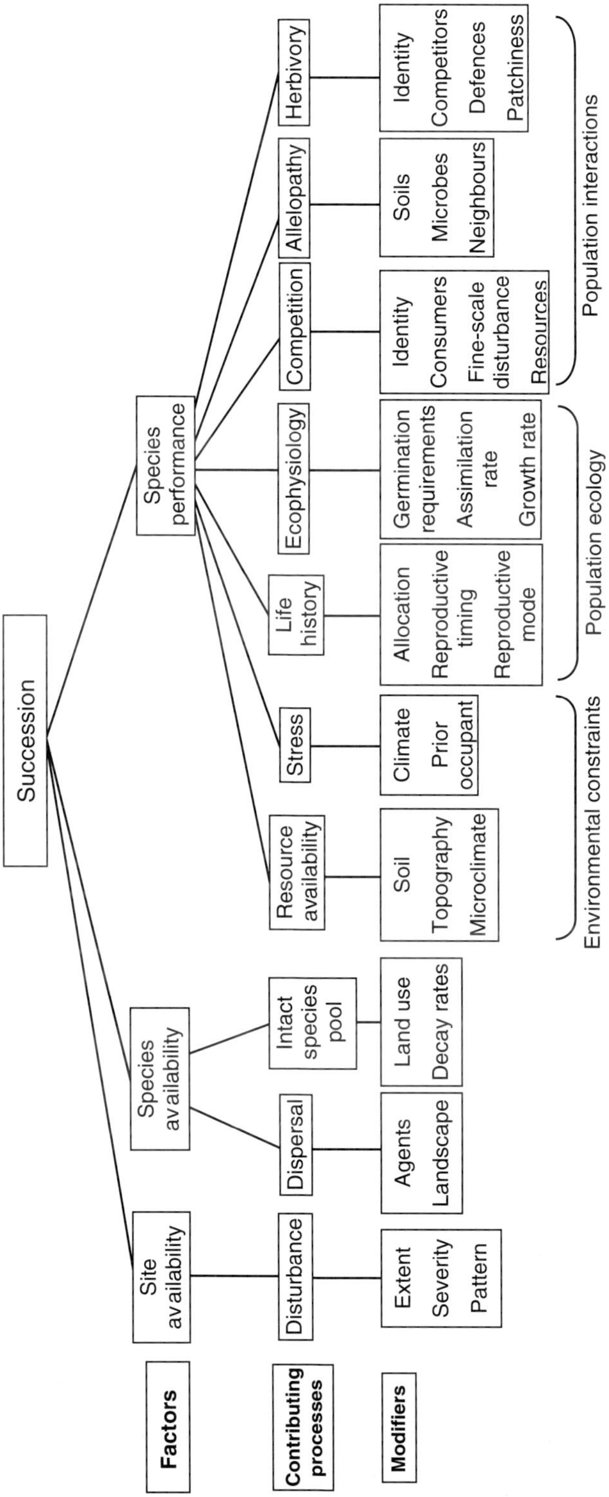

Fig. 12.8. Causes of succession and their contributing processes and modifying factors (redrawn from Pickett *et al.*, 1987a).

seeds will be limiting and the first species establishing at a newly exposed environment can determine what subsequent species will be successful. Conversely, buried seeds and propagules will be more important in secondary succession because these will determine early species composition. Facilitation is more important in nutrient poor environments, but this decreases over the course of succession because vegetation will moderate the environment. Mycorrhizal associations are important throughout succession when nutrients are lacking. When nutrients are available, mycorrhizae will not be important early in succession because there will be little competition, but their importance will increase as the competition for resources increases.

Hierarchy of successional processes

When Connell and Slatyer (1977) developed their three models, they did not look at interactions among processes. Therefore, Pickett *et al.* (1987a,b) proposed a hierarchy of successional processes. This model has also been applied to agricultural weed management (Swanton *et al.*, 1993; Sheley *et al.*, 1996). At the largest scale, Pickett *et al.* said that succession is determined by three factors: site availability, species availability and species performance (Fig. 12.8). These in turn, are influenced by contributing processes and their modifiers.

Site availability

The process of succession is initiated when a disturbance creates or alters a site. Disturbances can be characterized by their extent (area affected), frequency (probability of occurrence over a time period) and magnitude (Walker and Willig, 1999). Magnitude is a combination of intensity (physical force) and severity (impact on vegetation), but the terms are often used interchangeably. The extent of a disturbance determines the environmental conditions and the heterogeneity of the patch, whereas the magnitude will determine its openness and the number and types of propagules available for regeneration (Pickett *et al.*, 1987a,b).

Generally, anthropogenic disturbances are greater in extent and severity than natural disturbances, but they have similar frequencies. Human activities can accentuate or prevent natural disturbances (Luken 1990, p. 12); for example, fire suppression can lead to more extensive and intense fires in the long run (Bond and van Wilgen, 1996). Turner *et al.* (1997) showed that the extent and pattern of fire influenced the successional pathway following the 1988 fires in Yellowstone National Park.

Species availability

The ability of a species to colonize a newly disturbed site will depend on:

- whether it is present in an intact species pool, such as vegetation or in the seed bank;
- its ability to disperse into the site.

Succession following an intense and extensive disturbance will be more dependent on long-distance dispersal, whereas the seed bank and remnant vegetation will be the source of propagules following less intense disturbance (Walker and Chapin, 1987). Del Moral and Bliss (1993) found that proximity to vegetation influenced succession on volcanoes. Lupins and other large-seeded species were early colonists because there were remnant populations serving as seed sources. Sites further from remnant vegetation were composed mostly of small-seeded wind-dispersed species.

Species performance

A species' performance is dependent on its population ecology, how it interacts with other populations, and on environmental factors such as resource level and abiotic stress. A species' performance can be influenced by idiosyncratic events that occur during succession. For example, the timing of the Mount St Helens eruption, in May 1980, may have changed which species were successful (del Moral and Bliss, 1993). When the eruption occurred, much of the ground was still covered in thick snow, and this allowed some plants to persist. Also, ice blocks left crevices when the ice melted and this created moist habitats suitable for

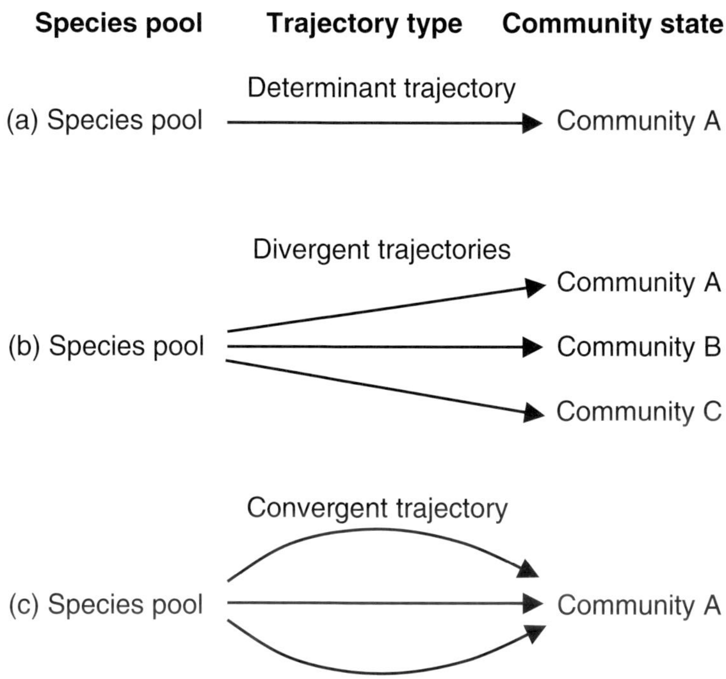

Fig. 12.9. Examples of how different community configurations may be produced from one species pool. (a) Assembly trajectories are deterministic when a species pool consistently produces the same extant community. These deterministic trajectories are relatively immune to historical influences such as invasion sequence. This is the classical view of succession. (b) Different communities may be produced from one species pool when trajectories diverge. These indeterminant trajectories are more sensitive to historical influences such as invasion sequence and changes in the assembly environment. (c) One community type may be produced when assembly trajectories converge (adapted from Drake, 1990).

seedling establishment. Furthermore, the favourable conditions during the three summers following the eruption probably increased seedling establishment allowing new recruits to get a 'foothold'. If conditions had been dry during these first summers, succession might have been delayed or altered. Slight changes in the abiotic environment can alter the competitive balance, or other interactions among colonizing species.

Community Assembly: a Broader Interpretation of Succession

Communities are complex entities and as a result, the thought of looking at 'whole communities' is daunting because there are so many interacting parts. Some scientists deal with complexity by trying to understand all of the community parts individually, and then trying to put the parts together. Another way to deal with complexity is to look at the community as a whole entity and to ignore the details. As the large-scale processes are understood, then more detail can be added. Community assembly theory allows us to do this (Booth and Swanton, 2002).

In recent years the ideas of succession have been reworked and reworded under the theory of community assembly. While there is extensive overlap of these theories, many think of succession as a process of individual replacement whereas community assembly takes a broader approach to community dynamics. Young *et al.* (2001) attempted to summarize the differences between succession and assembly and reconcile the two

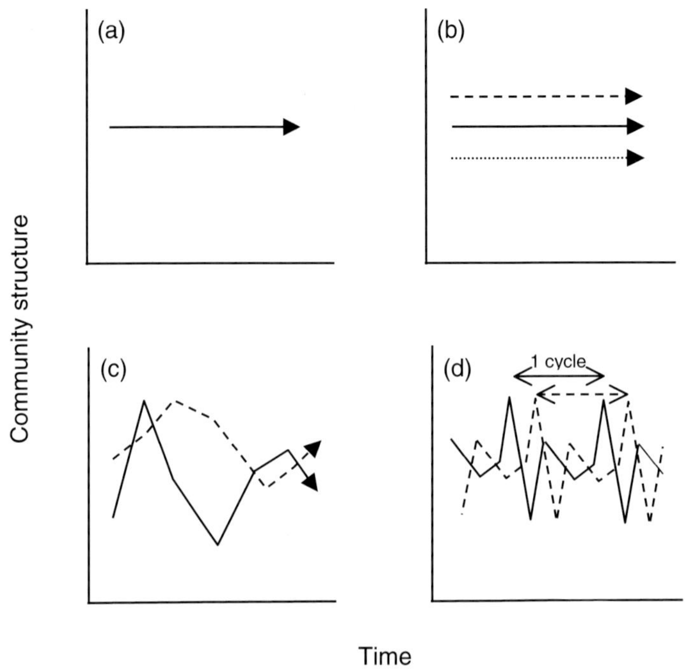

Fig. 12.10. Examples of the divergence and convergence of community trajectories over time. In: (a) communities will converge to a persistent state determined by environmental conditions; in (b) trajectories diverge, and multiple stable states arise out of the same species pool; in (c) communities are random or chaotic and trajectories will continue to converge and diverge, and in (d) one trajectory is reached, but the two communities are cycling out of sync (adapted from Samuels and Drake 1997).

theories. Unfortunately, succession and assembly theories were developed in isolation and only recently have people begun to consider where they overlap and diverge. Really, assembly is the broadening out of the ideas of community dynamics. The basic premise of community assembly is that all biological communities are assembled over time as they follow trajectories and that the membership of a community is limited by filters acting on the species pool. We will first consider how communities follow trajectories through time.

Communities follow trajectories through time

To help us think about communities as dynamic entities we can think of a community as following a trajectory through time (Drake *et al.*, 1999). A trajectory is a path through a series of community states. In a traditional Clementsian view of succession, community development follows a single trajectory: that is, given a species pool, only one community type develops along a deterministic trajectory (Fig. 12.9a). Alternatively, many types of communities can result from a species pool if trajectories are divergent (Fig. 12.9b). Sometimes divergent trajectories may converge producing identical community states (Fig. 12.9c).

It may be difficult to distinguish between a deterministic and an indeterministic trajectory. Over time, trajectories can repeatedly diverge and converge (Fig. 12.10) and this may or may not be predictable (Rodriguez, 1994; Samuels and Drake, 1997). In field experiments, for example, Inouye and Tilman (1988, 1995) found that old-field communities converged after 4 years of nitrogen addition, but after 11 years, these communities had diverged. Sometimes a community reaches a state of predictable cycling (i.e. similar to a population's stable limit cycle) and if viewed over too short a time span, the trajectory will appear indeterministic (Fig. 12.10). Furthermore, two communities following the same trajectory but offset in time will appear different, even though their dynamics are the same. Such communities must be observed over a long enough time scale to distinguish them from random or chaotic trajectories.

Species pools

Only some members of the available species pool will become part of a community. Other species will be removed or filtered out by biotic and abiotic processes at various life stages. Every community is composed of a subset of the total species pool – that is, the group of species available to colonize (Fig. 12.11). Belyea and Lancaster (1999) differentiated among five types of species pools.

- total species pool – large-scale species pool determined by landscape scale ecological and evolutionary processes;
- habitat species pool – the subset of the total species pool that could establish and survive in the habitat;
- geographical species pool – species able to disperse into the habitat;
- ecological species pool – overlap of species present in both the geographical and habitat species pool;
- actual species pool – species present in the above-ground community.

Thus, we have a series of species pools containing subsets of the total species pool. The only time when all types of species pools would be the same is in a closed and stable community where there is no dispersal into the community.

Ecological filters

The processes that remove species from a community are commonly called filters (or constraints). Thus filters limit membership to each species pool, and different types of filters will operate under different conditions. Dispersal constraints determine the geographical species pool, environmental constraints determine the habitat species pool and biotic constraints determine the actual species pool.

Dispersal filters – arriving at the party

Dispersal filters determine what species arrive at a site (see Chapter 6). Communities

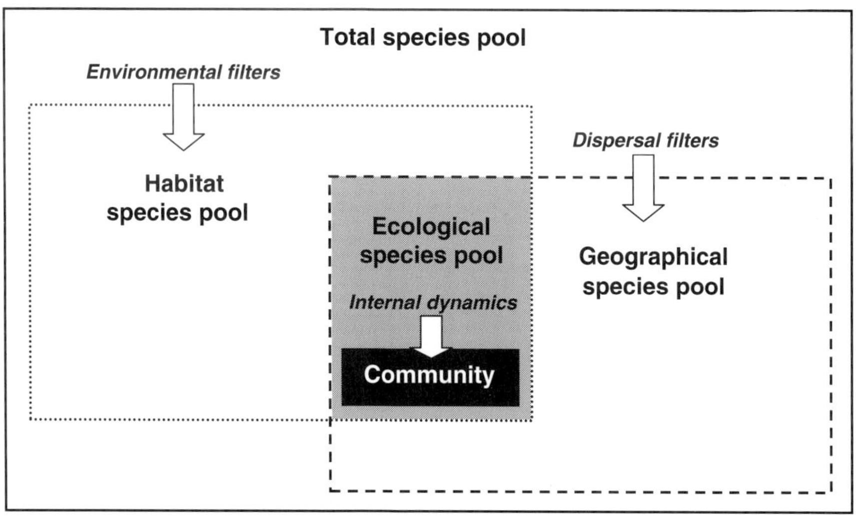

Fig. 12.11. Types of species pools (redrawn from Belyea and Lancaster, 1999).

do not have an unlimited and continuous supply of propagules (Belyea and Lancaster, 1999) because propagules are not produced at a constant rate, nor do they disperse evenly over space. Earlier in this chapter, we discussed the importance of proximity to seed sources (del Moral and Bliss, 1993). As well, seed characteristics, plant phenology and abiotic conditions determine when and whether a propagule can arrive at a site. There has to be the right combination of seed type, dispersal agent and environmental conditions for dispersal to occur.

The timing, sequence, frequency (number of times a species' invasion is repeated) and rate (how quickly invasions are repeated) of species introductions into a community can alter trajectories. The effect of invasion sequence is the most well understood of these. Numerous studies using natural communities (Abrams *et al.*, 1985; McCune and Allen, 1985) and microcosms (Robinson and Dickerson, 1987; Drake, 1991; Drake *et al.*, 1993) have shown that the order in which species are introduced can influence the community trajectory. Early invaders may have the advantage simply because they occupy biological space, inhibiting the invasion of late species, but this is not always the case. Later invaders may drive early ones to extinction by direct or indirect means. They may directly out-compete early invaders, or they may change the abiotic environment, making it inhospitable to the earlier species.

Early species may have a significant role in the assembly process even if the species is not dominant or if its presence is ephemeral. For example, a 'chaperone' species (Kelly, 1994) may facilitate the invasion of another species either by directly assisting it or by inhibiting a third species. Even though the chaperone species is not a dominant species, or is present only temporarily, it could alter significantly the trajectory and have cascading effects on community structure. These 'indirect effects' have also been demonstrated in natural systems (reviewed by Strauss, 1991) and may continue even after a chaperone species has become extinct.

Less studied are the effects of invasion rate and frequency on the trajectory. In experimental studies, species introductions are done singly and at a constant rate, but this is not how invasions occur in natural situations. Generally, increasing the invasion rate and/or frequency increases a community's richness and decreases the likelihood of there being a single stable trajectory because different species will be favoured over time (Hraber and Milne, 1997; Lockwood *et al.*, 1997). A high invasion rate and frequency minimize the influence of historical events (Lockwood *et al.*, 1997). Communities will be more persistent when the invasion rate and frequency are low because the assembly process is not disrupted.

Environmental constraints – crashing the party

After a species is dispersed into a community, it must be able to survive in the physical environment. We can get some indication of a plant's suitability to an environment; for example, by looking at a plant's growth rate versus average temperature and rainfall (Chapter 2). However, the environment can have subtle persistent effects on a community (Chesson, 1986). When we consider only average or typical environmental conditions, we neglect occasional environmental extremes which could have long-term persistent effects on a community. For example, the distribution of the saguaro cactus in Arizona is limited by periodic frosts that kill seedlings, rather than by the cactus's physiological response to average temperature (Hastings and Turner, 1965). Periods of stress, or environmental fluctuation or extremes, may in fact, have a greater impact on the long-term community dynamics than average, relatively predictable environmental conditions. An extreme event may cause some species to go extinct or severely reduce their abundance allowing other species to gain an edge. Environmental variation will alter a community's susceptibility to invasion. Species not usually able to establish may gain an advantage during a period of unusual environmental conditions.

Internal dynamics – being the life of the party

Seeds or propagules can arrive at a site and may be able to survive the abiotic conditions, but not all species will become part of the extant community. Internal dynamics act on the ecological species pool. Population interactions (Chapters 8 and 9) drive internal dynamics. This is possibly the most complicated aspect of community assembly because we cannot possibly understand all interactions. We cannot predict the outcome of all interactions between all species under all dispersal and environmental constraints, nor would this have any predictive value. To make this approach possible and useful, some researchers have used plant characteristic or traits to examine assembly dynamics (Box, 1981; Keddy, 1992a,b; Díaz *et al.*, 1999a,b; McIntyre *et al.*, 1999a,b; Weiher *et al.*, 1999). We will examine this more fully in the next section.

Plant Traits and Functional Groups

We have been taught to classify plants according to their phylogeny – that is, we place them in their family, genus and species. But it might also be useful to classify plants by the traits. Traits are the physical and physiological characteristics that determine a species' ecological function.

Environmental filters remove species from a species pool. In reality, filters act to remove species that lack specific traits (Fig. 12.12) and thus, *traits* rather than species are filtered (Weiher and Keddy, 1999; Booth and Swanton, 2002). Species without the suite of requisite traits will not be able to pass through the series of environmental filters. Plants respond at scales from the climate to disturbance to biotic interactions (Woodward and Diament, 1991). Each type of filter selects against a unique set of traits; therefore, the ability of a species to pass through one filter will not necessarily affect its ability to pass through another because different traits may be required for each. When we work with traits, we first have to decide what traits are important.

Selecting plant traits

How do we determine what traits are biologically relevant? We need to select traits

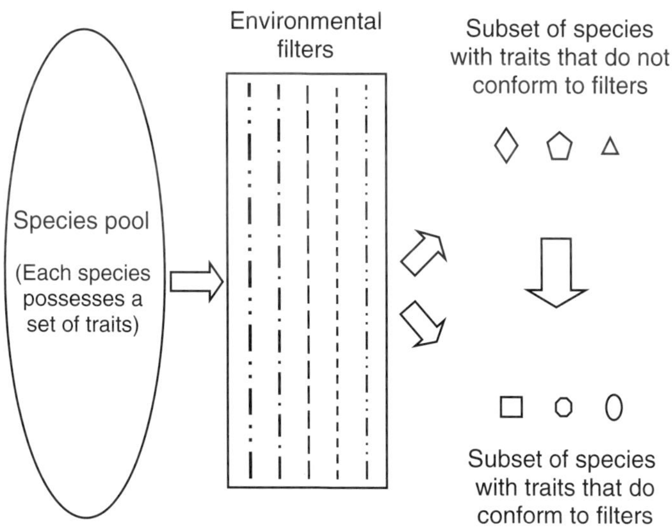

Fig. 12.12. A conceptual model of the trait-based approach to community assembly. A large pool of species is available but they must pass through a series of biotic and abiotic filters that remove species that do not possess specific traits (adapted from Weiher and Keddy, 1999).

Table 12.4. List of traits that could be used as a starting point for a trait-based approach to community assembly and possible ways to quantify them. Some or all of the traits could be selected to record, and of these only some traits would be ecologically important. Based on lists in Díaz *et al.* (1999b), Díaz, Barradas *et al.* (1999), Kleyer (1999) and McIntyre *et al.* (1999b).

Vegetative traits	Classification of trait
Vegetative traits	
Size	<10 cm; 11–20 cm; 21–50 cm; 51–100 cm; >100 cm
Height:width ratio	h:w>1; h:w/1; h:w<1
Specific leaf area	Aphyllous; <1 cm^2; 1.1–2 cm^2; 2.1–3 cm^2; 3.1–5 cm^2; >5 cm^2
Life cycle	Summer annual; winter annual; biennial; perennial monocarpic; perennial polycarpic
Life span	<1 year; 2–5 years; 5–10 years; 11–20 years; >20 years
General form	Prostrate; rosette; erect; tussock; vine; shrub
Leaf form	Aphyllous; evergreen; deciduous
Leaf angle	<90°; >90°
Leaf size	Aphyllous; <1 cm^2; 2–5 cm^2; 6–10 cm^2; 10–25 cm^2; 25–50 cm^2; >50 cm^2
Leaf shape (length:width)	l:w>1; l:w/1; l:w<1
Photosynthetic metabolism	CAM; C_3; C_4
Potential relative growth rate	Low; medium; high
Drought avoidance	None; succulent stem, taproot or other storage organ
Palatability	Unpalatable; low or just at juvenile stage; moderate; high
Leaf texture	Smooth; hairy; spines
Root morphology	Taproot; mostly horizontal; mostly vertical
Maximum rooting depth	<10 cm; 10–25 cm; 25–50 cm; 50–100cm; >100 cm
Extent of clonal expansion	None; some (dm); high degree (m)
Resprouting ability	None; moderate (daughter plants remain attached to parent plant for some period of time); high (daughter plants rapidly become independent)
Mycorrhizal associations	None; ectomycorrhizal; vesicular-arbuscular
Storage organs	None; tubers; bulbs; rhizomes
Reproductive traits	
Seed size (max. length)	<1 mm; 1–2 mm; 3–5 mm; 6–10 mm; >10 mm
Seed shape (variance of seed length, width and depth)	<0.15; 0.15–<1; 1–5; >5
Seed number (per plant)	<100; 100–999; 1000–5000; >5000
Weight of dispersal unit (fruit or seed)	<0.2 mg; 0.3–0.5 mg; 0.6–1 mg; 1–2 mg; >2 mg;
Seed dispersal	No mechanism; wind; by highly mobile animals (birds, bats); by low mobility animals (ants, rodents)
Fruit type	Dry indehiscent; dry dehiscent; fleshy
Season of germination	Plastic; early spring; late spring; summer; autumn
Age of first reproduction	<3 months; 3 months–1 year; 1–3 year; >3 year
Peak period of flower and fruit production	None; autumn to early spring; spring; late spring to late summer; late summer to autumn
Pollination mode	Wind; specialized animals; unspecialized animals
Position of dormant buds (physiognomic types)	thero-; geo-; hemi-; crypto-; chamaephyte
Agricultural specific traits	
Herbicide tolerance	
Weed size relative to crop	Smaller; same; larger

that address the multiple scales of environmental filters. Traits associated with both growth and reproduction should be included (Díaz *et al.*, 1999a). The former influences resource acquisition and storage, and the latter, recolonization and regeneration. Furthermore, both physiological and morphological traits are important. One trait may be important to several processes. For example, seed size affects dispersal, germination, risk of predation and seedling competitive ability. The selection of traits will depend on the habitat type, regional flora and on the goals of the study. Table 12.4 is a list of potential traits to consider.

Forming functional groups

We have reduced the complexity of our community from a list of species to a list of traits. We can simplify it still further by constructing functional groups from the traits. A functional group contains species with a similar set of traits. They serve similar ecological functions in a community and are therefore filtered from species pools in a similar manner, but how do we divide our traits into functional groups? There are a number of ways to do this (Smith *et al.*, 1997); we present two examples.

Nobel and Slatyer (1980) formed functional groups based on their ecological understanding of traits important to the community type. They identified a number of functional groups (Table 12.5) based on three sets of traits:

- arrival and persistence after disturbance;
- their ability to grow following a disturbance;
- the time to reach important life stages.

Using these functional groups they were able to describe vegetation changes of a wet sclerophyll forest in Tasmania following various disturbances (Fig. 12.13). The authors concluded that grouping plants by functional group 'could be applied to man-modified communities, and to such phenomena as weed invasion or suppression'.

Kleyer (1999) used another approach to

Table 12.5. List of nine functional groups derived by Nobel and Slatyer (1980) based on traits required to arrive or persist following a disturbance, ability to establish following a disturbance, and time to reach important life stages. Species types, their life stage parameters, and a general description of their functional group (modified from the original 15 groups presented in Nobel and Slatyer, 1980). Critical life history events indicated are: m, reproductive maturity, l, longevity of the population, e, longevity of the seed pool.

Type	time since disturbance 0 - - - - - - - - - - - - - - - - - ∞	Characteristics of species
1	- - m - - - - - - - - - - - - - - le	Tolerant of disturbance, and able to establish quickly and continue to establish following disturbance by various means
2	m - - - - - - - - - - - - - - le	Tolerant of disturbance, but propagules only available if vegetation is mature and can release seeds
3	- - m - - - l - - - - - - - - - - - e	Pioneer species, with propagules capable of long-distance dispersal into disturbed site
4	- - m - - - l - - - - e	Pioneer species, with propagules always available (usually in seed pool)
5	m - - - l - - - - e	Pioneer species, with propagules available as long as disturbance occurred after the juvenile stage of growth
6	m - - - le	Pioneer species, with short-lived seed pool available
7	- - m - - - le	Pioneer species, able to regrow vegetatively
8	- - m - - - - - - - - - - - - - - le	Require mature vegetation to re-establish, and propagules from dispersal or species pool
9	m - - - - - le	Pioneer species, capable of vegetative reproduction

(a)

Symbol	Functional Type	Species	Life stage parameters 0 10 30 70 400 500 ~
4A	4	*Acacia*	- - m - - - - - l - - - e
6E	6	*Eucalyptus*	- - m - - - - - - - - - le
6N	6	*Nothofagus*	m - - - - - - - - - le
6A	1	*Antherosperma*	- - - - - m - - - - - - - - - - - le

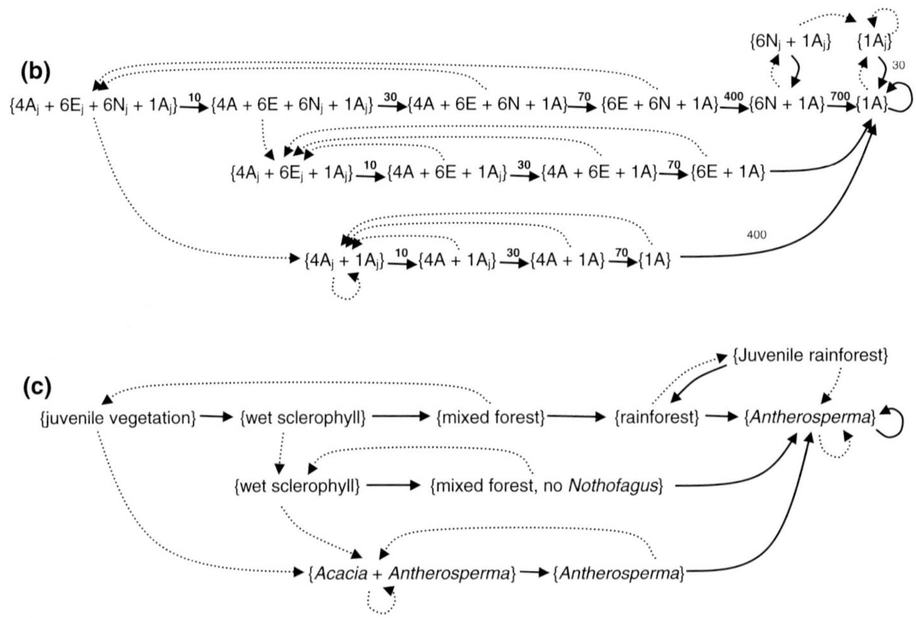

Fig. 12.13. Vegetation replacement series for a wet sclerophyll rainforest in Tasmania (Nobel and Slatyer, 1980). Part (a) describes the four dominant species, their functional types and life history parameters; (b) shows the replacement sequence; (c) summarizes the transition sequences by showing community types. Solid arrows indicate transitions where no disturbance occurs, whereas dotted arrows indicate transitions following disturbance. Numbers are years and subscript j ($_j$) indicates plants in the juvenile stage (redrawn from Nobel and Slatyer, 1980).

delineate functional groups. Kleyer sampled 129 agricultural and disturbed sites in Germany and noted three types of features:

- level of disturbance (from frequently disturbed field to 60-year-old undisturbed field);
- resource supply (level of nitrogen, phosphorus and soil moisture);
- plants species composition.

Species were then characterized using many of the traits listed in Table 12.4. Clustering techniques (see Chapter 15) were used and these identified 30 functional groups containing species with similar traits. Table 12.6 shows six examples, and lists representative species and associated traits. Kleyer was able to predict what functional groups were likely to occur under various combinations of disturbance and resource level (Fig. 12.14). For example, species in type I are more likely to be found in either frequently disturbed sites, or in less disturbed but resource-rich sites.

Table 12.6. Examples of functional groups with similar biology as resulting from the cluster analysis (based on information in Kleyer, 1999).

Functional groups and their representative species	Traits associated with functional group
I. Lambsquarters type *Amaranthus retroflexus, Atriplex patula, Avena fatua, Chenopodium album, Echinochloa crus-galli, Fallopia convolvulus, Polygonum lapathifolium, P. persicaria, Setaria verticillata*	Summer annual. Seed weight > 0.2 mg, seed number 1–10,000. Seed bank long-term persistent, long-range dispersal, plant height mostly 0.3–0.9 m, leaves regularly distributed
II. Chickweed type *Capsella bursa-pastori, Cardamine hirsuta, Matricaria discoidea, Poa annua, Senecio vulgaris, Stellaria media, Veronica persica*	Several generations per vegetation period. Seed weight > 0.2 mg, seed number 1000–10,000. Seed bank short-term to long-term persistent, long-range dispersal, plant height < 0.3 m, leaves regularly distributed. Regeneration from detached shoots reported for *C. hirsuta, S. media, V. persica*
III. Yarrow type *Achillea millefolium, Campanula persicifolia, C. rapunculoides, C. rotundifolia, Cerastium holosteoides, Poa compressa, Sedum acre*	First reproduction at 2–3 years (except *C. holosteoides, P. compressa*). Seed weight mostly < 0.2 mg, seed number 1000–10,000. Seed bank longevity various and dispersal mostly short-range. Plant height 0.3–0.9 m (or < 0.3 m), leaves regularly distributed. Lateral clonal extension some dm or m, vegetative growth mostly slow
IV. Yellow toadflax type *Artemisia vulgaris, Hypericum hirsutum, H. perforatum, Linaria vulgaris, Origanum vulgare, Solidago canadensis, Tanacetum vulgare*	Seed weight mostly < 0.2 mg, seed number mostly > 10,000. Seed bank longevity and dispersal various, plant height 0.3–0.9 m or > 0.9 m, leaves regularly distributed. Lateral clonal extension some m (except *A. vulgaris, H. hirsutum*), vegetative multiplication facultative or slow
V. Creeping buttercup type *Fragaria vesca, Glechoma hederacea, Poa subcoerulea, Potentilla anserina, Ranunculus repens, Rumex acetosella, Trifolium repens*	First reproduction at 1 year for *R. repens*, 2 years for the rest. Seed weight > 0.2 mg, seed number mostly low. Seed bank mostly long-term persistent and long-range dispersal. Plant height low, leaves semirosette or rosette. Lateral clonal extension some dm to m, vegetative multiplication mostly rapid.
VI. Blackthorn type *Corylus avellana, Malus domestica, Prunus spinosa, Salix caprea*	Woody, age at first reproduction 6–15 years. Seed weight > 0.2 mg, seed number 1000–10,000. Seed bank transient and medium- to long-range dispersal. Plant height > 0.9 m, leaves regularly distributed. Lateral clonal extension some m, vegetative multiplication slow

Summary

If we can accurately characterize a community based on its structure, why go the next, more difficult step to describe or predict a community's dynamics? First, we must remember that communities are dynamic, and their present structure may not be representative of their future structure. Second, there are practical applications to be gained from understanding how a community will change in the future, especially for weed ecologists. Thus, we gain predictive knowledge from an understanding of community dynamics.

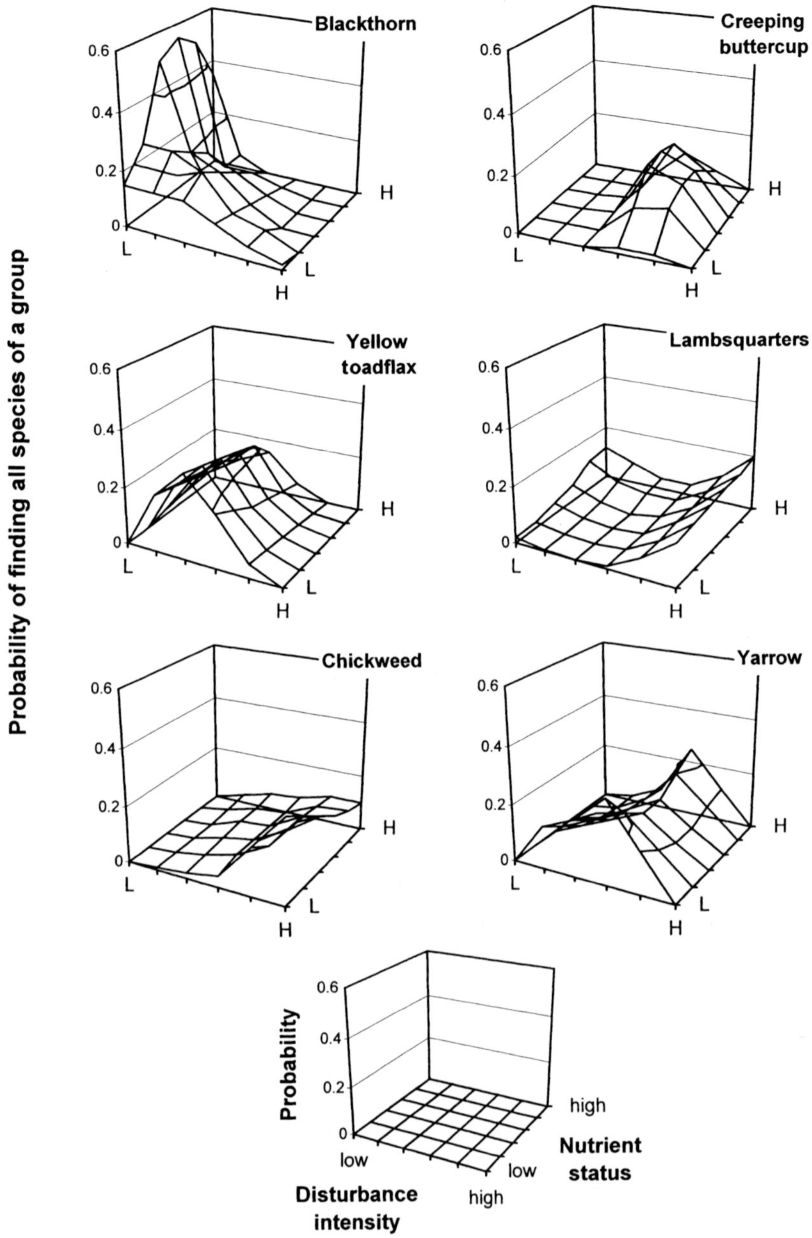

Fig. 12.14. Regression surfaces for six of Kleyer's (1999) functional groups. The probability of finding all of the species in a group is shown in relationship to disturbance intensity and resource supply. Disturbance intensities are (from low to high) 1, old field (disturbed 1×/60 years); 2, field balk (disturbed 1×/20 years); 3, field balk (disturbed 1×/ 10 years); 4, meadow (disturbed 1×/ year); 5, field road (disturbed 30×/year at soil surface); and 6, field (disturbed 6×/ year below soil surface) (redrawn from Kleyer, 1999).

Questions

1. What role does your species play in community dynamics? Is it an early-, mid- or late successional species?
2. Explain why more competitors (C) occur in Grime's model when there are higher levels of potential productivity.
3. Draw a successional trajectory of a no-till annual cropping system under the C-S-R model.
4. Why are the terms 'stability' and 'equilibrium' misleading when describing communities?
5. Refer to Fig. 12.7 (Walker and Chapin, 1987) and explain each of the ten patterns displayed.
6. Refer to Fig. 12.14 (Kleyer, 1999). In what type of community would you expect to find species of each of the six functional groups? Explain your answer.

General References

Booth, B.J. and Swanton, C.J. (2002) Assembly theory applied to weed communities. *Weed Science* 50, 2–13.

Luken, J.O. (1990) *Directing Succession.* Chapman and Hall, London.

Pickett, S.T.A., Collins, S.L. and Armesto, J.J. (1987) Models, mechanisms and pathways of succession. *Botanical Review* 53, 335–371.

Literature Cited

Abrams, M.C., Sprugel, D.G. and Dickmann, D.I. (1985) Multiple successional pathways on recently disturbed jack pine sites in Michigan. *Forest Ecology and Management* 10, 31–48.

Barbour, M.G., Burk, J.H., Pitts, W.D., Gillam, F.S. and Schwartz, M.W. (1999) *Terrestrial Plant Ecology*, 3rd edn. Benjamin/Cummings, Menlo Park, California.

Bazzaz, F.A. (1979) The physiological ecology of plant succession. *Annual Review of Ecology and Systematics* 10, 351–371

Belyea, L.R. and Lancaster, J. (1999) Assembly rules within a contingent ecology. *Oikos* 86, 401–416.

Bond, W.J. and van Wilgen, B.W. (1996) *Fire and Plants.* Chapman and Hall, London, 272 pp.

Booth, B.J. and Swanton, C.J. (2002) Assembly theory applied to weed communities. *Weed Science* 50, 2–13.

Booth, W.E. (1941) Algae as pioneers in plant succession and their importance in erosion control. *Ecology* 22, 38–46.

Box, E.O. (1981) *Macroclimate and Plant Form: an Introduction to Predictive Modeling in Phytogeography.* Junk, The Hague, pp. 11–20.

Brown, V.K. and Southwood, T.R.E. (1987) Secondary succession: patterns and strategies. In: Gray, A.J., Crawley, M.J. and Edwards, P.J. (eds) *Colonization, Succession and Stability.* Blackwell Scientific, California, pp. 315–337.

Chesson, P.L. (1986) Environmental variation and the coexistence of species. In: Diamond, J.M. and Case, T.J. (eds) *Community Ecology.* Harper and Row, New York, pp. 240–256

Clements, F.E. (1916) *Plant Succession.* Publication Number 242, The Carnegie Institution, Washington.

Clements, F.E. (1936) The nature and structure of the climax. *Journal of Ecology* 22, 9–68.

Connell, J.H. and Slatyer, R.O. (1977) Mechanisms of succession in natural communities and their role in community stability and organization. *American Naturalist* 111, 1119–1144.

Cowles, H.C. (1899) The ecological relationships of vegetation on sand dunes of Lake Michigan. *Botanical Gazette* 27, 95–117, 167–202, 281–308, 361–391.

Cowles, H.C. (1901) The physiographic ecology of Chicago and vicinity. *Botanical Gazette* 31, 73–108.

DeAngelis, D.L. and Waterhouse, J.C. (1987) Equilibrium and nonequilibrium concepts in ecological models. *Ecological Monographs* 57, 1–21.

del Moral, R. and Bliss, L.C. (1993) Mechanisms of primary succession: insights resulting from the eruption of Mount St Helens. *Advances in Ecological Research* 24, 1–66.

Díaz, S., Cabido, M. and Casanoves, F. (1999a). Functional implications of trait-environment linkages in plant communities. In: Weir, E. and Keddy, P.A. (eds) *Ecological Assembly Rules: Perspectives, Advances, Retreats.* Cambridge University Press, Cambridge, pp. 339–362.

Díaz, S., Cabido, M., Zak, M., Martínez Carretero, E. and Araníbar, J. (1999b) Plant functional traits, ecosystem structure and land-use history along a climatic gradient in central-western Argentina. *Journal of Vegetation Science* 10, 651–660.

Díaz Barradas, M.C., Zunzunegui, M., Tirada, R., Ain-Lhout, F. and Carcía Novo, F. (1999) Plant functional types and ecosystem function in Mediterranean shrubland. *Journal of Vegetation Science* 10, 709–716.

Drake, J.A. (1990) Communities as assembled structures: do rules govern pattern? *Trends in Ecology and Evolution* 5, 159–164.

Drake, J.A. (1991) Community-assembly mechanics and the structure of an experimental species ensemble. *American Naturalist* 137, 1–26.

Drake, J.A., Flum, T.E., Witteman, G.J., Voskuil, T., Hoylman, A.M., Creson, C., Kenny, D.A., Huxel, G.A. Larue, C.S., and Duncan, J.R. (1993) The construction and assembly of an ecological landscape. *Journal of Animal Ecology* 62, 117–130.

Drake, J.A., Zimmerman, C.R., Purucker, T. and Rojo, C. (1999) On the nature of the assembly trajectory. In: Weir, E. and Keddy, P.A. (eds) *Ecological Assembly Rules: Perspectives, Advances, Retreats.* Cambridge University Press, Cambridge, pp. 233–250.

Drury, W.H. and Nisbet, I.C.T. (1973) Succession. *The Arnold Arboretum Journal* 54, 331–368. Reprinted in: Golley, F.B. (ed.) *Benchmark Papers in Ecology* (1977) Dowden, Hutchison and Ross, Stroudburg, pp. 287–324.

Dublin, H.T., Sinclair, A.R.E. and McGlade, J. (1990) Elephants and fire as causes of multiple stable states in the Serengeti-Mara Woodlands. *Journal of Animal Ecology* 59, 1147–1164.

Ecke, F. and Rydin, H. (2000) Succession on a land uplift coast in relation to plant strategy theory. *Annals of Botanical Fennici* 37, 163–171.

Egler, F.E. (1954) Vegetation science concepts. I. Initial floristic composition, a factor in old field vegetation development. *Vegetatio* 4, 412–417.

Facelli, J.M. and Facelli, E. (1993) Interactions after death:plant litter control priority effects in a successional plant community. *Oecologia* 95, 277.

Fastie, C.L. (1995) Causes and ecosystem consequences of multiple pathways of primary succession at Glacier Bay, Alaska. *Ecology* 76, 1899–1912.

Frezen, P.M., Krasney, M.E. and Rigney, L.P. (1988) Thirty-three years of plant succession on the Kautz Creek mudflow, Mount Ranier National Park, Washington. *Canadian Journal of Botany* 66, 130-137.

Gleason, H.A. (1917) The structure and development of the plant association. *Bulletin of the Torrey Botany Club* 44, 463–481.

Gleason, H.A. (1926) The individualistic concept of the plant association. *Bulletin of the Torrey Botany Club* 53, 7–26.

Griggs, R.F. (1933) The colonization of the Katmai ash, a new and inorganic 'soil'. *American Journal of Botany* 20, 92–111.

Grime, J.P. (1977) Evidence for the existence of three primary strategies in plants and its relevance to ecological and evolutionary theory. *American Naturalist* 11, 1169–1194.

Grime, J.P. (1987) Dominant and subordinate components of plant communities: implications for succession, stability and diversity. In: Gray, A.J., Crawley, M.J. and Edwards, P.J. (eds) *Colonization, Succession and Stability.* Blackwell Scientific, Oxford, pp. 413–428.

Grubb, P.J. (1987) Some generalizing ideas about colonization and succession in green plants and fungi. In: Gray, A.J., Crawley, M.J. and Edwards, P.J. (eds) *Colonization, Succession and Stability.* Blackwell Scientific, Oxford, pp. 81–102.

Hastings, J.R. and Turner, R.M. (1965) *The Changing Mile.* University of Arizona Press, Tucson.

Holling, C.B., Schindler, D.W., Walker, B.H. and Roughgarden, J. (1995) Biodiversity in the functioning of ecosystems: an ecological synthesis. In: Perrings, C.A., Mäler, K.-G., Folke, C., Holling, C.S. and Jansson, B.-O. (eds) *Biodiversity Loss: Economic and Ecological Issues.* Cambridge University Press, Cambridge.

Hraber, P.T. and Milne, B.T. (1997) Community assembly in a model ecosystem. *Ecological Modeling* 104, 267–285.

Huston, M. and Smith, T. (1987) Plant succession: life history and competition. *American Naturalist* 130, 168–198.

Inouye, R.S. and Tilman D. (1988) Convergence and divergence of old-field plant communities along experimental nitrogen gradients. *Ecology* 69, 995–1004.

Inouye, R.S. and Tilman, D. (1995) Convergence and divergence of old-field vegetation after 11 years of nitrogen addition. *Ecology* 76, 1872–1887.

Keddy, P.A. (1992a) A pragmatic approach to functional ecology. *Functional Ecology* 6, 621–626.

Keddy, P.A. (1992b) Assembly and response rules: two goals for predictive community ecology. *Journal of Vegetation Science* 3, 157–165

Kelly, K. (1994) *Out of Control: The New Biology of Machines.* Fourth Estate, London.

Kleyer, M. (1999) Distribution of plant functional types along a gradient of disturbance intensity and resource supply in an agricultural landscape. *Journal of Vegetation Science* 10, 697–708.

Lĕps, J., Osbornová-Kosinová, J. and Rejmánek, M. (1982) Community stability, complexity and species life history strategies. *Vegetatio* 50, 53–63.

Lockwood, J.L., Powell, R.D., Nott, M.P. and Pimm, S.L. (1997) Assembling ecological communities in time and space. *Oikos* 80, 549–553.

Luken, J.O. (1990) *Directing Succession.* Chapman and Hall, London.

McCune, B. and Allen, T.F.H. (1985) Will similar forests develop on similar sites? *Canadian Journal of Botany* 63, 367–376.

McIntosh, R.P. (1999) The succession of succession: a lexical chronology. *Bulletin of the Ecological Society of America* 80, 256–265.

McIntyre, S., Díaz, S., Lavorel, S. and Cramer, W. (1999a). Plant functional types and disturbance dynamics – Introduction. *Journal of Vegetation Science* 10, 604–608.

McIntyre, S., Lavorel, S., Landsberg, J. and Forbes T.D.A. (1999b) Disturbance response in vegetation – towards a global perspective on functional traits. *Journal of Vegetation Science* 10, 621–630.

Miles, J. (1987) Vegetation succession: Past and present perceptions. In: Gray, A.J., Crawley, M.J. and Edwards, P.J. (eds) *Colonization, Succession and Stability.* Blackwell Scientific, Oxford, pp. 1–29.

Nobel, I.R. and Slatyer, R.O. (1980) The use of vital attributes to predict successional changes in plant communities subject to recurrent disturbances. *Vegetatio* 43, 5–21.

Perrings, C.A. and Walker, B.H. (1995) Biodiversity loss and the economics of discontinuous change in semi-arid rangelands. In: Perrings, C.A., Mäler, K.-G., Folke, C., Holling, C.S. and Jansson, B.-O. (eds) *Biodiversity Loss: Economic and Ecological Issues.* Cambridge University Press, Cambridge, pp. 190–210.

Pianka, E.R. (1970) On r and K selection. *American Naturalist* 104, 592–597.

Pickett, S.T.A., Parker, V.T. and Fiedler, P. (1992) The new paradigm in ecology: implications for conservation biology above the species level. In: Fiedler, P. and Fain, S. (eds) *Conservation Biology: The Theory and Practice of Nature Conservation, Preservation and Management.* Chapman and Hall, New York, pp. 65-88.

Pickett, S.T., Kolasa, J. and Jones, C.G. (1994) *Ecological Understanding.* Academic Press, Orlando.

Pickett, S.T.A., Collins, S.L. and Armesto, J.J. (1987a) A hierarchical consideration of causes and mechanisms of succession. *Vegetatio* 69, 109–114.

Pickett, S.T.A., Collins, S.L. and Armesto, J.J. (1987b) Models, mechanisms and pathways of succession. *Botanical Review* 53, 335–371.

Pimm, S.L. (1991) *The Balance of Nature? Ecological Issues in the Conservation of Species and Communities.* University of Chicago Press, Chicago.

Putman, R. (1994) *Community Ecology.* Chapman and Hall, London.

Robinson, J.V. and Dickerson, J.E. Jr (1987) Does invasion sequence affect community structure? *Ecology* 68, 587–595.

Rodriguez, M.A. (1994) Succession, environmental fluctuations, and stability in experimentally manipulated microalgal communities. *Oikos* 70, 107–120.

Samuels, C.L. and Drake, J.A. (1997) Divergent properties on community convergence. *Trends in Ecology and Evolution* 12, 427–432.

Sheley, R.L., Svejcar, T.J. and Maxwell, B.D. (1996) A theoretical framework for developing successional weed management strategies on rangeland. *Weed Technology* 10, 766.

Smith, T.M., Shugart, H.H. and Woodward, F.I. (eds) (1997) *Plant Functional Types.* Cambridge University Press, Cambridge, p. 369

Steneck, R.S. and Dethier, M.N. (1994) A functional group approach to the structure of algal dominated communities. *Oikos* 69, 476–498.

Strauss, S.Y. (1991) Indirect effects in community ecology: their definition, study and importance. *Trends in Ecology and Evolution* 6, 206–210.

Swanton, C.J., Clements, D. and Derksen, D.A. (1993) Weed succession under conservation tillage: a hierarchical framework for research and management. *Weed Technology* 7, 286–297.

Tansely, A.G. (1935) The use and abuse of vegetational concepts and terms. *Ecology* 16, 284–307.

Turner, M.G., Romme, W.H. and Hargrove, W.W. (1997) Effects of fire size and pattern on early succession in Yellowstone National Park. *Ecological Monographs* 67, 411.

Walker, B.H. (1981) Is succession a viable concept in African savanna ecosystems? In: West, D.C., Shugart, H.H. and Botkin, D.B. (eds) *Forest Succession: Concepts and Applications.* Springer-Verlag, New York, pp. 431–447.

Walker, L.R. (1999) Patterns and processes in primary succession. In: Walker, L.R. (ed.) *Ecosystems of Disturbed Ground.* Elsevier, Amsterdam, pp. 585–610.

Walker, L.R. and Chapin F.S. III (1987) Interactions among processes controlling successional change. *Oikos* 50, 131–135.

Walker, L.R. and Willig, M.R. (1999) An introduction to terrestrial disturbance. In: Walker, L.R. (ed.) *Ecosystems of Disturbed Ground.* Elsevier, Amsterdam, pp. 1–16.

Watt, A.S. (1947) Pattern and process in the plant community. *Journal of Ecology* 35, 1–22.

Weiher, E. and Keddy P.A. (1999) Assembly rules as general constraints on community composition. In: Weiher, E. and Keddy, P.A. (eds) *Ecological Assembly Rules: Perspectives, Advances, Retreats.* Cambridge University Press, Cambridge, pp. 251–271.

Weiher, E., van der Werf, A., Thompson, K., Roderick, M., Garnier, E. and Eriksson, O. (1999) Challenging theophrastus: a common core list of plant traits for functional ecology. *Journal of Vegetation Science* 10, 609–620.

Woodward, F.I. and Diament, A.D. (1991) Functional approaches to predicting the ecological effects of global change. *Functional Ecology* 5, 202–212.

Young, T.P., Chase, J.M. and Huddleston, R.T. (2001) Community succession and assembly: comparing, contrasting and combining paradigms in the context of ecological restoration. *Ecological Restoration* 19, 5–18.

13
Plant Invasions

Concepts

- An invasion is the geographical expansion of a species into an area not previously occupied by it. Native and non-native species can be invasive.
- Most invasions fail because the species do not possess the suite of traits necessary to disperse, establish, reproduce and spread into a habitat.
- Three habitat characteristics that are generally thought to encourage invasions are disturbances, low species richness and high resource availability.
- The impact of an invasion depends on the area covered by the species, its abundance and the effect per individual.
- Invasion meltdown is the acceleration of impacts on native ecosystems due to synergistic interactions.
- While we are becoming proficient at explaining invasions, we are not very good at predicting them.

Introduction

An invasion is 'the geographical expansion of a species into an area not previously occupied by that species' (Vermeij, 1996). This definition captures the idea that, although most invasive species tend to be non-natives, this is not a necessary condition. For example, in the northeastern USA, the native hay-scented fern (*Dennstaedtia punctilobula*) becomes invasive under conditions of intensive browsing by white-tailed deer and overstorey tree canopy thinning or removal. The hay-scented fern is able to take advantage of increased light conditions to form a thick canopy which prevents the regeneration of other species (de la Cretaz and Kelty, 1999).

Plant invasions have high ecological and economic costs associated with them (Parker *et al.*, 1999; Pimentel *et al.*, 2000, 2001). Some economic costs are easily quantifiable (the cost of weed control, yield loss), whereas other are not (damage to ecosystems, loss of recreational land, aesthetics). Pimentel *et al.* (2000) estimated the costs of invasive plants to crop and pasture land in the USA as well over $34 billion annually. In India, the cost is even higher at $38 billion per year (Pimentel *et al.*, 2001). While the general concepts of evaluating costs and benefits in ecosystems have been criticized (Gatto and de Leo, 2000; Nunes and van den Bergh, 2001), Pimentel *et al.*'s (2000, 2001) studies do show the context of the huge economic impact of weeds. The ecological costs of invasion are much harder to understand and quantify. What is the 'cost' associated with garlic mustard invading a forest? If an invasion causes another species to become extinct, what is the lost value?

Plant invasions are not, by any means, caused exclusively by human actions. Species' ranges have always fluctuated on small to large spatial and temporal scales in response to natural phenomena, and communities have always had to 'react' to these changes (Chapter 2). What has changed is the rate at which invasions are now occurring (Lodge, 1993). Of course, no community is truly safe from invaders, nor should it be, since succession is the normal process of species invading and replacing other species. The problem is that some species tend to be better at invading than others, and that some habitats appear to be more invasible than others. Invasions are commonplace at all scales: from invasions of weeds into a field or forest, to the introduction of a non-native plant into a continent and its subsequent invasion. In this chapter, we look at the process of plant invasion, the characteristics of invasive species and invasible habitats, and the ecological impacts of invasions.

Most Invasions Fail or Have No Effect, but Some Will Succeed

Most invasions fail. That is, most species when introduced into a community will not survive. Surprisingly, our knowledge of why invasions fail is quite limited (Rejmánek, 1999). Successful invaders are obvious, but failed invaders are not; therefore, it is difficult to study the process of a failed invasion. The principal reasons that invasions fail are (Crawley, 1987; Lodge, 1993):

- inappropriate abiotic conditions;
- introduced species are outcompeted by native (or non-native) species;
- presence of generalist natural enemies such as herbivores and diseases;
- lack of mutualists to pollinate, disperse or facilitate the invader;
- low-density effects such as difficulty in finding mates.

Sometimes a species can be introduced into a community, but have no effect. If we think of this from an assembly perspective, an introduced species may have no effect on the trajectory. When a species does become established, it will not necessarily change the community function: it may become part of the community and increase species richness, but cause no other changes.

If most invasions fail, then what are we

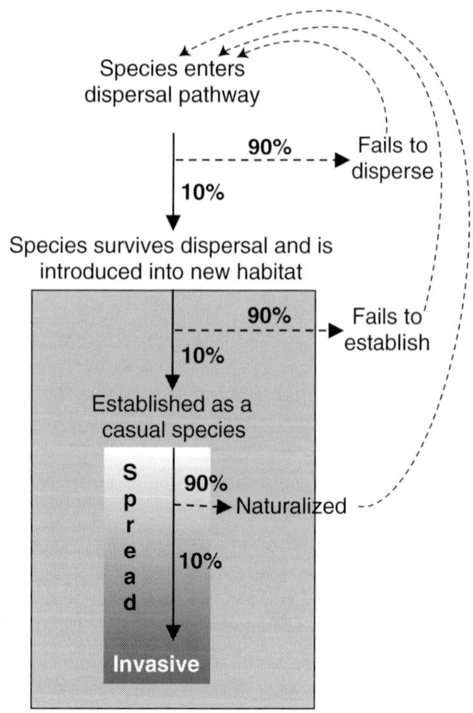

Fig. 13.1. Transitions that a species must overcome to become invasive. Approximately 90% (80–95%) of species are lost at each transition. While a species may fail at some stage, it may re-enter the invasion cycle numerous times, thus increasing its likelihood of success (adapted and redrawn from Kolar and Lodge, 2001).

worried about? Well, the small proportion of invasions that *do* succeed can have drastic effects on populations and communities. An invasion may simply cause a single species to become extinct, or alternatively could have community-wide effects if it changes the assembly trajectory. When the trajectory path is altered, the community is ruled by a new set of assembly processes (Drake *et al.*, 1999). Species invasions may alter the assembly process in an unpredictable manner. For example, adding a species that functions as a keystone species may alter community dynamics drastically. Introducing a non-native species may have a greater effect on the assembly trajectory because the community has no 'experience' of it (Drake *et al.*, 1996; Vitousek, 1990).

The Process of Invasion

A successful invasion is a rare event. How rare? Williamson proposed the 'tens rule' to describe how approximately 10% of species pass through each transition from being imported (dispersal) to becoming casual (introduced to wild) to becoming established, and finally becoming a weed (Williamson and Brown 1986; Williamson, 1993, 1996) (Fig. 13.1). This is a rough rule, but has been shown to apply around the world in many situations (Williamson and Fitter, 1996). What is evident from this rule is that successful invasions are rare (Williamson, 1996).

A successful invasion requires that a species arrive, establish, reproduce, spread and integrate with other members of a community (Vermeij, 1996; Williamson, 1996; Richardson *et al.*, 2000). A species faces a series of filters (barriers) to accomplish each of these. A variety of terms and schemes have been used to describe the process of invasion – often in conflicting fashion (Vermeij, 1996; Williamson, 1996). Here, we follow the terminology suggested by Richardson *et al.* (2000) (Table 13.1).

To invade, a species' propagules must first disperse to a recipient site. Dispersal provides the opportunity for invasion. Long-distance dispersal mechanisms are no longer an important feature of invasive species because human actions assist the movement of propagules (Williamson, 1989). In effect, human activity is expanding the total species pool (Hobbs, 2000). Most introductions fail because species die in transit or shortly after they arrive (Lodge, 1993).

Species that can persist and reproduce for some time after arrival, but rely on continued introductions to maintain their population are termed casuals (or transients). For example, of the 348 'wool aliens' introduced and established around the wool mills of Galashiels in Britain, none remained

Table 13.1. Terminology associated with the process of ecological invasions (adapted from Richardson *et al.*, 2000).

Term	Explanation
Non-native plants	Plant taxa in a given area whose presence is due to intentional or accidental introduction as a result of human activity Synonyms: exotic, alien, non-indigenous
Casual non-native plants	Non-native plants that do not form self-replacing populations, and which depend on repeated introductions to persist, although individuals may flourish and occasionally reproduce. Synonyms: transient, waif, occasional escape
Naturalized plants	Non-native plants that form self-sustaining populations but do not necessarily invade natural, managed or human-made ecosystems
Invasive plants	Naturalized plants that produce reproductive offspring, often in high numbers and at considerable distance from the parent plants, and thus have the potential to spread

after the mills were closed down (Crawley, 1987). Similarly, corn cockle (*Grostema githago*) was a weed in Britain only as long as it was reintroduced during sowing of maize (Firbank, 1988). With the advent of seed cleaning, the species virtually disappeared.

Naturalized species are those that are able to persist solely through reproduction, but are not necessarily increasing in distribution. Vermeij (1996) referred to an established population as one where the population is self-sustaining from local reproduction and recruitment.

To become invasive, a naturalized plant must be able to disperse and withstand the environmental conditions of its new habitat. The time when a naturalized species changes from non-invasive to invasive is not necessarily distinct, and a species may revert back over time. For example, recall the controversy over whether or not purple loosestrife is a serious weed problem (Chapter 2); this may be an argument over a species that is naturalized to some extent. Most may argue that it is and always will be highly invasive. Others may believe it is reverting to non-invasive status. Still others can argue that purple loosestrife was just a visible but relatively low-impact weed. What complicates the status of purple loosestrife is disagreement over when to measure the ultimate impact of an invading species, i.e. sound the alarm early while there is time to do something or wait and see if other species adapt to its presence. This essentially reflects the fact that there are many stages to an invasion.

During the many invasion stages, different species will be lost at the transition of one stage to another. However we define

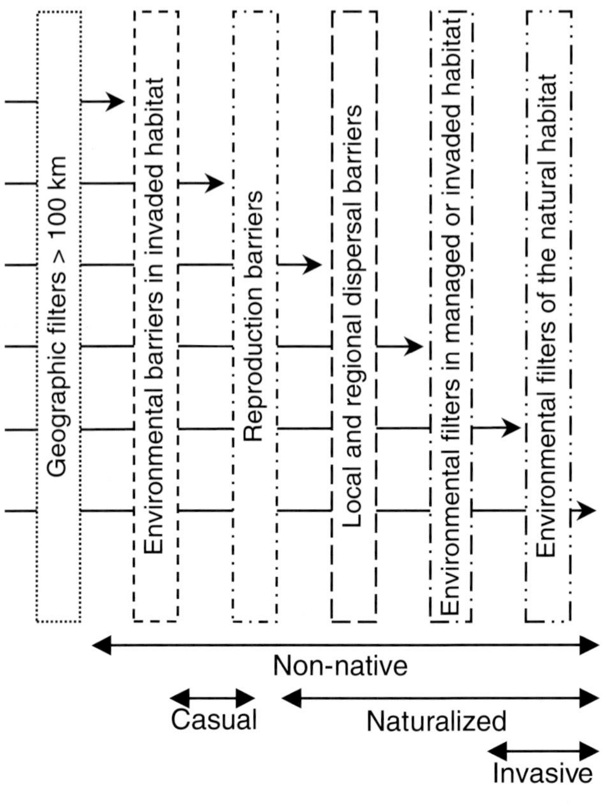

Fig. 13.2. A schematic representation of major filters limiting the spread of non-native plants (redrawn from Richardson *et al.*, 2000).

them, the limiting filters will be different at each transition. Over the course of an invasion, the filters change and the traits necessary to succeed are different. In Fig. 13.2 each filter is composed of a unique set of filters and therefore different traits (and therefore species) will be removed.

Time lags

A striking characteristic of invasions is that there is usually a lag between the time when a species is introduced and when its population growth explodes (Hobbs and Humphries, 1995; Kowarik, 1995; Crooks and Soulé, 1999). Time lags can sometimes be quite long; for example, white pine (*Pinus strobus*) was not considered invasive in Central Europe until more than 250 years after it was introduced for forestry (Rejmánek, 1996). In Britain, wild lettuces (*Lactuca virosa* and *Lactuca scariola*) were considered rare from the 1960s, when they were introduced, to the mid-1900s when their abundance increased as they spread into gravel pits (Crooks and Soulé, 1999). The lag phase commonly lasts from 20 to 100 years (Hobbs and Humphries, 1995; Wade, 1997).

The potential invasiveness of a species cannot be measured by its lack of a time lag. For example, of three non-native *Impatiens* introduced to Britain, Himalayan balsam (*I. glandulifera*) had the longest time lag, but was the most invasive (Fig. 13.3) (Perrins *et al.*, 1993). Jewelweed (*I. parviflora*) and spotted jewelweed (*I. capensis*) increased faster initially, but did not spread as far. Himalayan balsam is more tolerant to frost and therefore can establish early, forming thick monospecific stands. It is also more likely to be dispersed over a long distance via human plantings in gardens, and this forms foci for spreading to new regions. The spread of jewelweed is limited because it has more specific environmental requirements and therefore seed are less likely to land in a suitable habitat. Therefore, species that spread rapidly after introduction may not be the most invasive in the long-term. A species' ability to persist during a lag phase will also vary depending on the habitat type, and geographical location. The lag phase for Himalayan balsam was 40 years in Germany and 80 years in Britain (Wade, 1997).

Part of this time lag is expected because population growth follows an exponential growth curve (Chapter 3). Small populations will appear to grow very slowly ini-

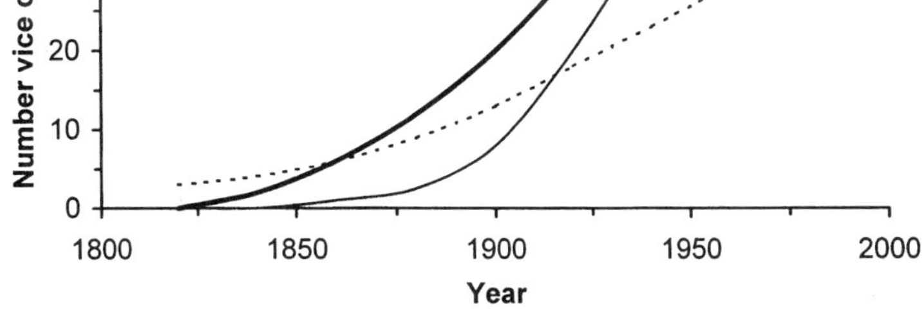

Fig. 13.3. Invasion process showing time lag of three *Impatiens* species introduced to Britain. Spread of the species is measured as the number of vice counties where it is present. Species are: Himalayan balsam (*I. glandulifera*), jewelweed (*I. parviflora*) and spotted jewelweed (*I. capensis*) (redrawn from Perrins *et al.*, 1993).

tially. The length of a time lag is dependent on the physical and biotic environment, and on any genetic changes that occur (Crooks and Soulé, 1999). Also, small populations will have low seed set if pollinators are hard to attract or because little wind-dispersed pollen reaches plant stigmas. Therefore, we can expect population expansion to be slow at first.

Time lags can also occur when species persist in small isolated pockets until a disturbance or a certain set of environmental conditions occurs that facilitates its rapid expansion. This happened in the case of Oxford ragwort (*Senecio squalidus*). Its abundance increased after World War II because it invaded railway lines and disturbed areas created by bombings (Baker, 1965; Kowarik, 1995; Crooks and Soulé, 1999). Populations of agricultural weeds also change in response to farming practices; for example, the expansion of Canada horseweed (*Conyza canadensis*) was facilitated by the increase in reduced tillage.

A final reason why small populations may have a sudden rapid increase in population size is that they may undergo some genetic change that increases the population's fitness. Hybridization and introgression (hybrid back-crossing with one of the parent populations and therefore new genes are introduced back into the species genome) are mechanisms by which genetic change can occur. For example, the introduced smooth cordgrass (*Spartina alterniflora*) has hybridized with the native California cordgrass (*Spartina foliosa*) and the hybrid is now spreading more rapidly than the native species (Daehler and Strong, 1997). Ellstrand and Schierenbeck (2000) list 28 well-documented examples of invasiveness occurring following hybridization.

Populations can have other types of genetic change. Weed populations that become resistant to prevalent herbicides can experience rapid population growth. For example, Canada horseweed that has become resistant to glyphosate (herbicide) has increased in Delaware, USA (vanGessel, 2001).

Invasion Success Depends on a Species' Invasive Ability and Habitat Invasibility

Invasions fail when the species does not possess traits necessary to survive and reproduce in a new habitat. From this it is clear that habitat characteristics as well as species traits determine the progress of an invasion. In the chapter introduction, we mentioned the native hay-scented fern which becomes invasive only when intensive deer browsing occurs (de la Cretaz and Kelty, 1999). In this case, the plant was not inherently invasive nor is the habitat inherently invasible. It was the combination of plant species traits and a change in the habitat characteristics (increased disturbance) that led to the invasion. While some plants are inherently 'weedy', others require specific environmental conditions to become invasive. Conversely, some habitats are more invasible, while others are more invasion resistant. In the case of hay-scented fern, only forest sites with high levels of grazing are invaded. Therefore, we must examine both species traits and habitat characteristics to understand how and why invasions occur (Lodge, 1993).

Species' invasive ability

It would be useful if we could list the traits of a plant species and from that list determine when and where a species will invade. Unfortunately this is not possible, but there are some ways that we can get a general idea whether a species is likely to become invasive.

Life history traits

Much of the early work on invasions tended to list traits likely to increase a species' invasive ability. Baker (1974) produced his classic list of traits of the ideal weed (Box 1.1). Lodge (1993) summarized traits commonly associated with weediness but noted that when tested, many of these are statistically rejected, or else there are too many exceptions to make them useful. A high r, for example, does not necessarily increase invasiveness (Lawton and Brown, 1986). Trait lists only provide a few indicators that can help explain a weed's invasiveness, but they are not helpful in predicting what species will become invasive (Perrins *et al.*, 1992). Himalayan balsam, for example, is a problem weed in Britain and yet possesses only two of Baker's characteristics. Other weeds such as common field-speedwell (*Veronica persica*) and common chickweed (*Stellaria media*) possess many of Baker's traits but are not as problematic (Perrins *et al.*, 1993). Native and non-native species with expanding distribution in England, Scotland, the Republic of Ireland and the Netherlands were almost indistinguishable as far as trait lists (Thompson *et al.*, 1995).

The ability to reproduce clonally has also been cited as being important to invasion (Leakey, 1981; Baker, 1986; Lodge, 1993) and a number of weedy species rely on clonal reproduction (e.g. horsetail *Equisetum* spp., bracken fern *Pteridium aquilinum*). However, in central Europe, 69% of native species have the potential for clonal growth, whereas only 35% of the established non-native species are clonal. Clonal reproduction may be more advantageous to weeds of native habitats. For example, in the Czech Republic 62% of non-natives invading natural habitats were clonal, whereas 33% of non-natives invading managed habitats (e.g. agricultural land) were clonal (Pyšek, 1997). Pyšek (1997) suggested that while clonal growth is useful in some situations, it is not a requirement for successful invasion.

Is it possible, then, to predict whether a species will invade based on its traits? Under specific circumstances, there has been limited success. For example, Rejmánek and Richardson (1996) were able to predict the invasive ability of pines in the southern hemisphere. Using only three traits (seed mass, length of juvenile period and interval between seed mast years) they were able to create an invasion equation (Fig.

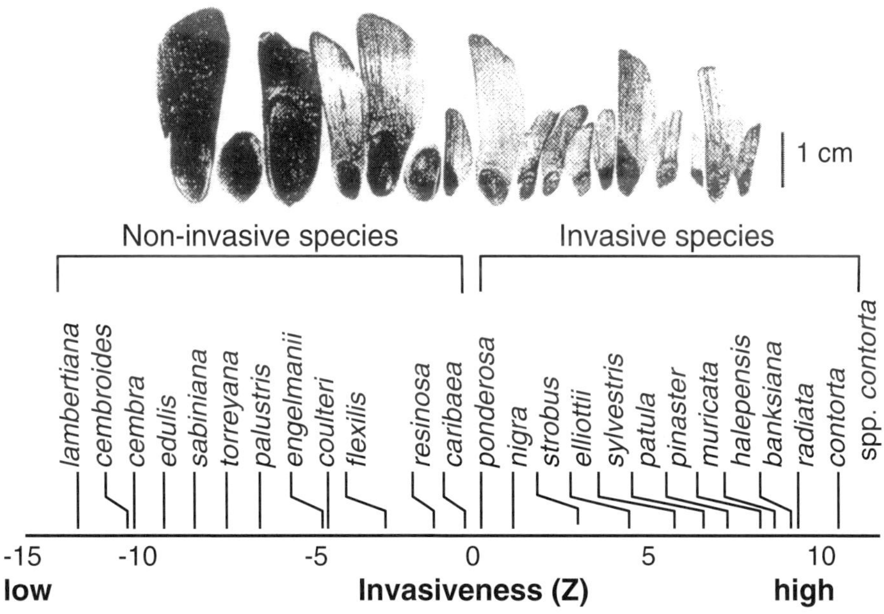

$$Z = 19.77 - 0.51\sqrt{M} - 3.14\sqrt{J} - 1.21S$$

Fig. 13.4. Placement of pines (*Pinus*) along an invasion gradient. The invasiveness (Z) equation was determined by using a discriminant analysis based on mean seed mass (M in mg), mean interval between large crop years (S in years), and minimum juvenile period (J in years) (Rejmánek, 1999; with permission of Kluwer Academic Publishers and the author).

Table 13.2. Rules for detecting invasive woody plants based on fruit and seed traits, the potential for vertebrate dispersal and a discriminant function '*Z*' (Fig. 13.4) (Rejmánek, 1999).

Discriminant function *Z**	Fruit and seed traits	Opportunities of dispersal by vertebrates: Absent	Present
Z>0	Dry fruit Large (>2 mg) seeds	Likely invasive	Very likely invasive
Z>0	Dry fruit Small (<2 mg) seeds	Likely invasive in wet habitats	
Z>0	Fleshy fruit	Unlikely invasive	Very likely invasive
Z<0		Non-invasive unless dispersed in water	Possibly invasive

* *Z* values >0 indicate invasiveness according to a discriminant function based on seed size, intervals between large seed crops and minimum juvenile period (see Fig. 13.4).

13.4). More recently, this equation has been found to apply to other woody plants (Table 13.2), and along with a decision tree (Fig. 13.5) can be used to decide whether to admit a woody species into North America (Reichard and Hamilton, 1997).

Large geographical native range

Species with a large native range often have the potential to be successful invaders. Goodwin *et al.* (1999) tested this by comparing pairs of species from the same genera where one species had successfully invaded New Brunswick, Canada, and the other had not. They found that the size of a species' native range was the single best predictor of invasiveness and could predict whether a species would be invasive or non-invasive for about 70% of the species tested. Other attributes such as plant height and length of flowering period were less useful.

Range seems to be important for two reasons (Scott and Panetta, 1993; Rejmánek, 1995; Goodwin *et al.*, 1999). Firstly, widespread species are more likely to be dispersed simply because they are in more locations and therefore are more likely to be picked up by dispersal agents. Secondly, widespread species are more likely to be adapted to a wider range of environmental conditions and therefore are more likely to survive in the abiotic environment of a new habitat following dispersal.

Native range, however, is not always a good indication of invasiveness. In cases where a species is controlled by biotic factors such as herbivory or disease in its native range, it may spread rapidly when released from these in a new habitat (Scott and Panetta, 1993). Monterey pine (*Pinus radiata*) is an example of a species with a small native range, but which became weedy when dispersed to other habitats (Chapter 2) (Richardson and Bond, 1991).

Taxonomic patterns

It is difficult to come up with taxonomic patterns related to invasiveness, because plant identification and taxonomy are not consistent around the world (Heywood, 1989; Pyšek, 1998). Also, we do not have the extensive data required to come to such conclusions. Nevertheless, we can make some generalizations. After examining non-native species in 26 regions of the world, Pyšek (1998) concluded that families with invasive species were dispersed throughout the phylogenetic tree, but that some families were concentrated together. For example, many of the *Asteridae* and *Caryophyllidae* subclasses had families with many invasive species in them. If we look at the number of invasive species as a proportion of the total species number in a family, then *Amaranthaceae*,

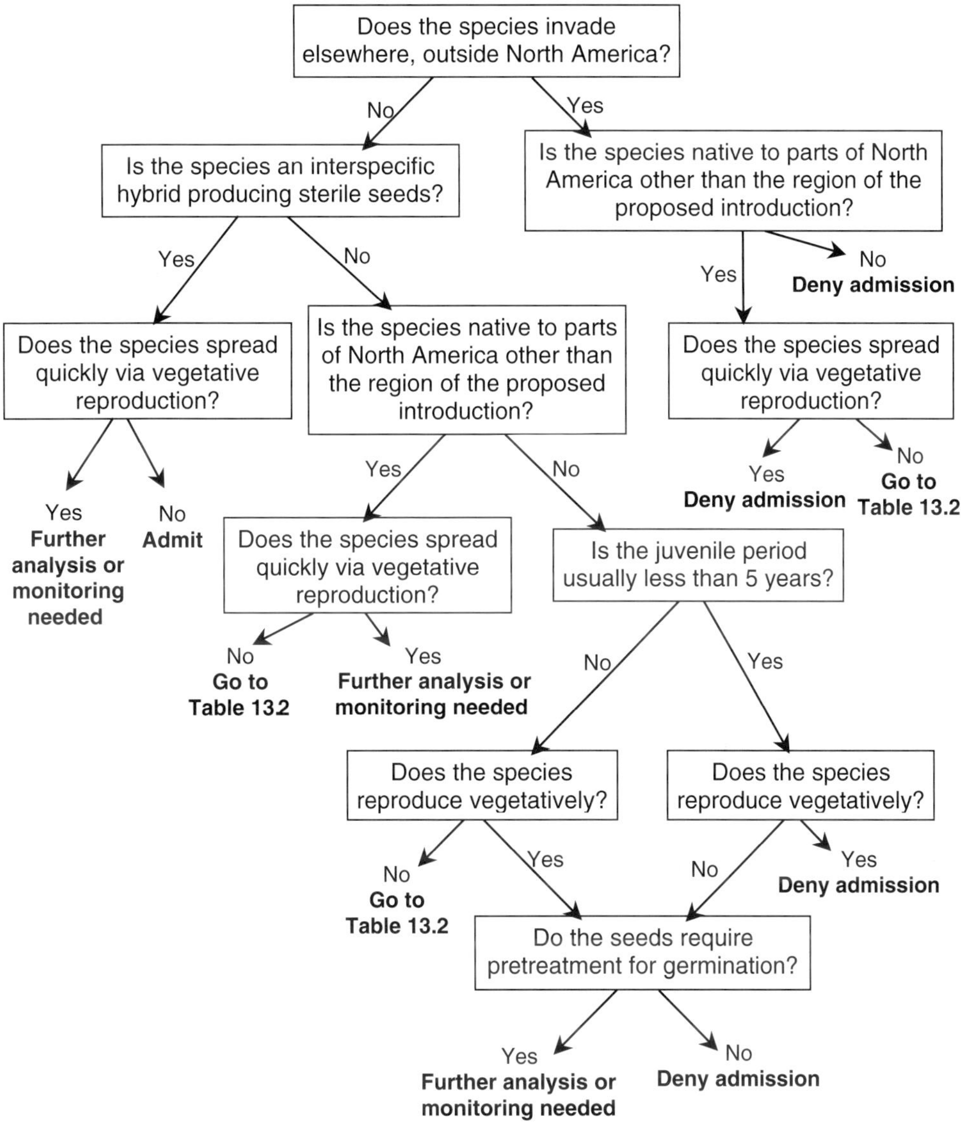

Fig. 13.5. Decision tree for admission of non-native trees into North America (redrawn from Reichard and Hamilton, 1997).

Chenopodiaceae, *Cruciferae*, *Poaceae*, *Papaveraceae* and *Polygonaceae* have the highest proportion of invasive species. Daehler (1998) concluded that based on family size, *Amaranthaceae*, *Convolvulaceae*, *Cyperaceae*, *Papaveraceae*, *Poaceae* and *Polygonaceae* had more agricultural terrestrial weeds than expected.

Habitat invasibility

Not all communities are equally invasible. Being able to characterize or predict what communities or habitat types are easier to invade would have obvious practical benefits. We could work on protecting vulnerable habitats and not worry about invasion resistant ones. Alas, we cannot do this. Three habitat characteristics that are generally thought to encourage invasions are disturbances, low species richness and resource availability.

Role of disturbance

Disturbance is often cited as a precursor to invasions (Baker, 1965; Fox and Fox, 1986; Orians, 1986; Crawley, 1987; Hobbs and Huenneke, 1992) because it provides weeds with a window of opportunity (Hobbs, 2000). In fact, any natural or managed ecosystem will experience disturbance and often specific types of disturbances may be required to maintain a community. Fire, for example, is a natural part of the boreal forest ecosystem. Disturbance may even inhibit invasion (Lodge, 1993); for example fire is used to control the invasion of yellow star thistle (*Centaurea solstitialis*) in California grassland (Hastings and DiTomaso, 1996).

While disturbance is a natural part of an ecosystem, human activities alter the disturbance regime and intensity (Hobbs, 2000). Disturbance can change or remove the filters acting on a community. Therefore, disturbance alters habitat characteristics such that they become more conducive to the spread of weeds (Vitousek *et al.*, 1997; Dukes and Mooney, 1999). A change in the type, frequency and extent of a disturbance in a community will create different invasion opportunities (Crawley *et al.*, 1999; Lonsdale 1999; Stohlgren *et al.*, 1999). The expansion of foxtail barley (*Hordeum jubatum*) and dandelion (*Taraxacum officinale*), for example, were facilitated by an increase in reduced tillage (Derksen *et al.*, 1993; Shrestha *et al.*, 2002). Plants invading from European Mediterranean climate into similar climates in Chile and California had more impact in Chile than California because the types of disturbances introduced were similar to those in California but quite novel to Chile (Holmgren *et al.*, 2000). Introducing a new species can alter the disturbance regime such that the extant species are no longer able to persist. Lehmanns lovegrass (*Eragrostis lehmanniana*) introduced into North America is more flammable than the native grasses, and has changed the fire regime of the community (Anable *et al.*, 1992).

Disturbance, however, is not always necessary for an invasion to occur, and even intact natural ecosystems can be invaded. For example, King and Grace (2000) found that cogongrass (*Imperata cylindrica*) did not require disturbance gaps to invade wet pine savanna. Similarly, melastone (*Tibouchina herbaceae*) is a small introduced perennial that invades undisturbed wet native forests of Hawaii and Maui (Almasi, 2000).

Role of diversity

Elton (1958) was the first to hypothesize that there was a negative relationship between native species richness (diversity) and community invasibility. Elton suggested that communities with many species would be invasion resistant, while species poor communities will be highly invasible. There is evidence both to support and refute this hypothesis. We will examine both.

A variety of experimental studies support the hypothesis that species poor communities are more invasible (Rejmánek, 1989; Case, 1990; Tilman, 1997, 1999; Doak *et al.*, 1998; Knops *et al.*, 1999). Tilman (1997, 1999) observed a negative relationship between species richness and invasibility showing that high species richness was related to a decrease in the invasion success in prairie grasslands. There was a similar negative relationship between native species richness and the percentage of species that were non-native in two shrubland and one heathland site in Australia (Fox and Fox, 1986).

Elton's idea does remain controversial (Crawley *et al.*, 1999; Levine and D'Antonio, 1999; Lonsdale, 1999; Mack *et al.*, 2000) because there is also evidence that areas with high species richness are *more* vulnerable to invasion than species-poor communities. Huston and DeAngelis (1994) highlighted that since species-rich communities have high levels of habitat heterogeneity, introduced species are likely to find suitable habitats to invade. The evidence supporting this relationship is typically derived from observational or correlation studies in natural habitats (Levine, 2000). Results typically show that species-rich communities have the highest number of non-native invasive species (Robinson *et al.*, 1995; Planty-Tabacchi *et al.*, 1996; Stohlgren *et al.*, 1998, 1999, 2001).

There may be no general relationship between richness and invasibility, or it may be that we are asking too general a question. Species richness may simply be too broad a factor to explain the relative invasibility of communities (Levine and D'Antonio, 1999) because it is more likely an aggregation of other variables that are determinants of invasibility. Diverse communities may have more invasive species simply because the environment is favourable to many species (Levine and D'Antonio, 1999; Lonsdale, 1999; Levine, 2000). Factors that allow more species to co-exist may also promote invasion (Levine, 2000) giving the appearance of cause-and-effect.

A community's composition can complicate how we interpret richness effects. Observed effects of richness on invasibility

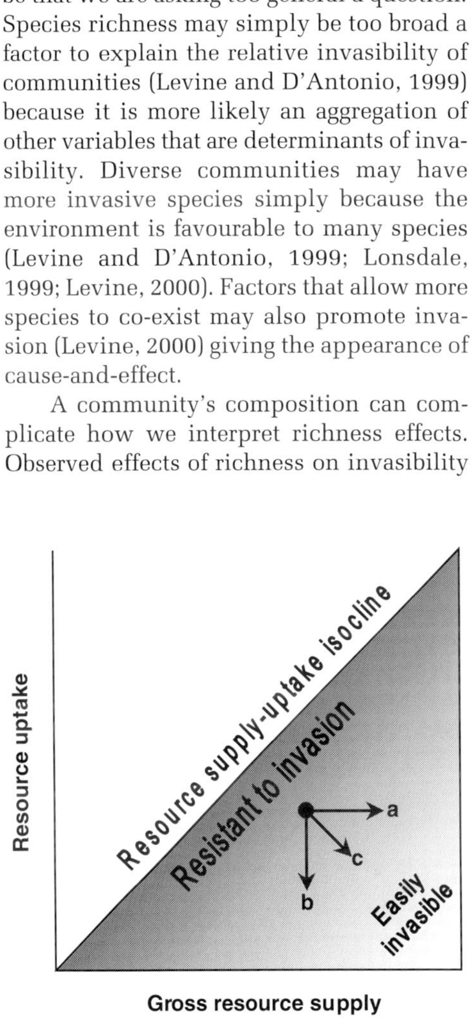

Fig. 13.6. Theory of fluctuating resource availability. The invasibility of a community is dependent on the ratio of gross resource supply to resource uptake. A community becomes more invasible if gross supply is increased (vector a), if resource uptake decreases (vector b) or if both occur (vector c) (redrawn from Davis *et al.*, 2000).

can be due to either the overriding effect of one or several keystone species or due to the effect of richness itself. Having more species increases the chance that one capable of excluding the introduced species will be present (Grime, 1997; Hooper and Vitousek, 1998; Lepš *et al.*, 2001). Therefore, species effects may appear to be richness effects.

Resource availability

Davis *et al.* (2000) proposed that resource availability may be used to explain the invasibility of a community. They compared gross resource supply with resource uptake and proposed that communities are invasible when resource supply is greater than resource uptake (Fig. 13.6). Therefore, a nutrient-rich community will not be invasible as long as the current community is using all the available nutrients. Conversely, a nutrient-poor community will be invasible if the community is not sequestering all of the nutrients.

This hypothesis incorporates ideas of both species richness and disturbance in explaining invasibility. For example, disturbance has often been associated with invasion because it reduces or eliminates competitors or because it increases the availability of resources (D'Antonio, 1993; Hobbs, 1989). According to Davis *et al.* (2000) disturbance is only one mechanism that alters resource availability. When the community is disturbed, or when a species is removed, then released resources become available to other invaders.

Phenomena other than disturbance can also release nutrients. For example, herbivory by a leaf beetle (*Trirhabda canadensis*) in experimental monocultures of prairie goldenrod (*Solidago missouriensis*) increased the availability of water, nitrate and light (Brown, 1994). Even though the relative growth rate of prairie goldenrod increased in response to herbivory, more species were able to invade compared to monocultures without the beetle.

Changing the abiotic environment can also alter the nutrient status of a community. The environment is never constant. It changes in regular predictable ways such as in the daily or seasonal fluctuations we observe. It also changes in unpredictable ways; for example, we may experience a particularly dry summer. Such natural fluctuations in the abiotic environment will affect whether a species can invade and/or persist in a community. For example, a drought may create a window of opportunity for invasion when it kills species and more light becomes available. Alternatively, a flood may bring in an influx of nutrients.

The fact that resource availability fluctuates over time helps to explain why a community's invasibility changes over time. A community may have regular or intermittent periods of vulnerability to invasion dependent on resource availability and use. This is why the question '*Which* communities are more invasible?' is misleading. Rather, we should be asking '*When* is a community invasible?' According to Davis *et al.* (2000) communities become invasible when resources are available in excess of what can be used by the extant community. Clearly, we cannot predict whether a community will be invasible based on simple habitat characteristics. A community's susceptibility to invasion is dependent on many factors and will change over time.

The Impact of Invasion

In natural ecosystems, species like buckthorns (*Rhamnus* spp.), canegrass (*Phragmites communis*), garlic mustard and kudzu (*Pueraria montana*) are obvious invaders because they can dominate an ecosystem (Anderson *et al.*, 1996; Reinartz, 1997; Galatowitsch *et al.*, 1999; Pappert *et al.*, 2000). However, even seemingly innocuous and familiar species like the Norway maple (*Acer plantanoides*), a tree commonly planted along urban roadways in North America, can devastate a natural ecosystem and severely reduce diversity of native species (Webb *et al.*, 2000). But just how do we quantify the effect of an invader?

In North America, we hear that purple loosestrife is having a major impact on wetlands, but just what do we mean by impact? Do we mean that purple loosestrife has a large and increasing range, that it is present in high abundance, or that each individual has a large effect on other species or on the community function? In fact the ecological impact is a combination of all of these (Parker *et al.*, 1999).

Range and abundance are fairly straightforward to quantify, and so we tend to use them synonymously with impact (Parker *et al.*, 1999), but they only tell part of the story. A plant could be widespread and abundant but still be fairly benign. Both oxeye daisy (*Chrysanthemum leucanthemum*) and common mullein (*Verbascum thapsus*) are examples of this. Sometimes the evidence of invasiveness is based solely on these criteria, with little knowledge of the effect per individual plant. For example, we know that the range of purple loosestrife is expanding and that it is highly abundant in some habitats, but there is little research on its effect, so we cannot quantify its impact (Hager and McCoy, 1998). Quantifying per capita effect is complicated because effects can be on a species' genetics, on individuals, population dynamics, communities or ecosystems (Table 13.3) (Parker *et al.*, 1999). This is a long list, and so far there is no solution to how we should measure effect in a consistent manner.

The effect of one species' invasion is further complicated by synergistic effects when more than one species invades. That is, the combined effect of two species may be greater than the effect of the two species individually. Synergistic effects may not be predictable. 'Invasion meltdown' is the acceleration of impacts on native ecosystems due to synergistic interactions (Simberloff and von Holle, 1999).

Invasion meltdown occurs in two ways (Parker *et al.*, 1999). First, established invaders may facilitate other invaders and therefore each successive invasion becomes easier. An example of this is the Japanese white-eye bird (*Zosterops japonicus*) introduced into Hawaii. It is the only bird to eat and subsequently disperse seeds of the invasive fire tree (*Myrica faya*) (Woodward *et al.*, 1990). Invasion of this plant was further facilitated by introduced pigs that also disperse seeds and create disturbance that

Table 13.3. Types of per capita effects that an invasive species can have on a species' genetics, or on individuals, population dynamics, communities or ecosystems (based on information in Parker *et al.*, 1999).

Effect on ...	Examples
Species' genetics	Hybridization Changing selection pressure Changing gene flow
Individual's population dynamics	Effect growth or reproduction of other species Abundance or dynamics of native population Alter species interactions
Community	Relative abundances Species richness Diversity
Ecosystems	Resource dynamics Disturbance type and frequency

favours the establishment of weeds over native species. Once the fire tree invades, it alters nutrient and water cycles facilitating the invasion of other non-native species (Vitousek *et al.*, 1987). Therefore, what is the effect of the Japanese white-eye bird?

The second way that invasion meltdown occurs is that each attempted invasion disrupts the abundances of native species to some extent, and thus the community becomes more and more invasible as the total number of invasion attempts increases. Therefore even unsuccessful invasions can influence long-term community dynamics. However, unsuccessful invaders still contribute to the invasion effect, but complicate our estimation of effect per individual.

Explaining vs. Predicting Invasion

We have become fairly proficient at explaining why a species has successfully invaded a habitat. Unfortunately such explanations come after the fact and so cannot be used to stop the invasion. Usually we use some type of *post hoc* analysis of how weeds fit with the trait list. Once a weed has successfully invaded, it is sometimes possible to look back and explain why it was successful.

Box 13.1. Ten general predictions that can be made about plant invasion (adapted from text in Rejmánek, 1999, 2000).

1. Invasive species are more likely to be able to maintain their fitness over a range of environments through phenotypic plasticity.
2. Invasive species are more likely to have a small genome size. This character is associated with small seed size and a high leaf area ratio.
3. Woody species are more likely to invade disturbed areas if they have small seeds, a short juvenile period and short time span between large seed crop years.
4. Seeds of woody species that invade disturbed and undisturbed habitats are more likely to be vertebrate dispersed.
5. Invasive herbaceous species often have large native (latitudinal) ranges.
6. Invasive species often reproduce vegetatively, especially at higher latitudes.
7. Non-native species are more likely to be successful if they are from a non-native genus that is not represented in the native flora.
8. Invasive species are less likely to be dependent on specific mutualists (pollinators, mycorrhizal fungi), and therefore will be able to overcome the abiotic and biotic filter in a new environment.
9. Tall plants often invade both disturbed and undisturbed herbaceous communities especially in mesic habitats.
10. Non-native species are often spread by human activities.

Harder though, is it possible to predict what species would likely be good invaders. 'The complexity of the interaction between species and community – each of which is difficult enough to characterize – is a central reason why predictions about specific introductions are so difficult to make' (Lodge, 1993). Some general predictions have been made. For example, Rejmánek (1999) summarized ten general predictions we can make (Box 13.1). While useful, such predictions cannot tell us whether a specific species will invade a specific habitat.

At this point, we should distinguish between *a posteriori* explanations of why a weed has invaded, and *a priori* predictions of what species are likely to invade. *A posteriori* explanations of invasive species are done after the invasion has occurred, and so can not be used to predict whether a species will become invasive. For example, after a weed has invaded, we can scan Baker's list of ideal weed characteristics and select the ones applicable to the species of interest. Proso millet (*Panicum miliaceum*), for example, is a successful invader of maize and soybean in Ontario, Canada, because it has dormant, hard-coated seeds with a prolonged emergence pattern and it is tolerant to several herbicides. When we look back, it is easy to tell why proso millet became a weed, but its weediness was not predicted beforehand. While adding to our knowledge of invasive species, such *a posteriori* explanations do not increase our power of prediction. A species with many 'weedy' traits will not necessarily be a weed and thus lists cannot be used to predict which species will become weeds. Furthermore, not all invasive species are likely to possess the same set of traits, making predictions impossible (Crawley, 1987).

Correct prediction has two parts (Daehler and Carino, 2000; Pheloung *et al.*, 1999);

- correctly identify the suite of traits that a weed will require to be a successful invaders;
- correctly identify which traits will make a species an *un*successful invader.

The first is useful to predict what species we want to prevent from invading or to minimize its spread. The second is useful to decide what species we do not need to worry about. Identifying both invaders and non-invaders will save money because it would allow us to focus management efforts.

In some cases, more specific predictions are possible. For example, Scott and Panetta (1993) were able to predict what southern African species were likely to become agricultural weeds in Australia. Species likely to become agricultural weeds in Australia:

- were already weeds in southern Africa;
- occurred in a wide range of climate types in their native range; and
- had congeneric species that were weeds.

Using these data, the authors looked at 55 newly introduced (since 1950) species and identified seven species likely to become agricultural weeds in Australia. They also identified 20 species in southern Africa that would likely become weeds if successfully introduced into Australia. The difficulty with making predictions is that we will not know if they are correct until the predictions have been tested.

Summary

Is it better to have one highly abundant invasive species, or many invaders that are each less abundant, but have the same total biomass? There is no correct answer to this, but it does highlight some important issues in invasive ecology. Understanding how and why invasions occur is complicated by the fact that most invasions fail. Furthermore, it is difficult to identify successful invaders during the important early stages of an invasion. Since we do know that both species traits and habitat characteristics have to be considered in order to understand why an invasion occurred, we can begin the next important step in weed ecology. This is to begin predicting when, how, and why invasions of weeds will occur, what their impact may be, and what (if anything) can and should be done to manage invasions.

Questions

1. What types of habitats does your selected weed species invade? Why is it able to invade so many (or so few) types of habitats?
2. Why are some communities more vulnerable to weed invasions?
3. What are some difficulties to determining the per capita effect of purple loosestrife (or another weed) on the invaded community.
4. Look at the decision tree by Reichard and Hamilton (1997). Explain the importance of each decision. Why would one option lead to an invasion?
5. Look at the predictions made by Rejmánek in Box 13.1. Explain why each is likely to be true.

General References

Williamson, M. (1999) Invasions. *Ecography* 22, 5–12.

Richardson, D.M., Pyšek, P., Rejmánek, M., Barbour, M.G., Panetta, F.D. and West, C.J. (2000) Naturalization and invasion of alien plants: concepts and definitions. *Diversity and Distributions* 6, 93–107.

Mack, R.N., Simberloff, D., Lonsdale, W.M., Evans, H., Clout, M. and Bazzaz, F.A. (2000) Biotic invasions: causes, epidemiology, global consequences, and control. *Ecological Applications* 10, 689–710.

Literature Cited

Almasi, K.N. (2000) A non-native perennial invades a native forest. *Biological Invasions* 2, 219–230.

Anable, M.E., McClaren, M.P. and Ruyle, G.B. (1992) Spread of introduced Lehmann lovegrass *Eragrostis lehmanniana* Nees. in Southern Arizona, USA. *Biological Conservation* 61, 181–188.

Anderson, R.C., Dhillion, S.S. and Kelley, T.M. (1996) Aspects of the ecology of an invasive plant, garlic mustard (*Alliaria petiolata*), in Central Illinois. *Restoration Ecology* 4, 181–191.

Baker, H.G. (1965) Characteristics and modes of origin of weeds. In: Baker, H.G. and Stebbins, G.L. (eds) *The Genetics of Colonizing Species*. Academic Press, New York, pp. 147–172.

Baker, H.G. (1974) The evolution of weeds. *Annual Review of Ecology and Systematics* 5, 1–24.

Baker, H.G. (1986) Characteristics and modes of origin of weeds. In: Baker, H.G. and Stebbins, G.L. (eds) *The Genetics of Colonizing Species*. Academic Press, New York, pp. 147–168.

Brown, D.G. (1994) Beetle folivory increases resource availability and alters plant invasion in monocultures of goldenrod. *Ecology* 75, 1673–1683.

Case, T.J. (1990) Invasion resistance arises in strongly interacting species-rich model competition communities. *Proceedings of the National Academy of Sciences USA* 87, 9610–9614.

Crawley, M.J. (1987) What makes a community invasible? In: Gray, A.J., Crawley, M.J. and Edwards, P.J. (eds) *Colonization, Succession and Stability*. Blackwell Scientific, Oxford, pp. 429–453.

Crawley, M.J., Brown, S.L., Heard, M.S. and Edwards, G.R. (1999) Invasion resistance in experimental grassland communities: species richness or species diversity. *Ecology Letters* 2, 140–148.

Crooks, J.A. and Soulé, M.E. (1999) Lag times in population explosions of invasive species: causes and implications. In: Sandlund, O.T., Schei, P.J. and Vilken, A. (eds) *Invasive Species and Biodiversity Management*. Kluwer Academic, Boston, pp. 103–125.

D'Antonio, C.M. (1993) Mechanisms controlling invasion of coastal plant communities by the alien succulent *Carpobrotus edulis*. *Ecology* 74, 83–95.

Daehler, C.C. (1998) The taxonomic distribution of invasive angiosperm plants: ecological insights and comparison to agricultural weeds. *Biological Conservation* 84, 167–180.

Daehler, C.C and Carino, D.A. (2000) Predicting invasive plants: prospects for a general screening system based on current regional models. *Biological Invasions* 2, 93–102.

Daehler, C.C. and Strong, D.R. (1997) Hybridization between introduced smooth cordgrass (*Spartina alterniflora*; Poaceae) and native California cordgrass (*S. foliosa*) in San Francisco Bay, California, USA. *American Journal of Botany* 84, 607–611.

Davis, M.A., Grime, J.P. and Thompson, K. (2000) Fluctuating resources in plant communities: a general theory of invasibility. *Journal of Ecology* 88, 528–534.

de la Cretaz, A.L. and Kelty, M.J. (1999) Establishment and control of hay-scented fern: a native invasive species. *Biological Invasions* 1, 223–236.

Derksen, D.A., Lafond, G.P., Thomas, A.G., Loeppky, H.A. and Swanton, C.J. (1993) Impact of agronomic practices on weed communities: tillage systems. *Weed Science* 41, 409–417.

Doak, D.F., Bigger, D., Harding, E.K., Marvier, M.A., O'Malley, R.E. and Thomson, D. (1998) The statistical inevitability of stability-diversity relationships in community ecology. *American Naturalist* 151, 264–276.

Drake, J.A., Huxel, G.R. and Hewitt, C.H. (1996) Microcosms as models for generating and testing community theory. *Ecology* 77, 670–677.

Drake, J.A., Zimmermann, C.R., Purucker, T. and Rojo, C. (1999) On the nature of the assembly trajectory. In: Weiher, E. and Keddy, P. (eds) *Ecological Assembly Rules: Perspectives: Advances and Retreats.* Cambridge University Press, Cambridge, pp. 233–250.

Dukes, J.S. and Mooney, H.A. (1999) Does global change increase the success of biological invaders. *Trends in Ecology and Evolution* 14, 135–139.

Ellstrand, N.C. and Schierenbeck, K.A. (2000) Hybridization as a stimulus for the evolution of invasiveness in plants? *Proceedings of the National Academy of Sciences USA* 97, 7043–7050.

Elton, C.S. (1958) *The Ecology of Invasions by Animals and Plants.* Methuen, London.

Firbank, L.G. (1988) Biological flora of the British Isles: *Agrostemma githago. Journal of Ecology* 76, 1232–1246.

Fox, M.D. and Fox, B.J. (1986) The susceptibility of natural communities to invasion. In: Groves, R.H. and Burdon, J.J. (eds) *Ecology of Biological Invasions.* Cambridge University Press, Cambridge, pp. 57–66.

Galatowitsch, S.M., Anderson, N.O. and Ascher, P. D. (1999) Invasiveness in wetland plants in temperate North America. *Wetlands* 19, 733–755.

Gatto, M. and de Leo, G.A. (2000) Pricing biodiversity and ecosystem services: the never-ending story. *BioScience* 50, 347–355.

Goodwin, B.J., McAllister, A.J. and Fahrig, L. (1999) Predicting invasiveness of plant species based on biological information. *Conservation Biology* 13, 422–426.

Grime, J.P. (1997) Biodiversity and ecosystem function: the debate deepens. *Science* 277, 1260–1261.

Hager, H.A. and McCoy, K.D. (1998) The implications of accepting untested hypotheses: a review of the effects of purple loosestrife (*Lythrum salicaria*) in North America. *Biodiversity and Conservation* 7, 1069–1079.

Hastings, M.S. and DiTomaso, J.M. (1996) Fire controls yellow star thistle in California grasslands. *Restoration Management Notes* 14, 124–128.

Heywood, V.H. (1989) Patterns, extents and modes of invasion by terrestrial plants. In: Drake, J.A., Mooney, H.A., di Castri, F., Groves, R.H., Kruger, F.J., Rejmánek, M. and Williamson, M. (eds) *Biological Invasions: a Global Perspective.* John Wiley & Sons, Chichester, pp. 31–60.

Hobbs, R.J. (1989) The nature and effects of disturbance relative to invasions. In: Drake, J.A., Mooney, H.A., di Castri, F., Groves, R.H., Kruger, F.J., Rejmánek, M. and Williamson, M. (eds) *Biological Invasions: a global Perspective.* John Wiley & Sons, Chichester, pp. 389–405.

Hobbs, R.J. (2000) Land-use changes and invasions. In: Mooney, H.A. and Hobbs, R.J. (eds) *Invasive Species in a Changing World.* Island Press, Washington, DC, pp. 55–64.

Hobbs, R.J. and Huenneke, L.F. (1992) Disturbance, diversity, and invasion: implications for conservation. *Conservation Biology* 6, 324–337.

Hobbs, R.J. and Humphries, S.E. (1995) An integrated approach to the ecology and management of plant invasions. *Conservation Biology* 9, 761–770.

Holmgren, M., Aviles, R., Sierralta, Segura, A.M. and Fuentes, E.R. (2000) Why have more European herbs so successfully invaded the Chilean matorral? Effects of herbivory, soil nutrients. *Oikos* 90, 67–78

Hooper, D.U. and Vitousek, P.M. (1998) Effects of plant composition and diversity on nutrient cycling. *Ecological Monographs* 68, 121–149.

Huston, M.A. and D'Angelis, D.L. (1994) Competition and coexistence: the effects of resource transport and supply rates. *American Naturalist* 144, 954–977.

King, S.E. and Grace, J.B. (2000) The effects of gap size and disturbance type on invasion of wet pine savanna by cogongrass, *Imperata cylindrica* (Poaceae). *American Journal of Botany* 87, 1279–1286.

Knops, J.M.H., Tilman, D., Haddad, N.M., Naeem, S., Mitchell, C.E., Haarstad, J., Ritchie, M.E., Howe, K.M., Reich, P.B., Siemann, E. and Groth, J. (1999) Effects of plant species richness on invasion dynamics, disease outbreaks, insect abundances and diversity. *Ecology Letters* 2, 286–293.

Kolar, C.S. and Lodge, D.M. (2001) Process in invasion biology: predicting invaders. *Trends in Ecology and Evolution* 16, 199–204.

Kowarik, I. (1995) Time lags in biological invasions with regard to the success and failure of alien species. In: Pyšek, P., Prach, K., Rejmánek, M. and Wade, M. (eds) *Plant Invasions: General Aspects and Special Problems*. SPB Academic, Amsterdam, pp. 15–38.

Lawton, J.H. and Brown, K.C. (1986) The population and community ecology of invading insects. *Philosophical Transactions of the Royal Society of London, Series B* 314, 607–617.

Leakey, R.R.B. (1981) Adaptive biology of vegetatively regenerating weeds. *Advances in Applied Biology* 6, 57–90.

Lepš, J., Brown, V.K., Diaz Len, T.A., Gormsen, D., Hedlund, K., Kailová, J., Korthals, G.W., Mortimer, S.R., Rodriguez-Barrueco, C., Roy, J., Santa Regina, I., van Dijk, C. and van der Putten, W.H. (2001) Separating the chance effect from other diversity effects in the functioning of plant communities. *Oikos* 92, 123–134.

Levine, J.M. (2000) Species diversity and biological invasions: relating local process to community pattern. *Science* 2888, 852–854.

Levine, J.M. and D'Antonio (1999) Elton revisited: a review of evidence linking diversity and invasibility. *Oikos* 87, 15–26.

Lodge, D.M. (1993) Biological invasions: lessons for ecology. *Trends in Ecology and Evolution* 8, 133–137.

Lonsdale, W.M. (1999) Global patterns of plant invasions and the concept of invasibility. *Ecology* 80, 1522–1536.

Mack, R.N., Simberloff, D., Lonsdale, W.M., Evans, H., Clout, M. and Bazzaz, F.A. (2000) Biotic invasions: causes, epidemiology, global consequences, and control. *Ecological Applications* 10, 689–710.

Nunes, P.A.L.D and van den Bergh, J.C.J.M. (2001) Economic valuation of biodiversity: sense or nonsense? *Ecological Economics* 39, 203–222.

Orians, G.H. (1986) Site characteristics favouring invasions. In: *Ecology of Biological Invasions of North America and Hawaii*. Springer-Verlag, New York, pp. 133–148.

Pappert, R.A., Hamrick, J.L. and Donovan, L.A. (2000) Genetic variation in *Pueraria lobata* (Fabaceae), an introduced, clonal, invasive plant of the southern United States. *American Journal of Botany* 87, 1240–1245.

Parker, I.M., Simberloff, D., Lonsdale, W.M., Goodell, K., Wonham, M., Kareiva, P.M., Williamson, M.H., Von Holle, B., Moyle, P.B., Byers, J.E. and Goldwasserr, L. (1999) Impact: toward a framework for understanding the ecological effects of invaders. *Biological Invasions* 1, 3–19.

Perrins, J., Williamson, M. and Fitter, A. (1992) Do annual weeds have predictable characters. *Acta Oecologia* 13, 516–533.

Perrins, J., Fitter, A. and Williamson, M. (1993) Population biology and rates of invasion of three introduced *Impatiens* species in the British Isles. *Journal of Biogeography* 20, 33–44.

Pheloung, P.C., Williams, P.A. and Halloy, S.R. (1999) A weed risk assessment model for use as a biosecurity tool in evaluating plant introductions. *Journal of Environmental Management* 57, 239–251.

Pimentel, D.A., Lach, L. Zuniga and R. Morrison, D. (2000) Environmental and economic costs of nonindigenous species in the United States. *BioScience* 50, 53–65.

Pimentel, D.A., McNair, S., Janecka, J., Wightman, J., Simmonds, C., O'Connell, C., Wong, E., Russel, L., Zern, J., Aquino, T. and Tsomondo, T. (2001) Economic and environmental threats of alien plant, animal, and microbe invasions. *Agriculture, Ecosystem and Environment* 84, 1–20.

Planty-Tabacchi, A., Tabacchi, E., Naiman, R.J., DeFerrari, C. and Décamps, H. (1996) Invasibility of species-rich communities in riparian zones. *Conservation Biology* 10, 598–607

Pyšek, P. (1997) Clonality and plant invasions: can a trait make a difference? In: de Kroom, H. and van Groenendael, J. (eds) *The Ecology and Evolution of Clonal Plants*. Backhuys Publishers, Leiden, The Netherlands, pp. 405–427.

Pyšek, P. (1998) Is there a taxonomic pattern to plant invasions? *Oikos* 82, 282–294.

Reichard, S.H. and Hamilton, C.W. (1997) Predicting invasions of woody plants introduced into North America. *Conservation Biology* 11, 193–203.
Reinartz, J.A. (1997) Controlling glossy buckthorn (*Rhamnus frangula* L.) with winter herbicide treatments of cut stumps. *Natural Areas Journal* 17, 38–41.
Rejmánek, M. (1989) Invasibility of plant communities. In: Drake, J.A., Mooney, H.A., di Castri, F., Groves, R.H., Kruger, F.J., Rejmánek, M. and Williamson, M. (eds) *Biological Invasions: a Global Perspective.* John Wiley & Sons, Chichester, pp. 369–388.
Rejmánek, M. (1995) What makes a species invasive? In: Pyšek, P., Prach, K., Rejmánek, M. and Wade, M. (eds) *Plant Invasions: General Aspects and Special Problems.* SPA Academic Publishing, Amsterdam, pp. 3–13.
Rejmánek, M. (1996) A theory of seed plant invasiveness: the first sketch. *Biological Conservation* 78, 171–181.
Rejmánek, M. (1999) Invasive plant species and invasible ecosystems. In: Sandlund, O.T., Schei, P.J. and Vilken, A. (eds) *Invasive Species and Biodiversity Management.* Kluwer Academic, Boston, Massachusetts, pp. 79–102.
Rejmánek, M. (2000) Invasive plants: approaches and predictions. *Austral Ecology* 25, 497–506.
Rejmánek, M. and Richardson, D.M. (1996) What attributes make some plant species more invasive? *Ecology* 77, 1655–1661.
Richardson, D.M. and Bond, W.J. (1991) Determinants of plant distribution: evidence from pine invasions. *American Naturalist* 137, 639–668.
Richardson, D.M., Pyšek, P., Rejmánek, M., Barbour, M.G., Panetta, F.D. and West, C.J. (2000) Naturalization and invasion of alien plants: concepts and definitions. *Diversity and Distributions* 6, 93–107.
Robinson, G.R., Quinn, J.F. and Stanton, M.L. (1995) Invasibility of experimental habitat in California winter annual grassland. *Ecology* 76, 786–794.
Scott, J.K. and Panetta, F.D. (1993) Predicting the Australian weed status of southern African plants. *Journal of Biogeography* 20, 87–93.
Shrestha, A., Knezevic, S.Z., Roy, R.C., Ball-Coelh, B.R. and Swanton, C.J. (2002) Effect of tillage, cover crop and crop rotation on the composition of weed flora in a sandy soil. *Weed Research* 42, 76–87.
Simberloff, D. and von Holle, B. (1999) Positive interactions of nonindigenous species: invasion meltdown? *Biological Invasions* 1, 21–32.
Stohlgren, T.J., Bull, K.A., Otsuki, Y., Villa, C. and Lee, M. (1998) Riparian zones as havens for exotic plant species in the central grasslands. *Plant Ecology* 138, 113–125.
Stohlgren, T.J., Binkley, D., Chong, G.W., Kalkham, M.A., Schell, L.D., Bull, K.A., Otsuki, Y. Newman, G., Bashkin, M. and Son, Y. (1999) Exotic plants invade hot spots of native plant diversity. *Ecological Monographs* 69, 25–46.
Stohlgren, T.J., Otsuki, Y., Villa, C., Lee, M. and Belnap, J. (2001) Patterns of invasions: a case example of native species hotspots and rare habitats. *Biological Invasions* 3, 37–50.
Thompson, K., Hodgson, J.G. and Rich, T.C.G. (1995) Native and alien invasive plants: more of the same? *Ecography* 18, 390–402
Tilman, D. (1997) Community invasibility, recruitment limitation, and grassland biodiversity. *Ecology* 78, 81–92.
Tilman, D. (1999) The ecological consequences of changes in biodiversity: a search for general principles. *Ecology* 80, 1455–1474.
vanGessel, M.J. (2001) Glyphosate-resistant horseweed from Delaware. *Weed Science* 49, 703–705.
Vermeij, G.J. (1996) An agenda for invasion biology. *Biological Conservation* 78, 3–9.
Vitousek, P.M. (1990) Biological invasions and ecosystem processes: towards an integration of population biology and ecosystem studies. *Oikos* 57, 7–13.
Vitousek, P.M., Aber, J.D., Howarth, R.W., Likens, G.E., Matson, P.A., Schindler, D.W., Schlesinger, W.H. and Tilman, D. (1997) Human alteration of the global nitrogen cycle: sources and consequences. *Ecological Applications* 7, 737–750.
Vitousek, P.M., Walker, L.R., Whiteaker, L.D., Mueller-Dombois, D. and Matson, P.A. (1987) Biological invasion by *Myrica faya* alters ecosystem development in Hawaii. *Science* 238, 802–804.
Wade, M. (1997) Predicting plant invasions: making a start. In: Brock, J.H., Wade, M., Pyšek, P. and Green, D. (eds) *Plant Invasions: Studies from North America and Europe.* Backhuys Publishers, The Netherlands, pp. 1–18.

Webb, S.L., Dwyer, M., Kaunzunger, C.K. and Wyckoff, P.H. (2000) The myth of the resilient forest: case study of the invasive Norway maple (*Acer platanoides*). *Rhodora* 102, 332–354.

Williamson, M. (1989) Mathematical models of invasion. In: Drake, J.A., Mooney, H.A., di Castri, F., Groves, R.H., Kruger, F.J., Rejmánek, M. and Williamson, M. (eds) *Biological Invasions: a Global Perspective*. John Wiley & Sons, Chichester, pp. 329–350.

Williamson, M.H. (1993) Invaders, weeds and the risk from genetically modified organisms (GMOs). *Experimentia* 49, 219–224.

Williamson, M. (1996) *Biological Invasions*. Chapman and Hall, New York.

Williamson, M. (1999) Invasions. *Ecography* 22, 5–12.

Williamson, M.H. and Brown, K.C. (1986) The analysis and modeling of British invasion. *Philosophical Transactions of the Royal Society of London, Series B* 314, 505–522.

Williamson, M.H. and Fitter, A. (1996) The characters of successful invaders. *Biological Conservation* 78, 163–170.

Woodward, S.A., Vitousek, P.M., Matson, K., Hughes, F., Benvenuto, K. and Matson, P.A. (1990) Use of the exotic tree *Myrica faya* by native and exotic birds in Hawai'i Volcanoes National Park. *Pacific Science* 44, 88–93.

14
Studying Community Structure and Dynamics

Concepts

- There are a number of indices available for measuring α-diversity. Their calculations are based on a number of variables including proportional abundance of each species (Shannon–Wiener index) and the probability of two individuals being from the same sample (Simpson's dominance index).
- Whittaker's statistic is used to measure β-diversity. It is based on species richness.
- Coefficients of similarity are used to determine whether the composition of two communities is similar. These indices are based on the number of species common to both communities (Jaccard's and Sørensen's coefficients) or on the species abundances (Steinhaus' coefficient).
- Succession can be studied using either long-term studies where one community is followed over time, or chronosequences where communities of different ages are compared.
- Multivariate analyses combine the data of a community's species' abundances into one analysis. Ordination is a way to examine how species abundances vary with the environment. Cluster analysis is a way of finding natural groups based on a number of variables.

Introduction

In Chapter 10 we discussed how to measure populations and their interactions. In this chapter we will present a few ways of measuring community structure and dynamics. The same general sampling methods (plot and plotless) are used in community ecology, except that the abundance of all populations in the community (i.e. all vascular plants) is measured, rather than just a few target species of interest. In this chapter we show how various indices of α- and β-diversity are calculated, show the two major ways to sample succession and give a general introduction to multivariate analyses.

Diversity

In Chapter 11 we presented the various types of indices used to measure diversity and we discussed their advantages and disadvantages. Diversity is a measure of the number of species (species richness) in a community and their relative abundances (evenness). There are also a number of indices that combine measures of richness and evenness. In general, density is used to calculate diversity indices rather than biomass or other measures of abundance. The timing and frequency of sampling is important when examining community-level phenomena because the appearance of a community changes in response to annual climate patterns. Ristau *et al.* (2001) examined how the frequency and timing of sampling could affect measures of diversity, and suggested that sampling understorey vegetation should be done both early and late in the growing season to ensure that species with different phenological patterns are included in the data. This was especially true for species richness, as it tended to increase over the season. This applies to other types of community sampling as well.

Throughout this section we present formulae for calculating various indices. We caution that different texts will use different notations for the equations. In our calculations, we use these notations:

- n = population density or number;
- n_i = density or number of the ith species (i.e. any particular species you choose);
- N = total number of individuals of all species in the community;
- S = species richness;
- 'Σ' means 'sum of all the following factors'.

α-Diversity

α-Diversity measures the amount of diversity within a community type. Here we show how to calculate two diversity indices: Shannon–Wiener diversity index and Simpson's dominance index. These indices differ in the types of measurements they incorporate.

Shannon–Wiener diversity index (H')

One of the nice things about this index is that if you calculate it for a large number of samples, the values will have a log-normal distribution. You may or may not be aware that most 'easy' statistics (e.g. *t*-test, analysis of variance (ANOVA)) only work if data have a log-normal distribution. Another nice thing about this index is that you can calculate evenness (not just richness). This index is based on the proportional abundance of each species (p_i). It is easiest to show how the indices are calculated using a table format (Table 14.1). The Shannon–Wiener diversity index is calculated using the following formula:

- $H' = -\Sigma\ [p_i\ (\ln p_i)]$, where
  - p_i = proportional abundance of a given species (call this species 'i').
    $p_i = n_i/N$.
  - 'ln' means 'natural logarithm' (ln was originally used because the numbers were so large that, in pre-computer days, it would have been difficult to do the calculations without spending several years at them. We continue to use the ln out of convention).

To interpret the value of this index, higher numbers indicate a more diverse community; however, it is an arbitrary scale. There is no predetermined value of H' that indicates whether a community is 'diverse' or not. However, as we will see below, there are ways to compare the diversity indices of two communities.

Example using the Shannon–Wiener diversity index

Menalled *et al.* (2001) conducted a 6-year study examining the effect of management system on the seed bank and above-ground vegetation of a maize–soybean–wheat rotation in Michigan. Their four treatments were:

- conventional – mouldboard ploughed, high external chemical inputs;
- no-till – no tillage, high external chemical inputs;

Table 14.1. An example below illustrates the calculations for the Shannon–Wiener diversity index (H'). This example compares the weed species found in two portions of the same habitat: sandy and wet areas of a meadow. Data were collected using a stratified random sample. In practice, to calculate all of this, it is easiest to construct tables (below)

	Sandy areas in the meadow				Wet areas in the meadow			
Weed taxa	n_i	p_i	$\ln p_i$	$-p_i(\ln p_i)$	n_i	p_i	$\ln p_i$	$-p_i(\ln p_i)$
1. Garlic mustard, *Alliaria petiolata*	32	0.101	−2.291	0.232	43	0.187	−1.677	0.314
2. Chicory, *Cichorium intybus*	41	0.130	−2.042	0.265	11	0.048	−3.040	0.145
3. Canada thistle, *Cirsium arvense*	58	0.184	−1.695	0.311	15	0.065	−2.730	0.178
4. Deptford pink, *Dianthus armeria*	48	0.152	−1.885	0.286	14	0.061	−2.799	0.170
5. Purple loosestrife, *Lythrum salicaria*	0	0.000			44	0.191	−1.654	0.316
6. Common reed, *Phragmites communis*	0	0.000			36	0.157	−1.855	0.290
7. Lady's-thumb, *Polygonum persicaria*	0	0.000			26	0.113	−2.180	0.246
8. Common purslane, *Portulaca oleracea*	59	0.187	−1.678	0.313	17	0.074	−2.605	0.193
9. Chickweed, *Stellaria media*	78	0.247	−1.399	0.345	24	0.104	−2.260	0.236
Σ (sum the columns)	316	1.000		H' = 1.753	230	1.000		H' =2.089

Using the table makes it easy to do the calculations.
For example, to calculate $-p_i$ in p_i for garlic mustard:

$$n_i = 32, \quad N = 316$$
$$p_i = n_i/N = 32/316 = 0.101$$
$$\ln p_i = \ln(0.101) = -2.291$$
$$-p_i(\ln p_i) = -(0.101)(-2.291) = -(-0.232) = 0.232.$$

Then H' is calculated by summing all the $-p_i(\ln p_i)$values.

$$H' = \Sigma -p_i(\ln p_i)$$
$$= 0.232 + 0.265 + 0.311 + 0.286 + 0.313 + 0.345$$
$$= 1.753.$$

We also can calculate evenness with these numbers.

$$\text{Evenness} = E = H' / \ln S$$
$$S = \text{number of species present} = 9$$
$$E_{(\text{wet meadow})} = H'_{(\text{wet meadow})} / \ln S_{(\text{wet meadow})}$$
$$= 2.089/\ln 9 = 2.089/2.197 = 0.951$$
$$E_{(\text{sandy meadow})} = H'_{(\text{sandy meadow})} / \ln S_{(\text{sandy meadow})}$$
$$= 1.753/\ln 6 = 1.753/1.792 = 0.978.$$

The calculations indicate that: $H'_{(\text{sandy meadow})} = 1.753$, $E_{(\text{sandy meadow})} = 0.978$; $H'_{(\text{wet meadow})} = 2.089$, $E_{(\text{wet meadow})} = 0.951$.

- low-input – mouldboard ploughed, low external chemical input;
- organic – mouldboard ploughed, no external chemical inputs.

Plots were 1 ha and there were six replicates per treatment. From 1993 to 1998, above-ground weed composition and biomass was measured using five 2 × 0.5 m quadrats per plot. Thus the researchers had data on species composition, density and biomass in the above-ground weed communities.

Among many analyses done by Menalled *et al.* (2001), they calculated weed biomass, species density (number of species per m^2) and Shannon–Wiener diversity index in each of the 6 years for each treatment. They found that, although there was year-to-year variation, organic and low-input systems tended to have higher weed biomass, species density and diversity than the conventional or high-input treatment (Table 14.2). In organic systems and low-input systems, you expect to have higher weed diversity and biomass, because herbicides (which reduce populations of susceptible species) are not used. These results are similar to those of other researchers (e.g. Moreby *et al.*, 1994; Leeson *et al.*, 2000; Murphy *et al.*, unpublished manuscript).

Table 14.2. Mean weed biomass, species density and Shannon–Wiener diversity index in a maize–soybean–wheat rotation with four management systems. Values are means (±1 SE) averaged over 6 years (from Menalled *et al.*, 2001).

Management system	Weed biomass ($g\ m^{-2}$)	Species density (no. species per m^2)	Shannon–Wiener diversity index
Conventional	18.2 ± 3.0	2.4 ± 0.1	0.18 ± 0.02
No-till	57.8 ± 8.5	2.9 ± 0.2	0.18 ± 0.02
Low-input	69.3 ± 9.3	4.7 ± 0.2	0.28 ± 0.13
Organic	108.5 ± 6.5	6.2 ± 0.2	0.32 ± 0.14

Evenness (*E*)

One advantage of the Shannon–Wiener index is that community evenness can also be calculated. The formula for evenness (*E*) is:

- $E = H' / \ln S$

To interpret evenness, values of 0 indicate that the habitat is extremely uneven (dominated by one species), whereas values approaching 1 indicate that the habitat is extremely even (maximum species diversity exists, no one species dominates).

The results in Table 14.1 indicate that the wet areas of the meadow have greater species richness but the sandy meadow has greater species evenness. Looking at the data set, the wet areas of the meadow do indeed have more species (S=9) but are dominated by a few species (mainly garlic mustard, purple loosestrife and common reed). In this example, we are fortunate because we can simply look at the data and verify that these

Table 14.3. Method to calculate whether the diversity of two communities is statistically different.

Sandy areas in the meadow						
Weed taxa	n_i	p_i	$\ln p_i$	$-p_i \ln p_i$	$p_i (\ln p_i)^2$	$[p_i (\ln p_i)]^2$
1. Garlic mustard	32	0.101	−2.290	0.232	0.531	0.054
2. Chicory	41	0.130	−2.042	0.265	0.541	0.070
3. Canada thistle	58	0.184	−1.695	0.311	0.528	0.097
4. Deptford pink	48	0.152	−1.885	0.286	0.539	0.082
5. Purple loosestrife	0	0.000				
6. Common reed	0	0.000				
7. Lady's thumb	0	0.000				
8. Common purslane	59	0.187	−1.678	0.313	0.526	0.098
9. Chickweed	78	0.247	−1.399	0.345	0.483	0.119
Σ (sum the columns)	316	1.000	−10.989	1.753	3.148	0.520

Wet areas in the meadow						
Weed taxa	n_i	p_i	$\ln p_i$	$-p_i \ln p_i$	$p_i (\ln p_i)^2$	$[p_i (\ln p_i)]^2$
1. Garlic mustard	43	0.187	−1.677	0.314	0.526	0.098
2. Chicory	11	0.048	−3.040	0.145	0.442	0.021
3. Canada thistle	15	0.065	−2.730	0.178	0.486	0.032
4. Deptford pink	14	0.061	−2.799	0.170	0.477	0.029
5. Purple loosestrife	44	0.191	−1.654	0.316	0.523	0.100
6. Common reed	36	0.157	−1.855	0.290	0.538	0.084
7. Lady's thumb	26	0.113	−2.180	0.246	0.537	0.061
8. Common purslane	17	0.074	−2.605	0.193	0.502	0.037
9. Chickweed	24	0.104	−2.260	0.236	0.533	0.056
Σ (sum the columns)	230	1.000	−20.799	2.089	4.564	0.518

Table 14.3, Continued.

Step 1: Calculate variance for each community

The formula to calculate variance is:

$$H'_{var} = 1/N * \{\Sigma p_i(\ln p_i)^2 - [\Sigma p_i(\ln p_i)]^2 - 1/(2N^2)[S\text{-}1]\}$$

Note: The last term of $1/(2N^2)[S\text{-}1]$ is usually so small (usually 10^{-6}) that it is omitted from the actual calculation; we will follow this convention here.
In effect the formula is:

$$H'_{var} = 1/N * \{\Sigma p_i(\ln p_i)^2 - [\Sigma p_i(\ln p_i)]^2\}$$

	Wet meadow	Sandy meadow
N	230	316
$1/N$	0.004	0.003
$\Sigma p_i(\ln p_i)^2$	4.564	3.148
$[\Sigma p_i(\ln p_i)]^2$	0.518	0.520
$H'_{var}=1/N*\{\Sigma p_i(\ln p_i)^2-[\Sigma p_i(\ln p_i)]^2\}$	$= 0.004 * \{4.564 - 0.518\}$	$= 0.003 * \{3.148 - 0.520\}$
	$= 0.004 * 4.046$ $= 0.016$	$=0.003 * 2.628$ $= 0.008$

Step 2: Calculate df and critical value of t (t_{crit})

$$\text{df}= \frac{[H'_{var(wet\ meadow)} + H'_{var(sandy\ meadow)}]}{\{[H'_{var(wet\ meadow)}]^2 / N_{wet\ meadow} + [H'_{var(sandy\ meadow)}]^2 / N_{sandy\ meadow}\}}$$

$$= \frac{[0.016 + 0.008]}{\{[0.016]^2/9 + [0.008]^2/6\}}$$

$$= \frac{[0.024]}{\{0.0003/9 + 0.00006/6\}} = \frac{0.024}{0.00003 + 0.00001}$$

$$= 0.024 / 0.000031 = 774.$$

The t_{crit} is determined by using a t-table (Table 14.3). The left column has the degrees of freedom and we have selected p=0.05.

We now know t_{crit} = 1.96 and df = 774.

Step 3: Calculate observed value of t (t_{obs})

In the case of the Shannon–Wiener comparison, the formula for calculating the observed t-statistic is:

$$t_{obs} = [H'_{(wet\ meadow)} - H'_{sandy\ meadow)}] / [H'_{var(wet\ meadow)} - H'_{var(sandy\ meadow)}]^{0.5}$$
$$= [2.089 - 1.753] / [0.016 + 0.008]^{0.5}$$
$$= 0.336 / [0.024]^{0.5}$$
$$= 0.336 / 0.155 = 2.168.$$

Step 4: Compare critical value of t with the observed value of t

Since t_{obs} (2.168) > t_{crit} (1.963), we conclude that the diversity of the two meadows is different.

Table 14.4. A *t*-table.

df	*P*		
	0.10	0.05	0.025
1	6.31	12.71	25.45
2	2.92	4.30	6.21
3	2.35	3.18	4.18
4	2.13	2.78	3.50
5	2.02	2.57	3.16
6	1.94	2.45	2.97
7	1.89	2.36	2.84
8	1.86	2.31	2.75
9	1.83	2.26	2.69
10	1.81	2.23	2.63
11	1.80	2.20	2.59
12	1.78	2.18	2.56
13	1.77	2.16	2.53
14	1.76	2.14	2.51
15	1.75	2.13	2.49
16	1.75	2.12	2.47
17	1.74	2.11	2.46
18	1.73	2.10	2.45
19	1.73	2.09	2.43
20	1.72	2.09	2.42
21	1.72	2.08	2.41
22	1.72	2.07	2.41
23	1.71	2.07	2.40
24	1.71	2.06	2.39
25	1.71	2.06	2.38
26	1.71	2.06	2.38
27	1.70	2.05	2.37
28	1.70	2.05	2.37
29	1.70	2.05	2.36
30	1.70	2.04	2.36
35	1.69	2.03	2.34
40	1.68	2.02	2.33
45	1.68	2.01	2.32
50	1.68	2.01	2.31
55	1.67	2.00	2.30
60	1.67	2.00	2.30
70	1.67	1.99	2.29
80	1.66	1.99	2.28
90	1.66	1.99	2.28
100	1.66	1.98	2.28
120	1.66	1.98	2.27
150	1.66	1.98	2.26
infin	1.64	1.96	2.24

calculations make sense; in reality, most data sets are too large to allow the luxury of visual inspection, hence diversity indices are necessary.

Comparing the diversity of two communities

If we want to compare the diversity of the two meadows, we need to use different statistics. By inspection, the higher value of the Shannon–Wiener index for the wet meadow implies that it is *more diverse* in terms of the number of species, but, we need some way to actually measure this. The first step of this procedure is to calculate the *variance in diversity* (H') at both sites. The formula looks a bit daunting, but we have actually already calculated most of the needed information (Table 14.1). To make it easy for the calculation, we can add two new columns (Table 14.3). We will use these variances to compare the diversity statistically between the two sites. Because the distribution of values of a Shannon–Wiener index is (assumed) log-normal, we can use common parametric statistics.

The second step is to calculate how many degrees of freedom (df) exist for our test. This measures how accurate our test will be; the more degrees of freedom, the more likely it is that we will get an accurate result. The number of degrees of freedom increases as the sample size increases. In our example, the resulting df of 774 is quite large.

Once we know the df, we can then determine the critical value of t (t_{crit}). To do this you look at a *t*-table (Table 14.4). You locate the df in the first column and then select the level of significance you desire (usually P=0.05) along the top. Since t_{crit} values increase very slowly after 150, you must interpolate to obtain the correct value of t_{crit}. In our case, with a df of 774 and a significance level of 0.05, the t_{crit}-value is 1.96. Note that on the *t*-table (Table 14.4), the t_{crit} values decrease as df increases and as the level of significance decreases.

The third step is to calculate the observed value of t from our data (t_{obs}). The calculation incorporates the values of the two diversity indices ($H'_{\text{wet meadow}}$ and $H'_{\text{sandy meadow}}$) and their variances ($H'_{\text{var(wet meadow)}}$ and $H'_{\text{var(sandy meadow)}}$).

We can use a *t*-test to determine whether $H'_{\text{wet meadow}}$ and $H'_{\text{sandy meadow}}$ are significantly different. The *t*-test compares the distribution of values derived from our observations to a statistical distribution (the *t*-distribution) and determines whether the values derived from the observations are significantly different than those expected from random. Therefore, the final step is to compare our observed *t* statistic (t_{obs}) with the critical *t* statistic (t_{crit}) obtained from the *t*-table. If species diversity of the two sites is the same, our t_{obs} value will be less than or equal to t_{crit}. If our t_{obs} is greater than t_{crit} value, we can conclude that the diversity of the two sites is different.

Since our t_{obs} (2.168) is greater than t_{crit} (1.96), we conclude that the wet and dry meadow are significantly different in terms of the Shannon–Wiener diversity index (at $p = 0.05$). This answer, however, does not explicitly tell you why there are differences between wet and sandy areas. You would need to examine the growth requirements of the weed species to determine why this occurs (essentially, wet areas provide enough water to allow species that are drought sensitive to survive – these perish in sandy areas).

Simpson's dominance index (D)

The Simpson's dominance index (D) is based on the probability of any two individuals sampled from a community being the same species. The more a community is dominated by any one (or a few) species, the less diverse it is. The advantage of Simpson's index is that it is simpler to calculate than the Shannon–Wiener index.

- $D = \Sigma \{[n_i(n_i-1)] / [N(N-1)]\}$
- for each species, calculate $n_i(n_i-1)/N(N-1)$ and add all these values together.

The calculations are relatively simple, despite the somewhat intimidating symbols used in the formula (Table 14.5). The value of $N(N-1)$ is constant for each species, making the calculation simple. By convention, Simpson's dominance index is usually written as the reciprocal value (D^{-1}). This means that as D^{-1} increases, so does diversity. In this example, the reciprocal of 0.177 (i.e. 1/0.177) is 5.65. By doing this, the higher the index value, the more diversity there is (though, more accurately, this actually means there is more species evenness). In the above example, no one species dominates, hence the value of D^{-1} is relatively high (it is a relatively even community). For

Table 14.5. Calculation of the Simpson's dominance index (D^{-1}) using the data from the sandy meadow.

	Sandy areas in the meadow						
Weed taxa	n_i	n_i-1	$n_i(n_i-1)$	N	N-1	$N(N-1)$	$n_i(n_i-1)/N(N-1)$
Garlic mustard	32	31	992	316	315	99540	0.010
Chicory	41	40	1640	316	315	99540	0.016
Canada thistle	58	57	3306	316	315	99540	0.033
Deptford pink	48	47	2256	316	315	99540	0.023
Common purslane	59	58	3422	316	315	99540	0.034
Chickweed	78	77	6006	316	315	99540	0.060
Σ (sum the columns)	316						D=0.177

Simpson's index = D

$$(D) = \Sigma\{[n_i(n_i-1)]/[N(N-1)]\}$$
$$= 0.010 + 0.016 + 0.033 + 0.023 + 0.034 + 0.060$$
$$= 0.177$$

n_i = number of individuals in any given species

N = the total number of individuals = 316.

Simpson's dominance index = D^{-1}

$$D^{-1} = 1/0.177 = 5.65.$$

our data set, this concurs with the results using the Shannon–Wiener diversity index and Evenness calculation.

β-Diversity

β-Diversity is a measure of change in diversity over one area. It is a measure independent of α-diversity. We use measures of β-diversity to determine whether there are separate communities within an area. We will show one method to calculate β-diversity: Whittaker's statistic (β_w).

Whittaker's statistic (β_w)

Whittaker's statistic (β_w)is calculated as:

- $\beta_w = (S/S_r)-1$ where:
  - S = the species richness in the sample (all quadrats);
  - S_r = the mean species richness/quadrat (the sum of all the total number of species per quadrat divided by the total number of species).

We will illustrate this by modifying our example from the Shannon–Wiener calculation (Table 14.6). The data for the wet meadow were obtained from seven quadrats located along a linear transect.

As β_w increases, the rate of species turnover increases. Our result (β_w = 1.7) indicates that species composition changes rather slowly along the transect. It probably means that, for example, soil moisture content in this wet area of the meadow is rather constant, at least where samples were recorded.

Similarity Indices

Similarity indices are used to compare what species two communities have in common, whereas diversity indices consider species number and abundance, but not species identity. Thus, two communities could have

Table 14.6. Method to calculate Whittaker's statistic (β_w) of β-diversity. Data are changed from numeric values into presence-absence data.

	Wet area of meadow							
	Did the species occur in this quadrat?							
	Quadrat number							
Weed taxa	1	2	3	4	5	6	7	Total
Garlic mustard	Y	Y	Y	Y	Y	Y	Y	
Chicory	Y	Y	Y	N	N	N	N	
Canada thistle	Y	Y	N	N	N	N	N	
Deptford pink	Y	N	N	N	N	N	N	
Purple loosestrife	N	Y	Y	Y	Y	Y	Y	
Common reed	N	N	Y	Y	Y	Y	Y	
Lady's thumb	N	N	N	N	Y	Y	Y	
Common purslane	Y	Y	N	N	N	N	N	
Chickweed	N	Y	N	N	N	N	N	
Number of species in this quadrat	5	6	4	3	4	4	4	30

Whittaker's statistic (β_w) = $(S/S_r)-1$, where:

S = total number of species in the sample (all quadrats) = 9 species

S_r = the mean species richness/quadrat = 30/9 = 3.33

= the sum of all the total number of species/quadrat divided by the total number of species.

$\beta_w = (S/S_r) - 1$

$= (9/3.33) - 1$

$= 2.7 - 1 = 1.7.$

the same diversity but be composed of different species. There are a number of similarity coefficients (Legendre and Legendre, 1998). We will present three.

Jaccard and Sørensen coefficients

The Jaccard coefficient (S_J) considers how many species are common to both communities (Jaccard, 1900). The common version of the formula is:

- $S_J = j / (a + b + j)$, where:
  - j = number of species found in both communities;
  - a = number of species found only in community a;
  - b = number of species found only in community b.

A variation of this index that emphasizes species common to both communities is sometimes called the Sørensen coefficient (Sørensen, 1948). It is calculated as:

- $S_S = 2j / (a + b + 2j)$

Note that these indices focus on what species are present in both communities and do not take into account their abundances. Therefore, a species that was rare in one community but common in another would have the same influence on the result. The advantage of these indices is that only

Table 14.7. Indices used to calculate similarity between communities: Jaccard's coefficient (S_J), Sørensen coefficient (S_S) and Steinhaus coefficient (S_A) .

Weed taxa	Sandy meadow (a) abundance		Wet meadow (b) abundance	Minimum
1. Garlic mustard	32		43	32
2. Chicory	41		11	11
3. Canada thistle	58		15	15
4. Deptford pink	48		14	14
5. Purple loosestrife	0		44	0
6. Common reed	0		36	0
7. Lady's thumb	0		26	0
8. Common purslane	59		17	17
9. Chickweed	78		24	24
Total no. of individuals per area	316		230	
No. species in each area	6		9	
No. species common to both areas		$j = 6$		
W				W =113

Jaccard's coefficient $S_J = j / (A + B - j)$

j = number of species found in both communities;

A and B = total number of species found in each community;

$$S_J = 6 / (6 + 9 - 6)$$
$$= 6 / 9 = 0.667.$$

Sørenson's coefficient $S_N = 2_j / (a + b + 2j)$

$$= 2 \times 6 / \{0 + 3 + (2 \times 6)\}$$
$$= 12 / 15 = 0.8$$

Steinhaus's coefficient $S_A = 2W / (A + B)$

W = sum of the lower of the two abundances of each species

A and B = the sum of abundances for each community

$$S_A = 2(113) / (316 + 230)$$
$$= 226 / (316 + 230)$$
$$= 226 / 546 = 0.414.$$

presence/absence data are required – not measures of abundance.

Steinhaus' coefficient

Some similarity indices do incorporate abundance data. For example, the Steinhaus coefficient (Motyka, 1947) is based on abundance data. This coefficient takes the smallest abundance for each species as a proportion of the average community abundance. The formula is:

- $S_A = W / \{(A + B) / 2\} = 2W / (A + B)$, where:
  - W = sum of the lower of the two abundances of each species in the community;
  - A and B = the sum of abundances of all species in each community.

Comparison of similarity coefficients

Again, we will use the relevant data from our example of the Shannon–Wiener calculation to illustrate how the Jaccard, Sørensen and Steinhaus coefficients are calculated (Table 14.7). For all three coefficients, we interpret their values on a scale from 0 (complete dissimilarity) to 1 (complete similarity). The Jaccard and Sørensen coefficients suggest that the communities are quite similar. The value of the Sørensen coefficient is higher than the Jaccard because it increases the value of species common to both, and every species in the sandy meadow is also present in the wet meadow. Therefore, there is substantial overlap in species composition. However, if you look at the data, the communities actually are very dissimilar and contradict these two coefficients. How can this occur? The communities may share many of the same species but the relative abundances of each species are very different. The Steinhaus coefficient accounts for differences in abundance and hence is more accurate than Jaccard's or Sørensen's coefficients.

Unfortunately, both similarity and diversity indices often have many names, and sometimes one name can apply to several indices (Legendre and Legendre, 1998). For example, the Steinhaus coefficient is sometimes incorrectly called the Czekanowski coefficient and the Shannon–Wiener diversity index is sometimes referred to as the Shannon–Weaver index. In addition, there are a number of different symbols used in the equations. We have chosen commonly used symbols in the equations, but different ones are used elsewhere. Thus, it is important to examine the equation being used to determine how the researcher is assessing similarity.

Using similarity indices to compare seed bank and seedling communities

We have said that the similarity indices are used to compare two communities. One interesting use of these indices is to compare the seed bank community to the aboveground vegetation community (Chikoye and Ekeleme, 2001; Grandin, 2001; Menalled *et al.*, 2001). This is often used by weed ecologists to examine whether weed seeds in the seed bank will be reflected in the aboveground weed community. For example, Dessaint *et al.* (1997) used the Jaccard's coefficient to test whether the similarity of the seed bank and the seedling flora changed over time following a 9-year chemical-free barley–pea–wheat rotation in France. Two tillage treatments were compared: shallow (15 cm) and deep (25–30 cm). They found that there was a high correspondence between species found in the seed bank and as seedlings, and that similarity increased over time in both treatments (Fig. 14.1). When seed and seedling densities were taken into account using the Steinhaus coefficient, there was a similarly high correspondence that also increased over time. These results show that repeated crop rotations tend to result in the seed bank and seedling communities becoming more similar over time because constant selection pressure selects for specific species.

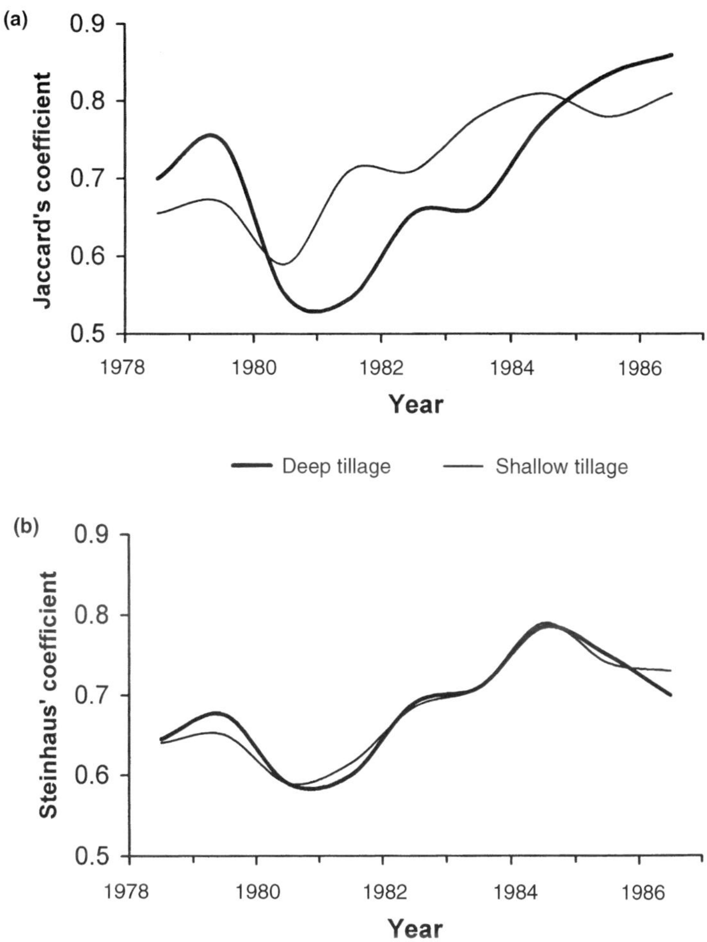

Fig. 14.1. Changes in the similarity ((a) Jaccard and (b) Steinhaus coefficient) between species found in the seed bank and as seedlings following a 9-year barley–pea–wheat rotation in France with two tillage treatments: shallow (15 cm) and deep (25–30 cm) (redrawn from data in Dessaint *et al.*, 1997).

Detecting Change in Vegetation Over Time

The change in vegetation over time is called succession (Chapter 12). There are two general ways to examine succession:

- use long-term studies to follow one community over time;
- compare community types of different ages (chronosequence).

Long-term studies

The most logical way to study how communities change over time is simply to watch and measure changes as they occur. You (the researcher) could regularly visit a community (say every year) and measure species composition, density and cover, as well as environmental factors such as soil type, temperature and chemistry, and light quality and quantity. Over time you might observe changes such as those discussed in Chapter 12.

An example of a long-term study is that described by Meiners *et al.* (2002). They conducted a study designed, in part, to determine:

- if species richness and cover of native and/or non-native species change during succession;
- if life history characteristics of native species differ from those of non-native species.

To conduct this study, they used 48 permanent plots in each of ten agricultural fields that were sequentially abandoned starting in 1958 (Small *et al.*, 1971). Plots were sampled yearly for 11 years and then every second year until year 40. On each sampling date, percentage cover of each species was recorded in permanent 0.5 × 2.0 m plots. From these data, the researchers could calculate proportional species richness and cover of native and non-native species.

Meiners *et al.* (2002) found that over time, richness of non-native species tended to decrease, while richness of native species increased (Fig. 14.2). The total and percentage cover of non-native species was highly variable among plots in the early stages of succession, but tended to decrease over time. To answer their second question, each species was characterized according to life history strategy and then expressed as proportional percentage cover for native and non-native species (Fig. 14.3). For native species, the proportion of total cover of trees increased while the proportion of annuals, biennials and herbaceous perennials decreased over time. A different pattern was observed for the cover of non-native species. Here, the proportional cover of shrubs increased over time and vines were more abundant during the mid-successional stages.

In the early stages of succession, non-native species outnumbered native species but this decreased over time. This was probably due to the increased cover of native trees shading the shade-intolerant non-native annuals. Not all non-native herbaceous species are shade-intolerant; for example, garlic mustard (*Alliaria petiolata*) has invaded some sites and is shown as the increase in biennial weeds in the later stages of succession.

Long-term studies are an excellent way to follow precise changes in community structure and function over time. However, as you will no doubt see, this type of long-term study requires, well, time and this is not always available. Also, it requires that a patch of land be preserved for the use of the researcher and protected from development (unless this is what is being studied).

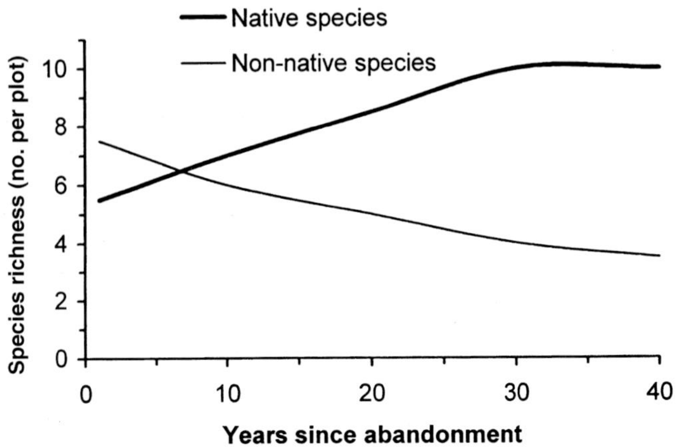

Fig. 14.2. Change in richness of native and non-native species in agricultural fields following abandonment (redrawn from data in Meiners *et al.*, 2002).

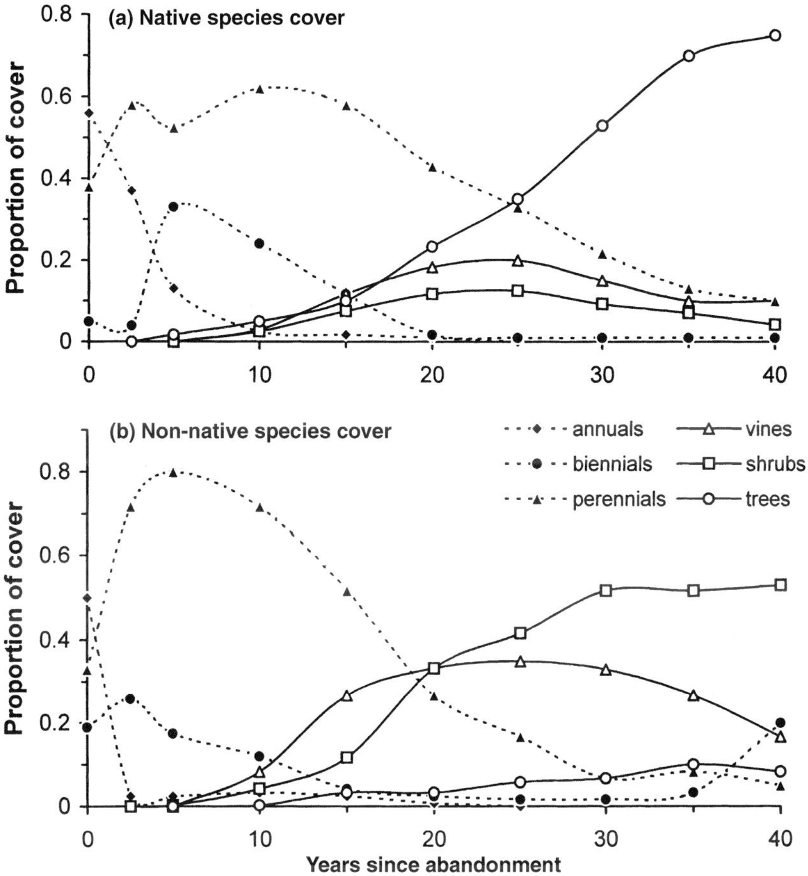

Fig. 14.3. Changes in percentage cover of: (a) native and (b) non-native species in agricultural fields following abandonment. Species were divided into six life history strategies and values are expressed as the proportion of total native and non-native cover (redrawn from data in Meiners *et al.*, 2002).

Chronosequences

To avoid the problems of long-term studies, some researchers use chronosequences to study changes in vegetation over time. To do this, plots of different successional stages are compared and the researcher recreates a chronological sequence (chronosequence) of the successional pathway. Pickett (1989) called this 'space for time substitution'.

Csecserits and Rédei (2001) used the chronosequence approach to study whether natural secondary succession was adequate to restore plant communities following field abandonment or whether active restoration efforts were required. They selected 54 abandoned fields in central Hungary that ranged in age from 1 to 33 years since abandonment. Fields were within a 60-ha area, had similar site characteristics, and were still grazed by sheep. Fields were divided into four age classes according to years since abandonment (1–5, 6–10, 11–23 and 24–33 years). Species were divided into four life history strategies (annual, biennial, perennial, woody) as well as five functional groups (Table 14.8).

The authors found that within each of the four life history strategies, species richness did not change over time but their relative abundance did change (Fig. 14.4). Over time, the relative abundance of annuals

Table 14.8. Plant functional groups used by Csecserits and Rédei (2001).

Type	Description
Weeds	Ruderal species
Sand generalists	Disturbance resistant, pioneer species of open sand steppe
Sand specialists	Species of open sand steppe that are less resistant to disturbance
Steppe generalists	Disturbance resistant species of closed sand steppe
Steppe specialists	Species of closed sand steppe that are less resistant to disturbance

tended to decrease, while the abundance of perennials and woody plants increased. In the functional groups, weed species richness decreased in the first ten years (between the first and second age group), while sand and steppe specialist groups increased during this interval. No other changes in species richness occurred. The patterns of abundance of functional groups were slightly different. From the first to third age group, weed abundance decreased while the abundance of sand specialists increased (Fig. 14.5). There was no change in other functional groups. Csecserits and Rédei (2001) concluded that there was no need to have active restoration efforts because weed abundance decreased over time and late successional species (sand and steppe specialists) had appeared after 10 years of abandonment. Therefore, the authors were able to determine that weeds would not cause persistent problems and that the process of natural succession was enough to return abandoned fields to semi-natural communities.

There are limitations to the use of chronosequences (Pickett, 1989; Bakker *et al.*, 1996; Foster and Tilman, 2000). First, this approach assumes that conclusions drawn from spatial relationships are the same as conclusions drawn from temporal ones. That is, if you sample sites of different ages (spatial), you will observe the same pat-

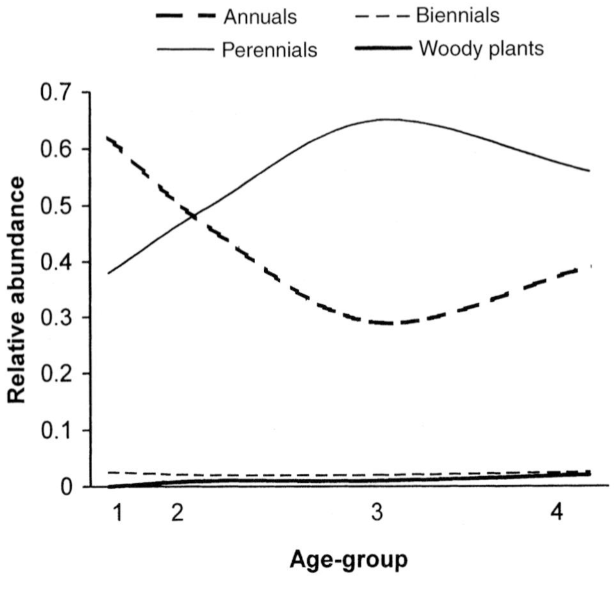

Fig. 14.4. Change in the relative abundance of annuals, biennials, perennials and woody plants following the abandonment of agricultural fields in central Hungary. Fields were divided into four age-classes: (1) 1–5 years, (2) 6–10 years, (3) 11–23 years, (4) 24–33 years (redrawn from data in Csecserits and Rédei, 2001).

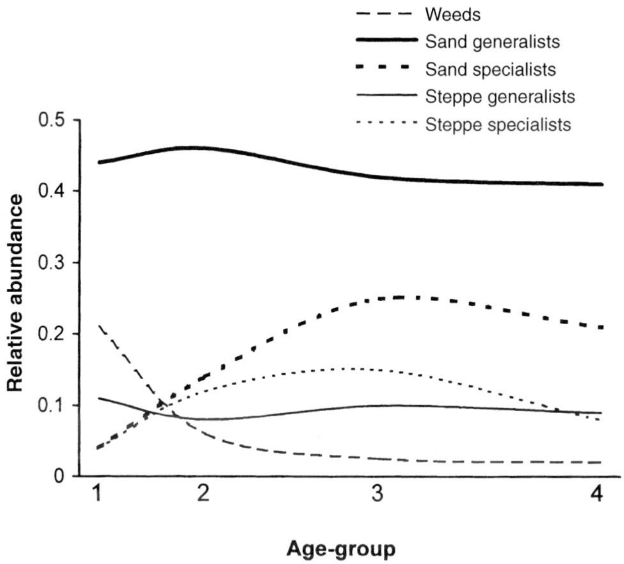

Fig. 14.5. Change in the relative abundance of weeds, sand generalists and specialists, and steppe generalists and specialists following the abandonment of agricultural fields in central Hungary. Fields were divided into four age-classes: (1) 1–5 years, (2) 6–10 years, (3) 11–23 years, (4) 24–33 years (redrawn from data in Csecserits and Rédei, 2001).

terns as you did if you observed them over time. In addition, because you are averaging the effects observed at different sites, you can only obtain a general level of detail to explain observations ('regional averaging'). Finally, since site-specific factors can confound results, correlations between species abundances and community attributes may not be directly correlated to the successional processes. Nevertheless, chronosequences are a useful and commonly used technique.

Multivariate Analyses of Community Data

In the previous section on succession, we used two datasets where the researchers examined the change in one variable (abundance or richness) at a time and considered how it changed through succession. Sometimes a researcher wants to look at more than one variable at a time and there are specific types of analyses for this – multivariate analyses. These types of analyses could be used to compare the plant community composition of two sites as related to specific factors such as moisture, wind or temperature.

Alternatively, we could use a series of *uni*variate analyses and compare the abundances of each individual species separately, but if we are asking a community-based question it is better to consider all the species together. Community patterns may differ from population patterns. If a researcher wants to know whether two communities are the same or different, then analyses based on populations might not answer the question. It is important to make sure that the methods of sampling and analysis match the question you are asking.

There are a number of types of multivariate statistics and their use depends on the types of question asked, the method of sampling and community type. The two main types we will address are ordination and cluster analysis. The statistics involved in analysing multivariate data are quite complicated, and we shall not go into them here

(Gauch, 1982; Digby and Kempton, 1987; Jongman *et al.*, 1995; Podani, 2000; Quinn and Keough, 2002). Kenkel *et al.* (2002) review the use of multivariate analysis in agricultural weed science.

Ordination

Ordination is used when a researcher wants to examine why species abundances vary in a community. When data are collected, species abundances and environmental variables are recorded. The first step of the process displays the species and sites along

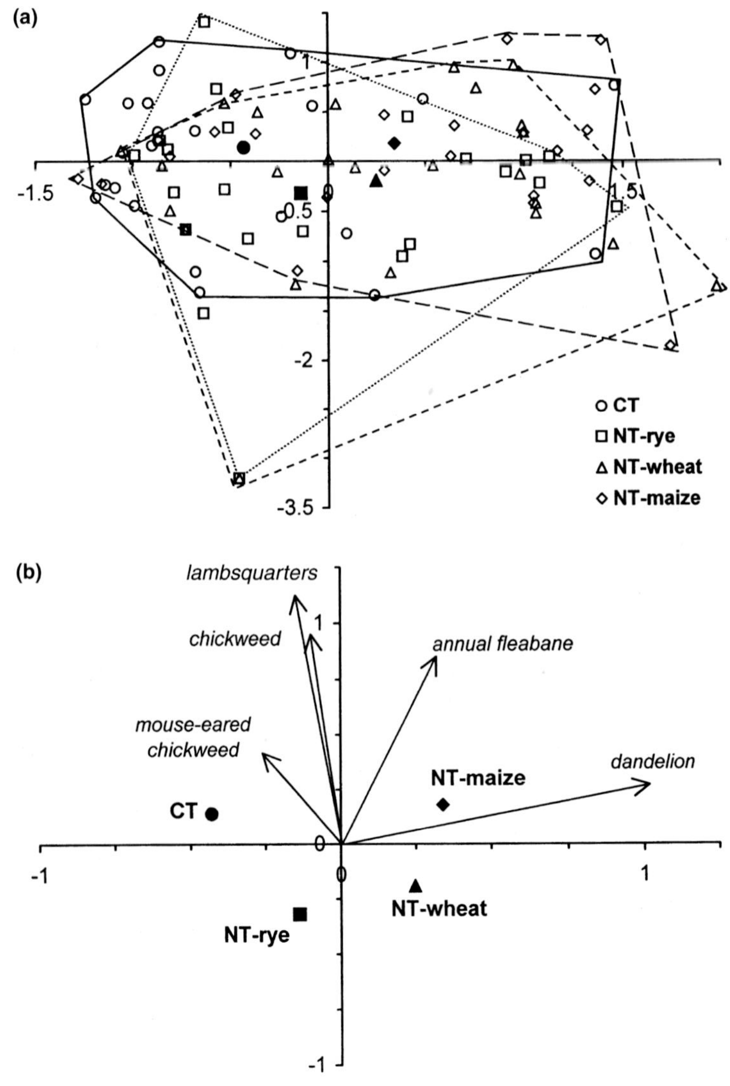

Fig. 14.6. Biplots illustrating the weed communities in a wheat crop under four tillage treatments: conventional tillage (CT), no-till with rye cover crop (NT rye), no-till with wheat stubble (NT wheat) and no-till with maize cover crop (NT maize). Part (a) shows the data points and the multivariate means (centroids) of each tillage treatment, (b) shows the associations between tillage treatments and weed species (redrawn from Shrestha *et al.*, 2002).

axes on ordination diagrams called biplots. The axes represent as yet unknown variables. Similar species will be located closer together on the biplot. From this type of analysis, we can see what species are grouped together, but we do not yet have a direct measure of what environmental variables influence species groupings. With the environmental data, we can then correlate environmental variables to the species patterns. However, this type of analysis does not show cause–effect relationships. Further experimental research will be required to determine the nature of the correlations.

Ordination is an exploratory technique that can be used to elicit patterns and is one way to reduce the complexity typical of ecological community data. 'The usual objective of ordination is to help generate hypotheses about the relationship between the species composition at a site and the underlying environmental gradients' (Digby and Kempton, 1987).

Example of ordination

Shrestha *et al.* (2002) used ordination to examine the effect of: (i) tillage and cover crop, and (ii) previous crop type on weed communities in winter wheat. Previous crop type consisted of soybean, white bean or kidney beans. Their tillage and cover crop treatments were:

- conventional tillage;
- no-till with rye cover crop;
- no-till in wheat stubble;
- no-till in maize cover crop.

The researchers used 6 × 10 m plots and each treatment was replicated four times. Weed densities were counted using eleven 0.09-m^2 quadrats per plot.

The results of ordination are often presented in biplots where the two axes separate treatments by placing more dissimilar communities further apart. Figure 14.6 represents the results of the effect of tillage and cover crop on weed communities in winter wheat. Here the data point for each replicate is shown (empty symbols) as well as the

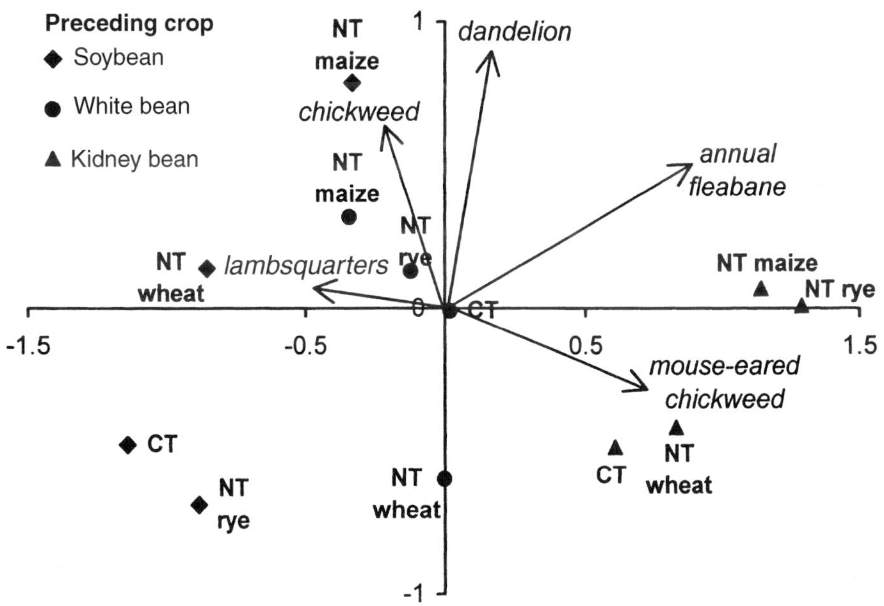

Fig. 14.7. Biplot of weed communities in a wheat crop which was preceded by either soybean, white bean or kidney beans. Treatments are: conventional tillage (CT), no-till with rye cover crop (NT rye), no-till with wheat stubble (NT wheat) and no-till with maize cover crop (NT maize). Centroids of treatments are shown (redrawn from Shrestha *et al.*, 2002).

treatment multivariate means (centroids represented by solid symbols) (Fig. 14.6a). In this case, the environmental variables were not measured directly. Instead, the environment will be influenced by management treatments. The centroids of the four treatments are dispersed along the two axes indicating that different weed communities were present. The arrows on the Fig. 14.6b point to species that are most associated with each management technique. Longer arrows indicate a stronger association. For example, the no-till wheat stubble and no-till maize cover treatments were more associated with dandelion than the other two treatments, while conventional tillage was associated with lambsquarters.

The researchers also found that weed communities present in wheat were influenced by the species of crop in the previous year (Fig. 14.7). For example, when wheat was preceded by kidney beans, it was associated with mouse-eared chickweed, and when preceded by soybean was associated with lambsquarters. This occurred for all tillage treatments.

Cluster analysis

Cluster analysis is a way of finding natural groupings of similar objects (e.g. two sites) based on a number of variables (e.g. many species abundances). Cluster analysis is useful when the researcher wishes to map or classify ecological communities.

The first step of a cluster analysis is to measure the level of similarity between the sites. This is done using similarity indices such as those mentioned previously. Next, this information is used to form clusters of similar sites. There are a variety of techniques used to do this. Some start with the entire data set and progressively divide it into smaller groups, whereas others start with small species groups and combine similar ones into larger species groups. The results of cluster analyses are often represented pictorially using a 'dendrogram'.

Example of cluster analysis

Jensen (1998) used cluster analysis to examine the relationship between the above-ground vegetation and the seed bank in wet meadows of northwest Germany. Jensen collected data seed bank and above-ground vegetation data from early- (I), mid- (II) and

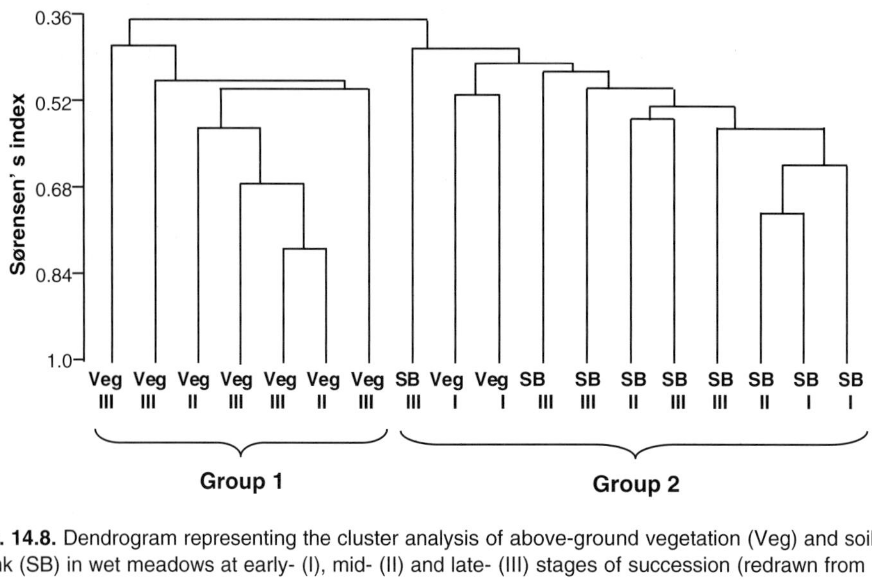

Fig. 14.8. Dendrogram representing the cluster analysis of above-ground vegetation (Veg) and soil seed bank (SB) in wet meadows at early- (I), mid- (II) and late- (III) stages of succession (redrawn from Jensen, 1998).

late-(III) successional communities. These represent stages following the abandonment of wet meadows.

The Sørensen coefficient was used to compare the similarity of the seed bank and above-ground vegetation and then a dendrogram of the results was created (Fig. 14.8). Looking at the top of the figure there are two major groups. Group 1 contains all the seed bank data and early successional communities (I), while group 2 contains all the mid- and late-successional above-ground communities. This means that the above-ground early-successional communities were more similar to the seed bank communities than to the mid- and late-successional above-ground communities. Does this make ecological sense? Yes it does. As succession proceeds, plant composition changes, but seeds of earlier stages of succession remain dormant in the seed bank. Thus, early successional above-ground communities are similar to all seed bank communities because seed banks contain seeds of their current community and all previous communities.

Summary

In this chapter, we have presented only a few ways that researchers study community patterns and processes. Some of the methods presented can be used in combination. For example, Carøe *et al.* (2000) used a combination of the Steinhaus similarity index, ordination and cluster analysis to examine the succession of ground vegetation in a Danish beech forest. There are many more ways to study communities. Sometimes the most convincing ecological evidence comes when several types of analyses point to the same conclusion. There is no one correct way to study community ecology and the type of analysis you do is dependent on your hypothesis and how you collected your data (Chapter 10).

Questions

1. Design a data sheet that could have been used by Meiners *et al.* (2002) to collect their field data.
2. Why is it important to calculate degrees of freedom (df)?
3. Why can you get different results when using the Jaccard and Sørensen coefficients of similarity?
4. Since β_w (Whittaker's statistic) can compare the richness of two communities: why is it better to use similarity indices?
5. What are the advantages and disadvantages of the long-term and chronosequence approaches to examining succession?

General References

Krebs, C.J. (1999) *Ecological Methodology,* 2nd edn. Benjamin/Cummings, Menlo Park, California.
Legendre, P. and Legendre, L. (1998) *Numerical Ecology,* 2nd English edn. Elsevier, Amsterdam.
Podani, J. (2000) *Introduction to the Exploration of Multivariate Biological Data.* Backhuys Press, Leiden.

Literature Cited

Bakker, J.P., Olff, H.J., Willems, J.H. and Zobel, M. (1996) Why do we need permanent plots in the study of long-term vegetation dynamics? *Journal of Vegetation Science* 7, 147–156.
Carøe, I., Barfod, A.S. and Lawesson, J.E. (2000) Temporal dynamics of the ground vegetation in a Danish beech forest. *Nordic Journal of Botany* 20, 585–597.
Chikoye, D. and Ekeleme, F. (2001) Weed flora and soil seedbanks in fields dominated by *Imperata cylindrica* in the moist savanna of West Africa. *Weed Research* 41, 476–491.

Csecserits, A and Rédei, T. (2001) Secondary succession on sandy old-fields in Hungary. *Applied Vegetation Science* 4, 63–74.

Dessaint, F., Chadoeuf, R. and Barralis, G. (1997) Nine years' soil seed bank and weed vegetation relationships in an arable field without control. *Journal of Applied Ecology* 34, 123–130.

Digby, P.G.N. and Kempton, R.A. (1987) *Multivariate Analysis of Ecological Communities.* Chapman and Hall, New York.

Foster, B.L. and Tilman, D. (2000) Dynamic and static views of succession: testing the descriptive power of the chronosequence approach. *Plant Ecology* 146, 1–10.

Gauch, H.G.J. (1982) *Multivariate Analysis in Community Ecology.* Cambridge University Press, Cambridge.

Grandin, U. (2001) Short-term and long-term variation in seed bank/vegetation relations along an environmental and successional gradient. *Ecography* 24, 731–741.

Jaccard, P. (1900) Contribution au problème de l'immigration post-glaciaire de la flore alpine. *Bulletin de Societé Vaudoise Science Natural* 36, 87–130.

Jensen, K. (1998) Species composition of soil seed bank and seed rain of abandoned wet meadows and their relation to aboveground vegetation. *Flora* 193, 345–359.

Jongman, R.H.G., ter Braak, C.J.F. and van Tongeren, O.F.R. (1995) *Data Analysis in Community and Landscape Ecology.* Cambridge University Press, Cambridge.

Kenkel, N.C., Derksen, D.A., Thomas, A.G. and Watson, P.R. (2002) Multivariate analysis in weed science research. *Weed Science* 50, 281–292.

Leeson, J.Y., Sheard, J.W. and Thomas, A.G. (2000) Weed communities associated with arable Saskatchewan farm management systems. *Canadian Journal of Plant Science* 80, 177–185.

Legendre, P. and Legendre, L. (1998) *Numerical Ecology,* 2nd English edn. Elsevier, Amsterdam.

Meiners, S.J., Pickett, S.T.A. and Cadenasso, M.L. (2002) Exotic plant invasions over 40 years of old field successions: community patterns and associations. *Ecography* 25, 215–223.

Menalled, F.D., Gross, K.L. and Hammond, M. (2001) Weed aboveground and seedbank community responses to agricultural management systems. *Ecological Applications* 11, 1586–1601.

Moreby, S.J., Aebischer, N.J., Southway, S.E. and Sotherton, N.W. (1994) A comparison of the flora and arthropod fauna of organically and conventionally grown winter wheat in southern England. *Annals of Applied Biology* 15, 13–27.

Motyka, J. (1947) O zadaniach i metodach badan geobotanicznych. Sur les buts et les méthodes des recherches géobotanique. *Annales Universitatis Mariae Curie-Sklodowsk, Sectio C Supplementum I.*

Pickett, S.T.A. (1989) Space for time substitution as an alternative to long-term studies. In: Likens, G.E. (ed.) *Long-term Studies in Ecology.* John Wiley & Sons, Chichester, pp. 71–88.

Podani, J. (2000) *Introduction to the Exploration of Multivariate Biological Data.* Backhuys Press, Leiden.

Quinn, G.P. and Keough, M.J. (2002) *Experimental Design and Data Analysis for Biologists.* Cambridge University Press, Cambridge, UK.

Ristau, T.E., Horsley, S.B. and McCormick, L.H. (2001) Sampling to assess species diversity of herbaceous layer vegetation in Allegheny hardwood forests. *Journal of the Torrey Botanical Society* 128, 150–164.

Shrestha, A., Knezevic, S.Z., Roy, R.C., Ball-Coelho, B.R. and Swanton, C.J. (2002) Effect of tillage, cover crop and crop rotation on the composition of weed flora in a sandy soil. *Weed Research* 42, 76–87.

Small, J.A., Buell, M.E. and Siccama, T.G. (1971) Old-field succession on the New Jersey Piedmont – the first year. *William L. Hutchinson Memorial Forest Bulletin* 2, 26–30.

Sørensen, T. (1948) A method of establishing groups of equal amplitude in plant sociology based on similarity of species content and its application to analysis of the vegetation on Danish commons. *Biological Skr* 5, 1–34.

Part IV

Conclusions

15
The Importance of Weed Ecology to Management

Introduction

We started out by noting that this book was designed to be an intermediate-level text that focused on weed ecology, and that we are interested primarily in the population and community ecology. If you revisit the various chapters and the examples we used in them, you might notice that our emphasis on population and community scales still incorporated ideas, processes and structures from other scales, e.g. genetic, physiology, morphology, ecosystem and landscape. This reflects the reality of ecology – it is a subject area that is integrative as it tries to explain how all these other scales influence population and community structure and dynamics. The focus on population and community scales is explained by the fact that these are easiest for humans to perceive and are at scales amenable to management of weeds in our everyday lives. Humans are also able to cope with scales smaller than this because we can through technology, control variables in experiments. At larger scales (ecosystem, landscapes), the processes and structures are more complicated and variables are difficult to control, hence we know less about the ecology of weeds at these larger scales.

Think about information that at different scales would apply to the weed you chose to study. What additional information did you or would you gain if you examined the literature on genetic, physiological, morphological, ecosystem and landscape scales? Does this information exist? If so, what are the implications for the economic and ecological impact of your weed? Importantly, information at all scales is critical to the subject that most weed textbooks focus on: management. How should the weed you chose be managed? To illustrate, we have included two examples from our experience: one weed colonizes most terrestrial habitats in North America, and one is a tropical weed.

North American Example: Garlic Mustard and Dame's Rocket

We have already used the two North American weeds (garlic mustard, *Alliaria petiolata*, and dame's rocket, *Hesperis matronalis*) as examples in several chapters. Now that you know many of the basic ecological concepts, we can draw a clear picture of the problems posed by these weeds and the contribution of ecology to their management. Both weeds are Eurasian species which have been invasive in many habitats including transportation and powerline corridors, riverbanks, forest edges, forest plantations and agricultural fields (Cavers *et al.*, 1979; Nuzzo, 1991, 1993a,b,c, 1999; Anderson *et al.*, 1996; Cruden *et al.*, 1996; Haber, 1996, 1998; Hoffman and Kearns, 1997; Beyers and Quinn, 1998; Drayton and Primack, 1999; Luken and Shea, 2000).

Of the two weeds, garlic mustard is more shade tolerant, possibly allelopathic, competitive for light and nutrients, and reproduces rapidly and repeatedly (Nuzzo, 1991, 1993a,b, 1999; Anderson *et al.*, 1996; Cruden *et al.*, 1996; Haber, 1996, 1998; Hoffman and Kearns, 1997; McCarthy, 1997; Beyers and Quinn, 1998; McCarthy and Hanson, 1998; Susko and Lovett Doust, 1998, 1999, 2000; Dhillon and Anderson, 1999; Drayton and Primack, 1999; Meekins and McCarthy, 1999; Luken and Shea, 2000). Dame's rocket has not yet made anyone's list of the 'world's worst weeds' because it does not (yet) cause a serious economic impact in any one agricultural or forest crop. Additionally, there is much less known about its genetics or physiology (Dvorak, 1982; Gohil and Raina, 1987; Conner and Sterling, 1995; Hoffman and Kearns, 1997; Davis *et al.*, 1998). However, dame's rocket is capable of colonizing most habitats and has the potential to be like its more established 'relative' in the mustard (Brassicaceae) family, garlic mustard (Haber, 1996, 1998).

What both dame's rocket and garlic mustard appear to share is the ability to self-pollinate, and produce many seeds that have a high rate of germination. Germination occurs in early spring (occasionally in late autumn) and flowering may be delayed for several years if conditions are not favourable (Cavers *et al.*, 1979; Baskin and Baskin, 1992, 1998; Anderson *et al.*, 1996; Beyers and Quinn, 1998; Dhillon and Anderson, 1999; Nuzzo, 1999). They are well adapted to disturbance, so much so that one of the worst management approaches is to pull out the adults or seedlings indiscriminately because the soil disturbance will cause hundreds more seeds to germinate. Most types of disturbance are helpful to garlic mustard – even flooding, wind-throw and prescribed burns (Nuzzo, 1999; Luken and Shea, 2000). The populations of both species will increase rapidly in disturbed areas and can dominate a community. For example, Murphy has recorded populations of both species increasing to the extent of occupying over 90% of the ground within 5 years of the arrival of the first dozen or so colonists.

Managing these weeds will depend on knowledge of their ecology. Herbicides and hand pulling may extirpate small populations, albeit temporarily (Drayton and Primack, 1999). However, neither strategy works well, since hand pulling causes more disturbance and herbicides normally eliminate most of the native species allowing weeds to recolonize preferentially. The short-term approach is to prevent flowering and exhaust the seedbank, e.g. by clipping the inflorescence just before seed set since at this time in their life cycle it will be too late for the weeds to compensate because their resources will be exhausted. A longer-term approach is to determine whether garlic mustard and dame's rocket are vulnerable to herbivores, parasites or pathogens in North America. So far, there appear to be none in North America but possibilities exist in Europe (Jones and Finch, 1987; Stobbs and Van Schagen, 1987; Ford *et al.*, 1988; Nielsen *et al.*, 1989, 1995; Larsen *et al.*, 1992; Chang *et al.*, 1996; Chen and Schwegman, 1996; Haribal and Renwick, 1998; Haribal *et al.*, 1999; Guglielmone *et al.*, 2000).

Ultimately, we probably need to manage the landscape, not the weeds, but this too relies on weed ecology. The success of garlic mustard and dame's rocket is attributed to fragmentation of forests and other habitats. This creates habitat characteristics that are conducive to invasion by weed species and hostile to native species. The edges are areas that experience constant shifts in temperature, light, moisture and wind – this is very suitable for species with a high degree of phenotypic plasticity, and both garlic mustard and dame's rocket appear to have this. Hence, if we reduce, halt or reverse fragmentation, then managing these weeds is easier because their ecological 'niche' is one that depends on small, fragmented areas (Matlack, 1993, 1994a,b, for general discussion).

Tropical Weed: Purple Nutsedge

Purple nutsedge (*Cyperus rotundus*) is widespread throughout tropical and subtropical areas of the world. Based on a worldwide survey (Holm *et al.*, 1977), purple nutsedge was found to be a serious weed problem in 52 crops in 92 countries. Genotypic differences, which enhance adaptability to environmental variation, have been reported within selected populations (Cheng and Thseng, 1990; Cheema *et al.*, 1992). In addition, success of this weed is also attributed to its ability to reproduce clonally. Both rhizomes and tubers are formed which are capable of producing multiple young sprouts (Siriwardana and Nishimoto, 1987). These rhizomes and tubers make control difficult. Manual control is rarely satisfactory. When shoots are removed by hoeing near the soil surface, new shoots will regrow at a rate of up to 3 cm per day (William, 1976; Komai and Ueki, 1982). Exhaustion of tubers did not occur even after 9 months of regular removal of shoots at bi-weekly intervals (Horowitz, 1972). Control by herbicides is only moderately successful (Hawton *et al.*, 1992) because of extensive clonal growth and the inability of a systemic herbicide to translocate throughout this underground structure.

In order to develop a more integrated approach to the management of purple nutsedge, Neeser *et al.* (1997, 1998, 1999) conducted several basic ecological studies. They discovered that purple nutsedge tubers can survive 16–42 months and that burial depth in the soil up to 23 cm had no effect on survival or dormancy. Tuber numbers increased with increasing levels of photosynthetically active radiation (PAR) received at the soil surface. As a result, the amount of light (PAR) reaching the soil surface could be used to rank the competitive ability of differing crops because they intercept light differently, therefore reducing the need for complete control with herbicides. In 1998 they introduced a mechanistic model of purple nutsedge population dynamics. This model combined ecological variables such as tuber age and tuber population density with variables that reflect crop competition for light and weed control levels, to calculate tuber fecundity and survival. When nutsedge was not controlled the model predicted that cumulated incident photosynthetic active radiation was a major determinant of population size whereas intraspecific competition was by far the major determinant of the rate of population increase '*r*'.

This same model also led to the formulation of several research hypotheses that could be tested through further field experimentation. For example, one hypothesis stated that 'a low level of weed control in competitive crops will not result in a significant increase of the purple nutsedge population'. The results of the model predicted that tuber populations would not increase if efforts to control purple nutsedge were reduced provided that the weed was present in a competitive crop such as maize or bean. In contrast to this, tuber numbers increased in non-competitive crops such as peppers if a high level of weed control was not maintained. These ecological studies on purple nutsedge are examples of how plant ecology, population modelling and weed management can be combined to develop effective weed management strategies.

Why Weed Management Needs Weed Ecology

An important concept is that without ecological information, weed management ultimately may fail, make weed problems worse, or have unintended and detrimental environmental and economic consequences. Humans often rely on single solutions to manage weeds. The problem with a single weed control strategy is that weeds adapt to management. Also, herbicides work very well and are excellent tools within a weed management strategy; however, misuse of this technology can lead to problems such as residual carry-over, cropping restrictions, ground water contamination and the development of genetically based herbicide resistance.

The introduction of herbicide resistant crops has had a major influence on cropping systems. In western Canada, herbicide-resistant canola (*Brassica napus* L.) is commercially available but the use of this technology has raised several weed management concerns. Volunteer herbicide-resistant canola has become a weed problem in succeeding crops. In addition, resistant herbicide genes may transfer to closely related weed species through hybridization. Finally, there exists the possibility for 'gene stacking', i.e. canola possessing resistant genes to two or more herbicides through hybridization from two or more different herbicide-tolerant canola cultivars (Hall *et al.*, 2000). Invariably, these types of problems can cause instability within a weed management programme, making control more difficult.

Our approach to weed management often reflects human nature. We are not completely comfortable with risk and as a result, we like the idea of control. Control of weeds with herbicides reduces the risk of weeds escaping and ultimately reducing yield and profitability. Weed management, however, is often focused on weeds as a series of individuals. Control measures often try to exploit a species' biological weakness. This approach may lead to instability in weed management, because once one species is removed, another will appear. This leads to a teeter–totter effect: as soon as one weed problem is solved, another will surely arise (Booth and Swanton, 2002). An ecologically based approach to weed management may lead to a more balanced and diverse weed community.

A diverse weed community may prove to be very important to weed management. If a diverse weed community could be managed successfully while maintaining acceptable crop yields, then perhaps some previously unrealized benefits of the presence of weeds on weed community dynamics could be seen. Although there is no simple relationship between diversity and stability, factors contributing to stability may yet be discovered that would show the benefits of maintaining a diverse weed community. Moreover, diversity in weed communities may be viewed in support of the goal of enhancing biodiversity in managed and natural ecosystems. For example, Murphy *et al.* (unpublished manuscript) found that in a three-crop rotation, no-tillage systems increased weed species diversity above ground and within the seed bank.

Weed Ecology and Weed Science: Building on This Book and Moving Forward

Management based on ecology should be the foundation upon which weed management strategies are designed and implemented within an integrated weed management approach. Management without ecology increases the likelihood of failure. An ecological approach broadens (increases) management options, thus decreasing the probability of failure. In the final analysis, weed ecology forces us to accept the fact that weeds are part of our plant communities. New species of plants colonize new habitats all the time; humans simply speed up the process by transporting weeds and creating habitats amenable to their existence. Thus, weeds need to be studied in the context of their populations and communities in which they survive. The only novel aspect to weeds is that they have a tremendous economic and ecological impact that would not exist if it were not for humans. Reducing this impact will require management approaches that are based first on ecological knowledge. Hopefully, you now have accomplished this.

Literature Cited

Anderson, R.C., Dhillon, S.S. and Kelley, T.M. (1996) Aspects of the ecology of an invasive plant, garlic mustard (*Alliaria petiolata*) in central Illinois. *Restoration Ecology* 4, 181–191.

Baskin, C.C. and Baskin, J.M. (1998) *Seeds: Ecology, Biogeography, and Evolution of Dormancy and Germination.* Academic Press, Boston, Massachusetts.

Baskin, J.M. and Baskin, C.C. (1992) Seed germination biology of the weedy biennial *Alliaria petiolata. Natural Areas Journal* 12, 191–197.

Beyers, D.L. and Quinn, J.A. (1998) Demographic variation in *Alliaria petiolata* (Brassicaceae) in four contrasting habitats. *Journal of the Torrey Botanical Club* 125, 138–149.

Booth, B.D. and Swanton, C.J. (2002) Assembly theory applied to weed communities. *Weed Science* 50, 2–13.

Cavers, P.B., Heagy, M.I. and Kokron, R.F. (1979) The biology of Canadian weeds. 35. *Alliaria petiolata* (M. Bieb.) Cavara and Grande. *Canadian Journal of Plant Science* 59, 217–229.

Chang, K.F., Hwang, S.F., Evans, I.R. and Howard, R.J. (1996) A new yellows disease of dame's violet (*Hesperis matronalis*) in Alberta Canada, caused by an unidentified phytoplasma. *Zeitschrift für Pflanzenkrankheiten und Pflanzenschutz* 103, 225–233

Cheema, P., Bir, S.S. and Sidhu, M.K. (1992) Chromosomal variation in *Cyperus rotundus* from northwest India. *Nucleus-Calcutta* 35, 83–86.

Chen, S.C. and Thseng, F.S. (1990) Studies on the intraspecific variation of the upland weed purple nutsedge (*Cyperus rotundus* L.) in Taiwan. 1. Variations of morphological characters in the population of field and field path. *Journal of Agriculture and Forestry* 39, 103–115.

Chen, W. and Schwegman, J.E. (1996) First report of *Fusarium* root rot of *Alliaria petiolata. Plant Disease* 80, 711–715.

Conner, J.K. and Sterling, A. (1995) Testing hypotheses of functional relationships: a comparative survey of correlation patterns among floral traits in five insect-pollinated plants. *American Journal of Botany* 82, 1399–1406.

Cruden, R.W., McClain, A.M. and Shrivastava, G.P. (1996) Pollination biology and breeding system of *Alliaria petiolata* (Brassicaceae). *Bulletin of the Torrey Botanical Club* 123, 273–280.

Davis, A.R., Pyaltuik, J.D., Paradis, J.J.C. and Low, N.H. (1998) Nectar-carbohydrate production and composition vary in relation to nectary anatomy and location within individual flowers of several species of Brassicaceae. *Planta* 295, 305–318

Dhillon, S.S. and Anderson, R.C. (1999) Growth and photosynthetic response of first-year garlic mustard (*Alliaria petiolata*) to varied irradiance. *Journal of the Torrey Botanical Society* 126, 9–14.

Drayton, B. and Primack, R.B. (1999) Experimental extinction of garlic mustard (*Alliaria petiolata*) populations: implications for weed science and conservation biology. *Biological Invasions* 1, 159–167.

Dvorak, F. (1982) Some results of the study of *Hesperis matronalis* L. from the Belanski Tatry Mts Czechoslovakia. *Biologia Seria A: Botanika* 37, 441–447.

Ford, R.E., Beczner, L. and Hamilton, R.I. (1988) Turnip, cucumber, and ribgrass mosaic virus isolated from *Hesperis matronalis* in British Columbia. *Plant Disease* 72, 101–106.

Gohil, R.N. and Raina, R. (1987) Polyploidy accompanied by structural alterations in the evolution of *Hesperis matronalis* L. *Cytologia* 52, 223–228.

Guglielmone, L., Jenner, C.E., Walsh, J.A., Ramasso, E., Marian, D. and Roggero, P. (2000) An unusual isolate of turnip mosaic virus potyvirus from *Abutilon theophrasti* in Piedmont Italy. *Phytoparasitica* 28, 149–152.

Haber, E. (1996) *Invasive Plants of Canada: 1996 National Survey Results.* Biodiversity Convention Office, Environment Canada, Ottawa, Canada.

Haber, E. (1998) *Invasive Plants Alert: 1997–1998 Report.* Environment Canada, Ottawa, Ontario.

Hall, L., Topinka, K., Huffman, J., Davis, L. and Good, A. (2000) Pollen flow between herbicide-resistant *Brassica napus* is the cause of multiple-resistant *B. napus* volunteers. *Weed Science* 48, 688–694.

Haribal, M. and Renwick, A.J.A. (1998) Isovitexin 6"-*O*-β-D-glucopyranoside: a feeding deterrent to *Pieris napioleracea* from *Alliaria petiolata. Phytochemistry* 47, 1237–1240.

Haribal, M., Renwick, A.J.A. and Attygalle, A. (1999) A new sinapoyl derivative of isovitexin 6"-*O*-b-D-glucopyranoside from *Alliaria petiolata. Journal of Natural Products* 62, 179–180.

Hawton, D., Howitt, C.J. and Johnson, I.D.G. (1992) A comparison of methods for the control of *Cyperus rotundus* L. *Tropical Pest Management* 38, 305–309.

Hoffman, R. and Kearns, K. (1997) *Wisconsin Manual for Control of Recreational Ecology Invasive Plants.* Wisconsin Department of Natural Resources, Madison, Wisconsin.

Horowitz, M. (1972) Effects of frequent clipping on three perennial weeds, *Cynodon dactylon* (L.) Pers., *Sorghum halepense* (L.) Pers. and *Cyperus rotundus* L. *Experimental Agriculture* 8, 225–234.

Jones, T.H. and Finch, S. (1987) The effect of a chemical deterrent, released from the frass of caterpillars of the garden pebble moth, on cabbage fly root oviposition. *Entomologia Experimentalis et Applicata* 45, 283–288.

Komai, K. and Ueki, K. (1982) Effect of shoot-cutting treatment on the regrowth ability and carbohydrate contents of purple nutsedge (*Cyperus rotundus* L.) tubers. *Weed Research – Japan* 27, 272–277.

Larsen, L.M., Nielsen, J.K. and Sørensen, H. (1992) Host plant recognition in monophagous weevils: specialization of *Centorhynchus inaffectatus* to glucosinolates from its host plant *Hesperis matronalis.* *Entomologia Experimentalis et Applicata* 64, 49–55.

Luken, J.O. and Shea, M. (2000) Repeated prescribed burning at Dinsmore Woods State Nature Preserve (Kentucky, USA): responses of the understorey community. *Natural Areas Journal* 20, 150–158.

Matlack, G.R. (1993) Microenvironment variation within and among forest edge sites in the eastern United States. *Biological Conservation* 66, 185–194.

Matlack, G.R. (1994a) Vegetation dynamics of the forest edge – trends in space and successional time. *Journal of Ecology* 82, 113–123.

Matlack, G.R. (1994b) Plant species migration in a mixed-history forest landscape in eastern North America. *Ecology* 75, 1491–1502.

McCarthy, B.C. (1997) Response of a forest understorey community to removal of an invasive nonindigenous plant (*Alliaria petiolata*, Brassicaceae). In: Luken, J.O. and Thieret, J.W. (eds) *Assessment and Management of Plant Invasions*, Springer-Verlag, New York, pp. 117–130.

McCarthy, B.C. and Hanson, S.L. (1998) As assessment of the allelopathic potential of the invasive weed *Alliaria petiolata* (Brassicaceae). *Castanea* 63, 68–73.

Meekins, J.F. and McCarthy, B.C. (1999) Competitive ability of *Alliaria petiolata* (garlic mustard, Brassicaceae), an invasive nonindigenous forest herb. *International Journal of Plant Sciences* 160, 743–752.

Neeser, C., Aguero, R. and Swanton, C.J. (1997) Incident photosynthetically active radiation as a basis for integrated management of purple nutsedge (*Cyperus rotundus*). *Weed Science* 45, 777–783.

Neeser, C., Aguero, R. and Swanton, C.J. (1998) A mechanistic model of purple nutsedge (*Cyperus rotundus*) population dynamics. *Weed Science* 46, 673–681.

Neeser, C., Aguero, R. and Swanton, C.J. (1997) Survival and dormancy of purple nutsedge (*Cyperus rotundus*) tubers. *Weed Science* 45, 784–790.

Nielsen, J.K., Jakobsen, H.B., Friis, P., Hansen, K., Møller, J. and Olsen, C.E. (1995) Asynchronous rhythms in the emission of volatiles from *Hesperis matronalis* flowers. *Phytochemistry* 38, 847–851.

Nielsen, J.K., Kirkenby-Thomsen, A.H. and Petersen, M.K. (1989) Host plant recognition in monophagous weevils: specificity in feeding responses *of Ceutorhynchus constrictus* and the variable effect of sinigrin. *Entomologia Experimentalis et Applicata* 53, 157–166.

Nuzzo, V.A. (1991) Experimental control of garlic mustard *Alliaria petiolata* [(Bieb.) Cavara and Grande] in northern Illinois using fire, herbicide, and cutting. *Natural Areas Journal* 11, 158–167.

Nuzzo, V.A. (1993a) Distribution and spread of the invasive biennial (*Alliaria petiolata* [Bieb.] Cavara and Grande) in North America. In: McKnight, B.L. (ed.) *Biological Pollution: Control and Impact of Invasive Exotic Species.* Indiana Academy of Science, Indianapolis, Indiana, pp. 115–124.

Nuzzo, V.A. (1993b) Current and historic distribution of garlic mustard *(Alliaria petiolata)* in Illinois. *Michigan Botanist* 23, 23–33.

Nuzzo, V.A. (1993c) Natural mortality of garlic mustard (*Alliaria petiolata* [Bieb.] Cavara and Grande). *Natural Areas Journal* 13, 132–133.

Nuzzo, V.A. (1999) Invasion pattern of the herb garlic mustard (*Alliaria petiolata*) in high quality forests. *Biological Invasions* 1, 169–179.

Siriwardana, G. and Nishimoto, R.K. (1987) Propagules of purple nutsedge (*Cyperus rotundus*) in soil. *Weed Technology* 1, 217–220.

Stobbs, L.W. and Van Schagen, J.G. (1987) Occurrence and characterization of a turnip mosaic virus isolate infecting *Alliaria petiolata* in Ontario, Canada. *Plant Disease* 71, 965–968.

Susko, D.J. and Lovett Doust, L. (1998) Variable patterns of seed maturation and abortion in *Alliaria petiolata* (Brassicaceae). *Canadian Journal of Botany* 76, 1677–1686.

Susko, D.J. and Lovett Doust, L. (1999) Effects of resource availability, and fruit and ovule position on components of fecundity in *Alliaria petiolata* (Brassicaceae). *New Phytologist* 144, 293–306.
Susko, D.J. and Lovett Doust, L. (2000) Patterns of seed mass variation and their effects on seedling traits in *Alliaria petiolata* (Brassicaceae). *American Journal of Botany* 87, 56–66.
William, R.D. (1976) Purple nutsedge: tropical scourge. *HortScience* 11, 357–364.

Part V

Appendix

Glossary

α- (alpha-) diversity – the total species diversity within one defined habitat or community
Abiotic – non-living, physical or chemical environment
Abundance – is a measure of a population's success in terms of numbers
Accuracy – describes how well data reflect the true value of the variable being estimated
Additive series design – type of competition experiment using a range of total densities and where the relative proportion of species changes
Agamospermy – the production of seeds without fertilization
Allelopathy – an interaction where one individual has a direct effect on another through the release of chemical compounds from roots, shoots, leaves or flowers
Amensalism – an interaction whereby only one individual is negatively affected and the other neither benefits nor is harmed
Apomixis – asexual reproduction
Apparent competition – an interaction that gives the appearance of being due to competition but is actually due to other factors
Asexual reproduction – the creation of offspring that are genetically identical to their parents, through a variety of mechanisms
β- (beta-) diversity – the comparison of diversity between habitats, communities and/or along an environmental gradient
Bias – describes how well the sampling procedure reflects the true value of the variable
Biological control – the management of weeds using introduced herbivores (often insects) as 'biological control agents'
Biomass – the weight of vegetation per area
Biotic – living
Casual plant – a non-native plant that does not form self-replacing populations
Clonal growth – the creation of new, potentially independent plants through vegetative growth (also vegetative reproduction)
Cluster analysis – a type of multivariate analysis used to find natural groups based on a number of variables
Cohort – a group of individuals born within the same age class
Community – a group of populations that co-occur in the same space and at the same time

Competition – a negative interaction where individuals make simultaneous demands that exceed limited resources
Competitive ability – a combination of competitive effect and competitive response
Competitive effect – the ability of an individual to suppress the growth or survival of another
Competitive response – the ability of an individual to avoid being suppressed by another
Complete additive design – type of competition experiment using all combinations of densities
Cover – the proportion of ground covered by a species when viewed from above.
δ- (delta-) diversity – the comparison of diversity between landscapes
Demography – the study of a population's size and structure and how it changes over time
Density – the number of individuals in a given area
Diversity – a measure of the number of taxa present (richness) and their relative abundances (evenness)
Dormant – in a resting state and unable to germinate
ε- (epsilon-) diversity – the total species diversity within a larger ecological landscape
Ecology – the study of organisms and their environment
Emergence – appearance of a shoot above the soil
Epiphyte – a parasite that is dependent on its host for physical support
Establishment – occurs once a seedling no longer depends on seed reserves (endosperm and cotyledons), i.e. it is photosynthetically independent
Evenness – a measure of how similar the relative abundances of each species in a community are
Exploitation competition – a type of competition where both individuals compete for the same resource, but do not interact directly
Facilitation – a successional process whereby early invading species ameliorate the environment for later invaders
Facultative interaction – an interaction where both species can survive independently, but both benefit when they are found together.
Fitness – a relative measure of how well an individual succeeds at continuing its lineage.
Frequency – the proportion of sampling units (e.g. quadrats) that contain the target species
Fruit – structure formed from the flower ovary or receptacle and containing one or many seeds
Functional group – a groups of species with a similar set of traits
γ- (gamma-) diversity – the total species diversity within an ecological landscape
Genet – an entire genetic individual, composed of ramets
Germination – the emergence of a root and/or shoot from a seed coat
Guerrilla-type growth – a type of clonal growth that results in loosely packed, often linear patches
Hemiparasite – parasite that relies on its host for only some resources
Herbivory – the consumption of plant tissue by animals
Holoparasite – parasite that is entirely dependent on its host
Hypothesis – an idea that can be tested using experiments
Inbreeding depression – reduced fitness that results from the accumulation of deleterious alleles caused by mating with a close relative
Indirect plant defences – plant uses another organism to defend itself against herbivory
Inhibition – a successional process whereby existing plants prevent or inhibit the establishment of subsequent species
Interference competition – a direct interaction between individuals, commonly where one individual is able to deny access to the resource to the other
Interspecific – between species
Intraspecific – within a species
Invasibility – the ease with which a habitat is invaded

Invasion – the expansion of a species into an area not previously occupied by it
Invasion meltdown – the acceleration of impacts on native ecosystems due to synergistic interactions of non-native species
Iteroparous – reproduction that occurs repeatedly through a plant's lifespan (often used as a synonym for polycarpic)
K – carrying capacity
Keystone species – a species that has a disproportionate effect on community function relative to its biomass
K-strategists – species that are large, have delayed reproduction, are long-lived and are found in stable environments
Life history – the general description of a plant's life cycle and the more specific aspects of life cycles within population (age, stage, size)
Life table – a table summarizing the survival data of a population
Metapopulation – a group of spatially isolated populations that interact through migration or distant pollination
Monocarpic – sexual reproduction that occurs only once in a plant's lifespan (often used as a synonym for iteroparous)
Multivariate analysis – a type of statistical analysis that combines species' abundance data into one analysis
Mutualism – an interaction that benefits both individuals
Naturalized plant – a non-native plant that forms self-sustaining populations but is not necessarily invasive
Neighbour manipulation experiment – type of competition experiment where the density of neighbour plants around a target individual is increased or decreased
Non-dormant – able to germinate
Non-native plant – a plant whose presence is due to intentional or accidental introduction as a result of human activity, also exotic, alien, non-indigenous
Null hypothesis – a hypothesis stating that there are no differences among observed populations
Obligate interaction – an interaction where both partners of the association require each other in order to survive
Observer effect – a phenomenon whereby the act of touching a plant while making observations or measurements induces a change in its growth, survival, etc.
Ordination – a type of multivariate analysis used to examine how species abundances vary with the environment
Overcompensation – a type of tolerance to herbivory where the effect of herbivory benefits the plant
Parasitism – an interaction where an individual obtains nutrients, shelter and/or support from its host
Partial additive design – type of competition experiment where one species is kept at a constant density with a second species at a range of densities
Persistence – a measure of how long a community remains the same
Phalanx-type growth – a type of clonal growth that results in slow growing, branched clones that form dense patches
Phenology – the study of life cycle events and the environmental conditions that influence them
Physiognomy – the general appearance of a community
Plot sampling – a method of sampling that uses quadrats to physically delineate the area being sampled
Plotless sample – a method of vegetation where quadrats are not used
Pollination – the transfer of pollen from an anther to a stigma
Polycarpic – sexual reproduction that occurs repeatedly throughout a plant's lifespan (often used a synonym for semelparous)

Polymorphism – when a structure produced has two or more morphology types – especially seeds
Population – a group of potentially inter-breeding individuals of the same species found in the same place at the same time
Post-dispersal seed predation – consumption of the seed after it has been dispersed
Potential distribution – the area in which a species is able to persist as determined by the abiotic environment (also physiological distribution or climatic range)
Precision – describes how close the values of replicated data are to each other
Pre-dispersal seed predation – consumption of the seed while it is still on the maternal parent plant
Primary dispersal – dispersal of seed from the parent plant to the ground
Primary dormancy – seeds that are unable to germinate when they first mature
Primary succession – succession that occurs on newly created land where no plants have grown previously or where there is no effective seed bank on site
Quadrats – physical sampling units that are placed over the vegetation to act as boundaries for sampling
Quiescent – seeds that are not dormant, but do not germinate because they have not encountered appropriate environmental conditions
'*r*' – the intrinsic rate of population growth
Ramet – an individual which is genetically identical to the parent plant and capable of physiologically independent growth
Replacement series design – type of competition experiment where the relative densities of both species vary while the total density is kept constant
Resilience – a measure of a community's ability to return to its original state following a disturbance
Resistance – a measure of whether a community resists stress or disturbance
Resource manipulation experiment – type of competition experiment where resources are increased or decreased
Richness – the number of taxa (i.e. species) present in an area or in a community
r-strategists – species that are small, annuals, have a rapid growth rate, reproduce early, and produce many small seeds and are therefore able to establish rapidly following a disturbance
Safe site – a site that provides all the conditions necessary for the seed to germinate and emerge from the soil
Secondary dispersal – movement of seed subsequent to primary dispersal
Secondary dormancy – dormancy that is imposed after seeds have dispersed
Secondary succession – succession that occurs on land previously vegetated, but disturbed by natural or human-caused factors
Seed – the embryonic plant that develops from the fertilized ovule
Seed bank – seeds that become incorporated into the soil
Seed dispersal – movement of a seed or fruit away from the maternal parent plant
Seedling – a young plant
Self-compatible – individuals that can successfully mate with themselves if pollen is transferred from stigma to style
Self-incompatible – an individual that is not able to mate with itself
Semelparous – reproduction that occurs only once in a plant's lifespan (often used as a synonym for monocarpic)
Sexual reproduction – the creation of offspring via fusion of two gametes (a sperm and ovum) to form a zygote
Stability – a measure of how communities resist change in response to disturbance or stress, comprised of persistence, resistance and resilience
Strategy – a group of similar characteristics which causes species to exhibit ecological similarities

Succession – the directional change in community composition
Tens rule – describes how approximately 10% of species pass through each transition from being imported to becoming casual to becoming established, and finally becoming a weed
Time lag – the time between when a species is introduced and when its population growth explodes
Tolerance – a successional process whereby existing species have no effect on subsequent ones
Traits – the physical and physiological characteristics of a plant that determine its ecological function
Trajectory – a path through a series of community states
Transects – lines used to help determine where to locate quadrats to test for changes along environmental gradients
Treatment – manipulations made as part of an experiment
Water use efficiency – the ability to minimize water use for a given amount of carbon assimilation
Weed – a native or introduced species that has a perceived negative ecological or economic effect on agricultural or natural ecosystems

Subject Index

Species Index